CRC Series in The Biochemistry and Molecular Biology of the Cell Nucleus

Editor-in-Chief

Lubomir S. Hnilica, Ph.D.
Professor of Biochemistry and Pathology
Department of Biochemistry
Vanderbilt University School
of Medicine
Nashville, Tennessee

The Structure and Biological Function of Histones

Author

Lubomir S. Hnilica, Ph.D.

Chromosomal Nonhistone Proteins

Volume I: Biology
Volume II: Immunology
Volume III: Biochemistry
Volume IV: Structural Associations

Editor

Lubomir S. Hnilica, Ph.D.

Enzymes of Nucleic Acid Synthesis and Modification

Volume I: DNA Enzymes
Volume II: RNA Enzymes

Editor

Samson T. Jacob, Ph.D.
Professor
Department of Pharmacology
The Pennsylvania State University
The Milton S. Hershey Medical Center
Hershey, Pennsylvania

Chromosomal Nonhistone Proteins

Volume III
Biochemistry

Editor

Lubomir S. Hnilica, Ph.D.
Professor of Biochemistry and Pathology
Department of Biochemistry
Vanderbilt University School of Medicine
Nashville, Tennessee

CRC Series in The Biochemistry and Molecular Biology of the Cell Nucleus

CRC Press, Inc.
Boca Raton, Florida

Library of Congress Cataloging in Publication Data

Main entry under title:

Chromosomal nonhistone proteins.

(CRC series in the biochemistry and molecular biology of the cell nucleus)
Includes bibliographies and indexes.
Contents: v. 1. Biology -- v. 2. Immunology.
1. Nucleoproteins. 2. Chromosomes. I. Hnilica,
Lubomir S. II. Series. [DNLM: 1. Chromosomal
proteins, Non-Histone. QU 58 C5576]
QP625.N87C47 1983 574.87'322 82-17742
ISBN 0-8493-5511-7 (v.1)
ISBN 0-8493-5512-5 (v.2)
ISBN 0-8493-5513-3 (v.3)
ISBN 0-8493-5514-1 (v.4)

This book represents information obtained from authentic and highly regarded sources. Reprinted material is quoted with permission, and sources are indicated. A wide variety of references are listed. Every reasonable effort has been made to give reliable data and information, but the author and the publisher cannot assume responsibility for the validity of all materials or for the consequences of their use.

Direct all inquiries to CRC Press, Inc., 2000 Corporate Blvd., N.W., Boca Raton, Florida, 33431.

International Standard Book Number 0-8493-5511-7 (Volume I)
International Standard Book Number 0-8493-5512-5 (Volume II)
International Standard Book Number 0-8493-5513-3 (Volume III)
International Standard Book Number 0-8493-5514-1 (Volume IV)
Library of Congress Card Number 82-17742
Printed in the United States

INTRODUCTION

The third volume of the *Chromosomal Nonhistone Proteins* treatise deals with the isolation and analysis of chromosomal proteins in general and with specific features of these proteins in particular. In addition to the high mobility group (HMG) proteins, postsynthetic modifications of chromosomal nonhistone proteins are discussed, together with some of their enzymatic properties and DNA binding.

As with all multiauthored volumes dealing with rapidly advancing and closely related aspects of a busy research area, the readers may see some overlaps between the individual chapters of this or the companion volumes. Unfortunately, this, despite the best intentions of all the contributing authors, is inevitable. Nevertheless, I hope that most will find this as well as the other volumes informative and helpful in their scientific endeavors.

THE EDITOR

Lubomir S. Hnilica is a professor of biochemistry and pathology at the Vanderbilt University School of Medicine, Nashville, Tennessee.

Dr. Hnilica received his degree in organic chemistry from Masaryk University, Brno, Czechoslovakia in 1952 and an advanced degree in biochemistry from the Czechoslovak Academy of Sciences. After postdoctoral training at the Czechoslovak National Cancer Institute in Bratislava, Dr. Hnilica continued his research as a WHO research fellow at the Chester Beatty Research Institute in London, England. After 2 years at the Baylor College of Medicine in Houston, Texas, Dr. Hnilica joined the staff of the University of Texas System Cancer Center, M.D. Anderson Hospital and Tumor Institute in Houston where he rose to the rank of Professor of Biochemistry and Chief of the Section of Biochemical Regulatory Mechanisms. In 1975 Dr. Hnilica joined the Vanderbilt University School of Medicine where he is Professor of Biochemistry and Pathology as well as director of the A. B. Hancock, Jr. Memorial Laboratory of the Vanderbilt University Cancer Center. His present research concerns proteins of the cell nucleus, their biochemistry and immunology, the interactions of nuclear proteins with DNA, and changes in gene expression during chemical carcinogenesis.

CONTRIBUTORS

Paul Byvoet, M.D., Ph.D.
Professor of Pathology
Chief, Clinical Chemistry
University of South Florida College of Medicine
J. A. Haley Veterans Administration Hospital
Tampa, Florida

Jen-Fu Chiu, Ph.D.
Associate Professor of Biochemistry
Department of Biochemistry
University of Vermont College of Medicine
Burlington, Vermont

G. H. Goodwin
Chester Beatty Laboratories
Institute of Cancer Research
Royal Cancer Hospital
London, England

Lubomir S. Hnilica, Ph.D.
Professor of Biochemistry and Pathology
Department of Biochemistry and the A. B. Hancock, Jr. Memorial Laboratory of the Vanderbilt University Cancer Center
Vanderbilt University School of Medicine
Nashville, Tennessee

E. W. Johns
Chester Beatty Laboratories
Institute of Cancer Research
Royal Cancer Hospital
London, England

Lewis J. Kleinsmith, Ph.D.
Professor of Biological Sciences
Division of Biological Sciences
University of Michigan
Ann Arbor, Michigan

Leokadia Klyszejko-Stefanowicz
Professor of Biochemistry
Department of Biochemistry
Institute of Biochemistry and Biophysics
University of Lodz
Lodz, Poland

Najma Malik
Department of Biochemistry
Schools of Medicine and Dentistry
Georgetown University
Washington, D.C.

Samuel J. Mitchell*
Division of Biological Sciences
University of Michigan
Ann Arbor, Michigan

Mark O. J. Olson, Ph.D.
Associate Professor
Department of Biochemistry
University of Mississippi Medical Center
Jackson, Mississippi

Mark E. Smulson, Ph.D.
Professor of Biochemistry
Department of Biochemistry
Georgetown University School of Medicine
Molecular Biology Head
Vincent T. Lombardi Cancer Center
Georgetown University Medical Center
Washington, D.C.

* Deceased.

TABLE OF CONTENTS

Volume III

Chapter 1

ISOLATION, FRACTIONATION, AND ANALYSIS OF NONHISTONE CHROMOSOMAL PROTEINS

Leokadia Klyszejko-Stefanowicz and Lubomir S. Hnilica

TABLE OF CONTENTS

Chromatin, the interphase chromosomal material of eukaryotic cells is a complex of DNA with a fixed level of histone proteins, variable amounts of nonhistone chromosomal proteins (NHCP), and small amounts of RNA. Of these principal chromatin components, the non-histone proteins have received considerable attention as a class of macromolecules which may be associated with the regulation of genetic transcription.[1-4,25]

I. SOME CONSIDERATIONS OF NHCP ISOLATION AND FRACTIONATION

The resolution of NHCP by conventional methods of protein chemistry has proved particularly difficult because of (1) affinity of these proteins for DNA, and their tendency to form insoluble complexes with histones and themselves, (2) general insolubility in aqueous buffered solutions at physiological values of ionic strength and pH, and (3) proteolytic degradation during prolonged isolation and fractionation procedures.

The same considerations which apply for the isolation of nuclei[6] and chromatin[7] are operative in evaluating procedures for isolation and fractionation of NHCP (for review see Reference 5). Cytoplasmic contamination is important as the probable source of both proteolytic activity and cytoplasmic proteins in the nonhistone pool. Moreover, some special requirements in the NHCP isolation and fractionation are due to very difficult solubilization of the whole and especially, dehistonized chromatin. The greatest barrier lies in the potential ionic, hydrogen, and hydrophobic interactions between chromatin components, and often leads to the use of harsh conditions such as extremes of pH and organic solvents, e.g., sodium dodecyl sulfate (SDS), urea, guanidine hydrochloride (GdnCl), and/or 2-mercaptoethanol (2-ME), which may irreversibly affect the native structure of some nonhistone proteins.

A. Proteolysis

Since the late 1960s several investigators have reported that, apart from proteases of cytoplasmic origin associated with subcellular fractions as microsomes,[8] lysosomes,[9-11] mitochondria[12-14] or outer nuclear membrane,[15] there occurs a neutral protease of low molecular weight (e.g., 24,000,[18] 15,400,[19] or 25,000[20]), tightly associated with chromatin.[16,17] This acid-soluble enzyme was purified from calf thymus chromatin[19,21] and characterized as a serine-type protease.[22] It also appears to be the major diisopropylfluorophosphate (DIFP)-binding protein in the chromatin from nuclei of several rat tissues (liver, thymus, kidney, lung, testis, and Morris hepatoma 9121).[20]

Contrary to cytoplasmic proteases active at lower ionic strength, the activity of the chromatin-bound protease increases in solutions of high ionic strength, being optimal at pH 7.8[23] to 8.5.[19] This enzyme functions in a variety of denaturing solvents such as 5 *M* urea,[24] 0.2 *M* HCl, or H_2SO_4,[16] but not in SDS.[17] It is considered as chromatin nonhistone protein,[25] which attacks histones and nonhistones. It is extractable in acid with the lysine-rich histones,[18] more exactly with histone H2B,[26,27] and its proteolytic activity could not be separated from histone H2B.[26]

Although the danger of proteolytic degradation of chromatin was recognized long ago, the use of protease inhibitors such as $NaHSO_3$ (5 m*M*,[28] 50 m*M*,[29] or even 1 m*M*[30]), DIFP (0.5 m*M*[31] or 1 m*M*[29]), and 0.1 m*M* phenylmethanesulfonyl fluoride (PMSF)[32] inspired by experiments of Fahrney and Gold[33] on esterase inhibition, was not common initially. Sodium bisulfite, being an exception, has been reported to inhibit endogenous chromatin proteolytic activity at a concentration of 50 m*M*,[29,34] generally in solutions of low ionic strength. Proteolytic degradation of calf thymus nucleohistone could be totally inhibited over a period of several days at 2°, either in the presence of $NaHSO_3$ or simply by storing at very low ionic strength (5×10^{-4}).[35] Chae and Carter[24] showed that histones and nonhistone chromatin proteins are intensively degraded when chromatin is exposed to 2 *M* NaCl-5 *M* urea (most often used for its dissociation and reconstitution) or to 5 *M* urea (pH 6 to 8). The report from Busch's laboratory[36] that degradation of chromosomal proteins, observed in chromatin dissociaedt in 3 *M* NaCl-7 *M* urea, was not prevented by 50 m*M* $NaHSO_3$ but only by 1 m*M* PMSF, is also important. To this time, little attention has been given to the effects of proteolysis on the interpretation of chromatin dissociation or reassociation experiments as well as on polypeptide composition of NHCP in their electrophoretical patterns. Also, controls to check for possible proteolytic activity were often not carried out, although long dialysis and centrifugation steps were generally required.

More precise studies on chromatin-bound protease, active in 2 *M* NaCl-5 *M* urea or 5 *M* urea alone, came from Carter and Chae.[37] The proteolytic activity was demonstrated by these authors in chromatin of six rat tissues (liver, thymus, kidney, testis, brain, Ehrlich ascites), rabbit bone marrow (but not in chicken reticulocyte chromatin), and examined in the presence of seven different protease inhibitors at several concentrations. It was shown that chromatin protease can be irreversibly and most efficiently inhibited by 1 m*M* concentration of two serine-specific inhibitors: water-soluble, but extremely toxic DIFP, and water-insoluble and less toxic PMSF. The protection of chromatin proteins was only partial when 1 m*M* PMSF was used without the presence of organic solvent (isopropyl alcohol, *p*-dioxan, or ethanol), in which it should be prepared as stock solution at 0.1 *M* concentration.[17] At 0.1 m*M* PMSF the protection reaches only 15% of complete proteolysis inhibition observed at final concentration of 1 m*M* PMSF-1% organic solvent.[17,37] Two principal protease inhibitors are used most frequently: PMSF at concentrations of 0.1 m*M*,[38-48] 0.2 m*M*,[49] 0.5 m*M*,[50-52] 1 m*M*,[53-60] 2 m*M*[61] or even 2.85 m*M*,[62] and DIFP at concentrations of 0.1 m*M*,[63] 0.5 m*M*,[64,65] or 1 m*M*.[66]

Among less-used protease inhibitors, the following should be mentioned: soybean trypsin inhibitor (1 μg/mℓ) without[44] or with 1 m*M* $NaHSO_3$;[67] *p*-chloromercuribenzoate at concentration of 2.5 m*M* (homogenate) and 1 m*M* (in other solutions),[68] or Trasylol.®[69] The $NaHSO_3$ at such relatively low concentrations as 1 m*M*,[70,71] 5 m*M*,[72] or 10 m*M*[40] alone or in combination with 1 m*M* PMSF[73] is also used in chromatin isolation. The use of 50 m*M* $NaHSO_3$ along with 40 m*M* NaF can also be quoted.[74] Some caveats should be observed with using protease inhibitors: PMSF may affect extraction of some proteins;[75] DIFP and $NaHSO_3$ may produce artifactual charge heterogeneity in electrophoretic patterns — DIFP by its nonspecific binding to proteins and $NaHSO_3$ by oxidizing conditions.[76]

Some authors failed to demonstrate autolytic activity of chromatin or of NHCP by differences in electrophoretic patterns of NHCP isolated with or without the use of serine-specific inhibitors.[40-50,77-79] Proteolytic degradation of NHCP has not been observed by Elgin,[79] who used SDS to dissolve dehistonized chromatin[80] or salt-formic acid-urea in the case of whole chromatin.[81] The experiments of Boffa and Allfrey[40] have pointed out the usefulness of [^3H] catalase, added as a marker of proteolysis to the nuclear extracts during isolation from dehistonized chromatin of phenol-soluble and urea-GdnCl-2-ME-Na_2HPO_4-soluble NHCP.

A neutral protease of high molecular weight (200,000) was isolated from rat liver chromatin and characterized by Bonner's group.[34,82] Most active at pH 7.0, sensitive to sulfhydryl and chelating reagents, inactive at salt concentrations higher than 1 *M* NaCl, this enzyme can be considered as an extranuclear protease,[37] inhibited by 50 m*M* $NaHSO_3$ and also by DIFP and PMSF.[82]

Another high molecular weight protease (103,000) was purified over 800-fold from chromatin isolated from fresh rat liver homogenate.[83] This enzyme has a similar substrate specificity as the protease isolated from rat liver mitochondria,[84] but it shows marked differences in molecular weight, in the effects of monovalent and divalent ions, and different sensitivity to inhibitors. For example, the soybean trypsin inhibitor, a strong inhibitor of the mitochondrial enzyme, is almost totally ineffective with this protease (17% of inhibition).[83] With other inhibitors tested, the following percentages of inhibition were obtained: 0.1% SDS (88%), 1 m*M* EDTA (54%), 50 m*M* $NaHSO_3$ (53%).

Three types of proteases have been described by Miyazaki et al.[66] in nuclei from various normal rat tissues and Rhodamine sarcoma, with the pH optima at 4.5, 7.5, and 10.0, all significantly inhibited by 1 m*M* DIFP.

In addition to neutral proteases, Hagiwara et al.[85] found an alkaline protease associated with chromatins of various normal and tumorous tissues of rats. This enzyme was not observed in other subcellular fractions. Perhaps it is functional in the hydrolysis of NHCP and H1 histones. Inhibited by DIFP, soybean trypsin inhibitor, and chymostatin, it is a chymotrypsin-like protease, with an approximate molecular weight of 18,000. Its activity was found high in rapidly proliferating cells: Yoshida sarcoma $>$ Rhodamine sarcoma $>$ AH130 $>$ thymus $\gg$ spleen $\gg$ kidney $>$ liver $\gg$ brain.

B. Cytoplasmic and Membrane Contamination

Since the experiments of Dounce and Lan[86] who used saponin (heterogeneous mixture of sapogenic glycosides) to isolate chicken erythrocyte nuclei, many efforts have been exerted to eliminate cytoplasmic and membrane contamination in nuclear preparations used as starting material for the isolation of chromatin and chromosomal proteins. To remove the outer nuclear membrane and the attached traces of endoplasmic reticulum, treatment with detergents is required. For this aim the use of SDS (0.4%)[87] and the following nonionic detergents should be noted: Tween® 80, i.e., polyoxyethylene sorbitan monooleate (0.1%);[88] Cemulsol NPT6 (0.15%) and Cemulsol® NPT12 (0.15%);[89,90] a mixture of sodium deoxycholate (0.43%) with Tween® 40 (0.86%) in Penman's[91] double detergent technique[50,92] or with Tween® 80;[93] Nonidet® P-40 (0.2%);[42,54,56,94-96] a mixture of Nonidet® P-40 (0.5%) and Triton® X-100 (0.5%);[97,98] Triton® N-101 (0.3%).[45,99] However, the most widely accepted is the use of Triton® X-100 alone at different concentrations, e.g., 0.05%,[100] 0.1%,[40] 0.2%,[101] 0.25%,[65,102] 0.5%;[46,55,72,97,103-107] 1%,[57,61,73,108-114] or even 2%.[115]

Triton® X-100, extensively used for the disruption and solubilization of cytoplasmic components, is a member of the octylphenoxyethanol series of the nonionic detergents, in which X indicates the average length of the hydrophobic polyoxyethylene chain. For Triton® X-100, X has the value of 9 to 10.[116] Rapid and satisfactory preparation of nuclei with the application of Triton® X-100 was first performed by Hymer and Kuff.[116] Similar products having an additional -CH_3 group on the hydrophobic end of the detergent molecules, i.e., Tritons® of nonylphenoxyethanol series[116] (Triton® N-101), are also effective in removing contaminating membranes. Usually concentration of Triton® X-100 (most often 0.25 to 1%) depends on the fragility of nuclei and must be individually determined for the particular tissue.[16] Nuclei treated with Triton® X-100 appear to have lost cytoplasmic and outer nuclear membranes as determined by electron microscopy, without fragmentations of the nuclei which has been observed by Blobel and Potter[117] in the case of the Penman's (DOC and Tween® 40) procedure.[91]

Both the amount and distribution of phospholipids, which were taken to indicate the presence of membrane (microsomal and nuclear) fragments by Tata et al.,[118] were extremely reduced in chromatin prepared from nuclei pretreated with Triton® X-100, as compared with chromatin preparations isolated from whole tissue[119] or from nuclei purified without the Triton® X-100 wash.[120]

Following reports indicating that Triton® X-100 washing of rat liver[121] or chicken erythrocyte nuclei[105] does not remove the outer nuclear membrane completely, Jackson[122] has shown that nonionic detergents (0.1 and 1% Triton® X-100, 0.1% Nonidet® P-40) are capable of extracting a majority of the membrane phospholipids but fail to remove membrane proteins to the same degree as with the Waring® blender procedure.[123] These data[122] suggest that the bulk of contaminating membranes in chromatin prepared from whole tissue is microsomal[118] while in chromatin prepared from the isolated nuclei, even pretreated with detergents, nuclear membrane is probably the major contaminant. Thus, the fraction of membrane proteins solubilized or contaminating the nonhistone protein preparations undoubtedly varies, depending upon the source of the membranes and the exact conditions of the extraction procedure.[124] It should be emphasized, however, that in the experiments of Jackson[122] the nuclear preparations pretreated with Triton® X-100 were not further purified by centrifugation through dense sucrose solution,[6,55] which is extremely helpful in removing the residual cytoplasmic and outer nuclear membranes.

The inclusion in the extraction buffers of divalent cations, required for the maintenance of morphologically intact nuclei, has been reported to be especially detrimental to complete membrane solubilization.[125] This may explain why the detergent extraction procedures, commonly employed in the preparation of detergent nuclei, remove so little membrane protein.

While considering the efficiency of Triton® X-100 in the removal of cytoplasmic contaminants from isolated cell nuclei, the experiments of Weiss and Braun[126] should be mentioned. According to these authors, the use of 1% Triton® X-100 for purifying nuclei from the cellular slime mold *Dictyostelium discoideum* resulted in a dramatic enrichment of two polypeptides: 44,000 mol wt (actin-like) and 180,000 mol wt (myosin-like), as compared with the preparation of nuclei purified without this detergent. This was presumably caused by artificial adsorption of an actin-like protein to chromatin. The authors suggested that studies demonstrating a large amount of actin and other contractile proteins as chromatin constituents deal with cytoplasmic contaminations since they all employ nonionic detergents for lysing cells and purifying nuclei.

C. Use of Organic Solvents

Many media used to solubilize chromatin, especially dehistonized chromatin, involve some organic solvents, among them strong denaturants such as: SDS, urea, GdnCl, and 2-ME.

1. Sodium Dodecyl Sulfate

Of all the synthetic detergents which may be classified as anionic, cationic, or nonionic, the SDS (principally the salt of alkyl sulfate half esters containing straight 12-hydrocarbon tail and a polar head on the same molecule) has found most application in the studies of NHCP. The dispersal effects of this amphiphile are produced by its two kinds of association with proteins: primarily through the salt-like interactions between the charged groups of protein and detergent, and secondly, by less specific weak forces between the nonpolar side chains of the proteins and hydrophobic moiety of the detergent.[127] The latter stabilize the ionic bonds and are responsible for the secondary nonpolar attraction of the additional detergent to that already electrostatically bound with proteins (Figure 1).

Apparently, SDS was first used by Sreenivasaya and Pirie[128] to disintegrate crystalline plant virus nucleoprotein and later by Mirsky and Pollister[129] to dissociate DNA from the isolated chromosomes of animal tissue. Since the experiments of Marco and Butler[130] and Kay et al.,[131] SDS solution (0.41%)[131] has been used as extracting and deproteinizing agent, improving the preparation of highly purified and polymerized DNA.[132] The high protein-denaturing potency of synthetic detergents, and their ability to keep the denatured protein in solution, was discoverea by Anson[133] and in the experiments of Bonner and associates,[80,81,134] SDS has been introduced to isolate and also to fractionate NHCP, i.e., using polyacrylamide gel electrophoresis (PAGE).

Like urea and GdnCl, SDS brings about, in addition to unfolding of the native protein molecule which assumes a more random orientation, the dissociation of oligomer structures. Profound studies by Tanford[137] and Tanford and Reynolds[135,136] on interactions of proteins with amphiphiles have shown that a variety of proteins bind ionically and hydrophobically identical amounts of SDS on a weight-to-weight basis with the eventual formation of a micelle with a high ratio of dodecyl sulfate to protein (1.5 to 2.3 g dodecyl sulfate per gram protein).[135,138] Saturated complex with a stoichiometry of 0.4 g of SDS per gram of protein is formed between 0.5 and 0.8 m*M* SDS monomer, and a second complex, which is saturated at 1.4 g of SDS per gram of protein, is observed above the 0.8 m*M* SDS monomer.[136] The binding of large amounts of SDS to protein is primarily hydrophobic in nature and independent of ionic strength; only the monomeric form of the amphiphile binds to protein, not the micelar form.[135,137] From hydrodynamic studies[124,136] the protein-SDS complex is believed to be a highly ordered, rod-like particle, the length of which varies uniquely with the molecular weight of the protein moiety. This explains the empirical observations that proteins dissolved in aqueous solutions containing high concentrations of SDS lose all their specificity

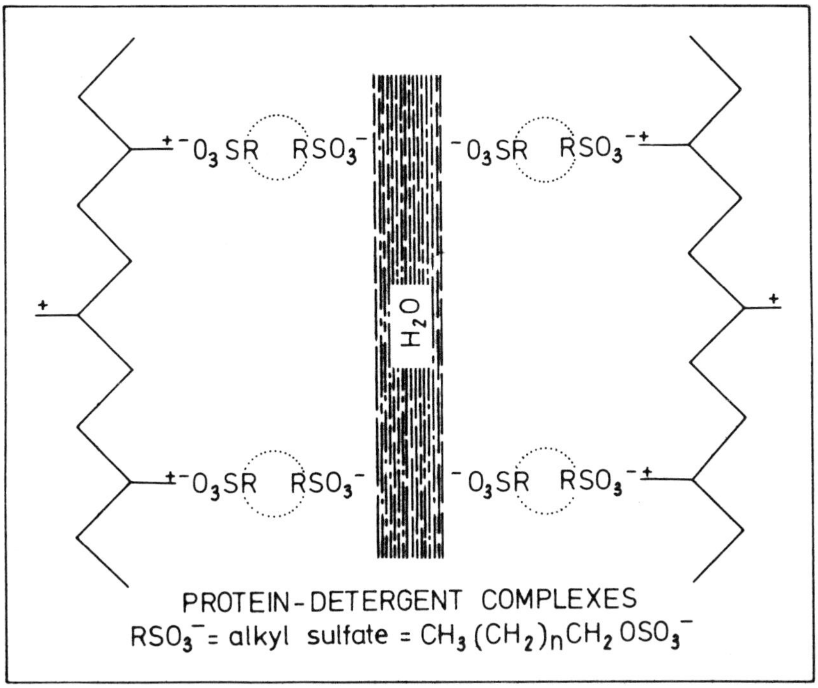

FIGURE 1. Proposed structure of protein-detergent complexes. Dotted lines indicate that extra bound detergent is attached by weak nonpolar forces to detergent stoichiometrically bound by electrostatic forces. This gives rise to distribution of similarly charged groups along the peptide, which repel similarly changed groups of an adjacent chain (From Putnam, F. W., *Advances in Protein Chemistry*, Vol. 4, Anson, M. L. and Edsall, J. T., Eds., Academic Press, New York, 1948, 79. With permission.)

and have electrophoretic mobilities in polyacrylamide gels which are the function of their molecular weight.[139,140]

The principal disadvantages of using SDS are protein denaturation caused by this detergent, which is known to avidly bind to proteins[135-137] and to markedly influence their physical state, and the difficulty of removing SDS from the NHCP without rendering them insoluble.[134,141] The majority of SDS can be removed from chromosomal protein preparations by the urea dialysis-$BaCl_2$ precipitation procedure described by Marushige et al.[134] and by Shirey and Huang.[141]

Knowing that the removal of denaturing agents like urea or GdnCl, which transform proteins into random coil conformation,[142] is generally accompanied by recovery of the native structure,[143] Weber and Kuter[144] have described a procedure for successful reactivation of several oligomeric enzymes after denaturation by exposure to SDS. To avoid disulfide formation, all solutions were made in 2-ME (10 m*M*). Proteins in SDS solutions were first incubated in concentrated urea (6 *M*) and then freed of the detergent by an anion exchange resin (Dowex® AG 1-X2, 200 to 400 mesh). The resulting dodecyl sulfate-free proteins could be renatured from the urea solution by standard procedures. This method has also been used for successful recovery by enzymatic[144] and immunological activity[43,48] of proteins after their electrophoresis on polyacrylamide gels in the presence of SDS.

2. Urea

For over two decades, multimolar concentrations of urea[145,146] have been used to solubilize

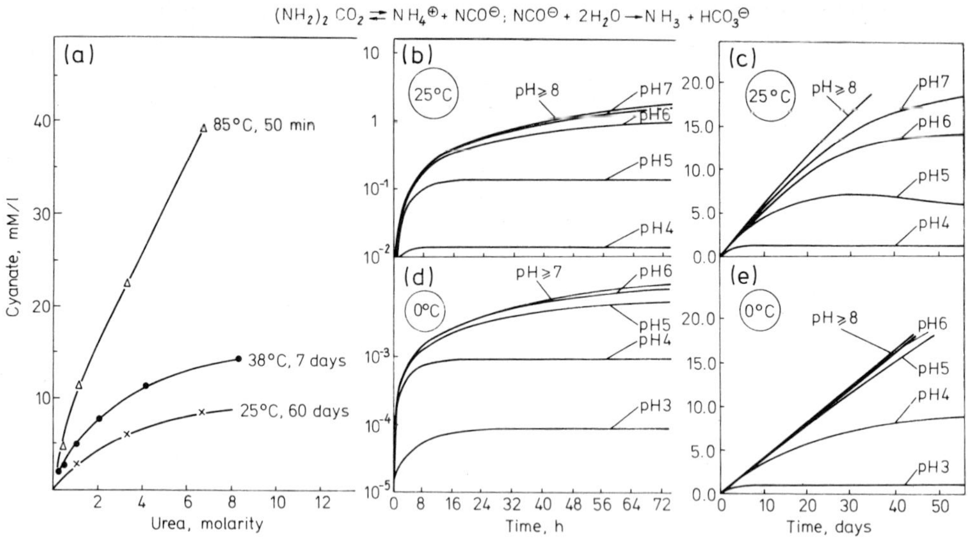

FIGURE 2. Effect of temperature and urea concentration on cyanate accumulation during equilibration of aqueous urea solutions. (a) Accumulation of cyanate ions in an aqueous solution of 8 *M* urea at indicated pH (b-e) during the first 3 days (b,d) and over a period of 2 months (c,e) at 25°C (b,c) and at 0°C (d,e). (Figure (a) from Marier, J. R. and Rose, D., *Anal. Biochem.*, 7, 304, 1964; Figures (b-e) from Hagel, P., Gerding, J. J. T., Fleggen, W., and Bloemenal, H., *Biochem. Biophys. Acta,* 243, 366, 1971. With permission.)

and dissociate chromosomal proteins from DNA as well as to isolate and fractionate NHCP on a preparative and analytical scale (for review see Reference 5). Unfortunately, the use of this reagent is not free from drawbacks. High concentrations of urea (5 to 9 *M*) destroy noncovalent and nonionic interactions in proteins and may cause complete protein denaturation. However, experiments on reconstitution of chromatin (for review see Reference 147), and on dissociation and reassociation of chromosomal proteins from DNA indicate that the effect of urea is not irreversible and under suitable conditions many of the proteins return to their native conformation.[148]

Urea is often assumed to be a reagent which brings about physical rather than chemical changes in protein molecules. This assumption is valid only if the urea is completely free of cyanate.[149] It is known[149-154] that the dissociation of proteins into subunits by the action of urea may be accompanied by the formation of ammonia, cyanate, and carbonate ions (Figure 2). The amount of cyanate formed in aqueous solutions of urea depends on urea concentration, duration of storage, and temperature and pH.[149,151,155] In the range from 1 to 9 *M* urea (the most widely used in protein chemistry) there exists a linear relationship between the cyanate and urea concentrations, especially evident at higher temperatures (Figure 2a). At a given urea concentration, the maximum levels of cyanate attainable by direct equilibration at 85, 38, and 25°C can be expressed by the proportions 4.3: 1.7: 1.0, respectively.[151] In other words, whereas only 50 min are required to attain a maximum cyanate level at 85°C, approximately 7 days are required at 38°C and 60 days are needed at 25°C.

Figure 2 (b to e) summarizes graphic representations of the accumulation of cyanate in 8 *M* urea solutions of different pH (3 to 8) at two temperatures (0°C [d,e] and 25°C [b,c]) over the period of 2 months (c,e) and in the range from 0 to 72 hr (b,d). Hence, it is clear that the accumulation of cyanate ions is most pronounced at a pH higher than 4 and increases with the duration of storage, especially at higher temperatures. The formation of the cyanate ions in urea solutions is extremely slow at low temperatures.[155]

It has been shown by comparative electrophoresis in polyacrylamide gels that a long contact of proteins (α-crystalline[155] and very low density lipoproteins[156]) with concentrated urea (e.g., 6 *M*), especially at room temperature and at higher pH (8.9 to 9.4), gives rise to a significant cyanate accumulation and, in consequence, to a high degree of protein carbamylation[149,154-156] at the α- and ε-amino as well as hydroxyl and –SH groups. Especially important is carbamylation of the ε-amino groups, which converts positively charged lysine to neutral homocitruline,[156] leading to changes in electrophoretic mobility (greater anodal migration) of the affected proteins. This can produce artifactual polymorphism and erroneous conclusions about the number of individual subunits or protein species.[154-156] The carbamylation of tyrosine or cysteine would not change the molecular charge at the given pH; moreover *O*-carbamyltyrosine and *S*-carbamylcysteine are decomposed in a slightly alkaline medium.[149,154]

The potential problems caused by the formation of cyanate and ammonium ions can be minimized or avoided by simple precautions during the preparation of urea buffers, e.g., according to Herbert et al.[156] urea should be prepared in a stock solution (10 *M*) using high-quality crystalline reagent,[76] without heating (at temperatures not exceeding 25°C), and stored at 5°C in stoppered glass bottles.[151] Immediately before the preparation of buffer, the urea solution should be deionized by passing through a column of mixed-bed resin (e.g., AG® 501-X8,[157,158] Rexyn® I-300,[156,159] or Amberlite® MB-3[160]). Fresh buffer should be rapidly cooled, if necessary, and used immediately after preparation. The urea buffers must be used within two days of preparation,[161] even in the case of deionized urea solutions.[157] To compete for the reaction of isocyanate with protein, lysine should be included in all urea buffers.[76] The addition of 2 m*M* TrisCl to buffers for chromatography on hydroxylapatite[162] has also been reported to protect the proteins against urea cyanate degradation.[155,156]

3. Guanidine Hydrochloride and 2-Mercaptoethanol

Interactions between DNA, RNA, histones, and nonhistones in chromatin involve ionic, hydrophobic, and hydrogen bonds. A few covalent disulfide linkages can also exist between chromatin proteins.

Compared with others salts, GdnCl can be considered to be an effective denaturant which brings about complete unfolding of globular proteins.[163] In a series of papers, Tanford et al.[164-167] presented the results of measurements of some hydrodynamic properties which are especially sensitive to conformation, e.g., intrinsic viscosity, sedimentation coefficients, optical rotatory dispersion, osmotic pressure, and titration studies of globular proteins in concentrated GdnCl solutions. Such proteins lose their characteristic native structure and appear to be devoid of any residual noncovalent structures, either in the presence or absence of disulfide bonds. In the presence of a reducing agent, e.g., 2-ME, they behave like randomly coiled simple linear organic polymers.

In addition to urea and GdnCl, some authors introduce moderate concentrations of a reducing agent, 2-ME, to break disulfide bonds and to prevent their formation when they are not present in the native proteins. The rupture of disulfide inter- and intra-chain cross-links facilitates protein unfolding[168] and was found to diminish aggregation between histone and nonhistone proteins, thereby facilitating the subsequent separation of these two protein groups.[169]

II. ISOLATION OF CHROMOSOMAL NONHISTONE PROTEINS

A. General Remarks

Since the discovery of nonhistone proteins in deoxyribonucleoprotein (DNP) complex,[170] numerous attempts have been made to isolate these proteins and to characterize their chemical and biological properties. For a long time, drastic procedures such as strong acid, alkaline

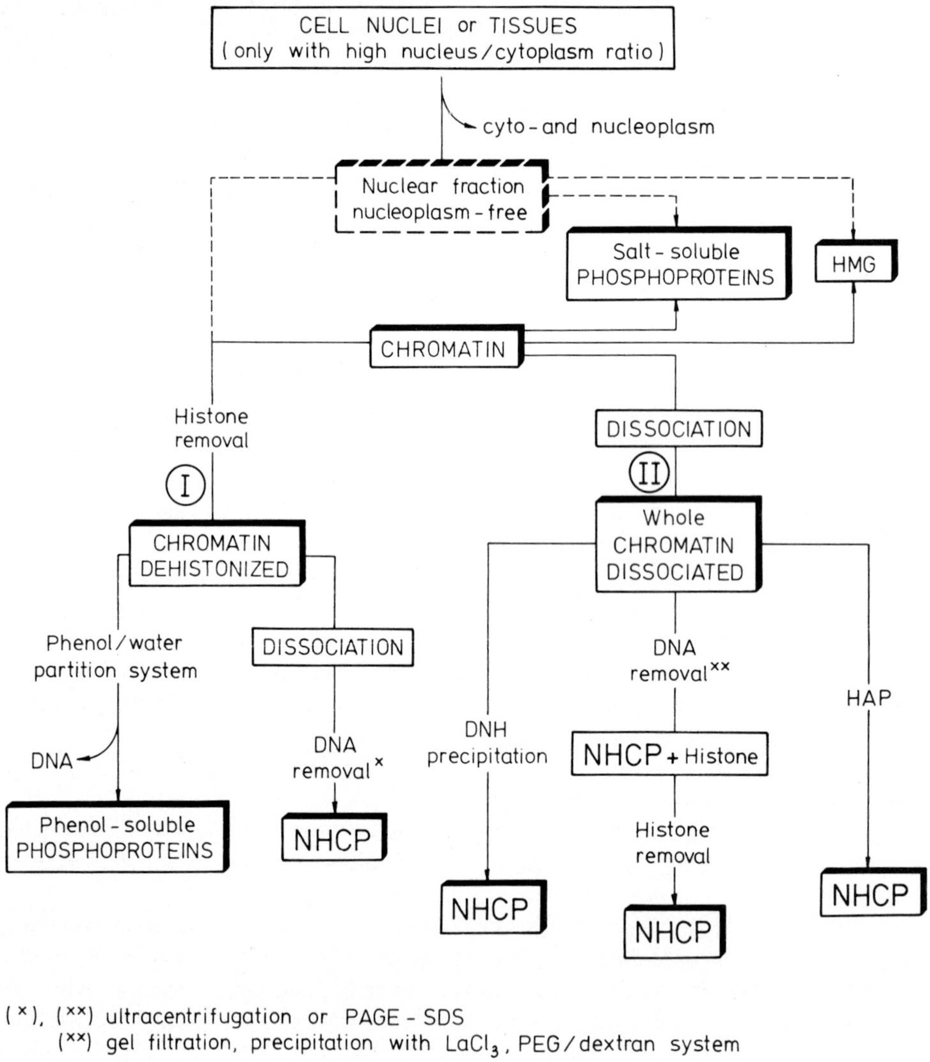

FIGURE 3. Scheme for principal procedures of NHCP isolation.[457]

or thioglycolate treatment,[171-173] extraction by ionic detergent,[80] urea,[174] GdnCl,[163] or phenol[175] have been used for NHCP isolation or solubilization. To study the biological phenomena involving NHCP it is essential to avoid denaturing conditions. Unfortunately, milder extraction procedures[176-178] frequently lead to only partial recovery of these proteins.

The isolation of NHCP generally follows one of two basic schemes. For isolation from dissociated dehistonized or whole chromatin the starting point is most frequently purified and nonsheared chromatin (for review of isolation methods see Reference 7) obtained from purified cytoplasm-free nuclei (for review of isolation methods see Reference 6) or even from total cell homogenate. This second procedure, described by Bonner et al.,[179] and based on the method of Zubay and Doty[180] may be applied only to tissues with a high ratio of nucleus to cytoplasm, such as thymus.[16] There are some NHCP groups which may be isolated both from chromatin and cell nuclei, e.g., nuclear phosphoproteins (for review see Reference 181 and Chapters 3 and 4) or HMG (High Mobility Group) proteins (for review see Reference 182 and Chapter 2). As can be seen in Figure 3, there are at least two principal approaches to the isolation of nonhistone proteins.

Methods of the first group, which use dehistonized chromatin to start with, involve the separation of NHCP by detergents, alkaline solutions of CsCl (Table 1) or, by phenol extraction in a two-phase partition system (Table 5a). In this last case, dehistonized and delipidated chromatin or nuclei are used as starting material.

In the second group (Figure 3), whole chromatin is dissociated in detergent or in various buffered salt solutions (NaCl or GdnCl) alone or with multimolar urea concentrations and with or without the addition of other organic reagents, e.g., 2-ME, ethylenediamine tetraacetate (EDTA) or dimethylsulfoxide (DMSO) (Tables 2 to 5). The DNA is removed by ultracentrifugation (Table 2), precipitation with histones at low ionic strength, i.e., at 0.14 *M* NaCl (Table 3a), or 0.4 *M* NaCl (Table 3b), with lanthanum chloride (Table 3c), gel filtration, HAP-chromatography (Table 4) or two-phase partition systems, using mainly polyethylene glycol (PEG)-dextran (Table 5b). The NHCP are obtained as such (Table 1, 3a, and 5a) or separated from histones by cation-exchange (Table 6) or anion-exchange chromatography (Table 7). Electrophoretic methods to separate NHCP from DNA in dehistonized or total chromatin were also used.

B. Dissociation and Solubilization of Dehistonized or Whole Chromatin

1. Solubilization of Acid-Treated Chromatin

Older methods often employed removal of histones by acid extraction followed by the separation of NHCP from DNA. In this approach, the NHCP suffered considerable damage as was indicated by: (1) extremely difficult solubilization of such dehistonized chromatin as well as incomplete separation of NHCP from histones, (2) protein denaturation and their extensive aggregation, (3) removal of acid-labile groups, and (4) changed structural and functional properties of these proteins, making them suitable only for chemical and some physicochemical studies. It has also been reported that acid treatment of chromatin can lead to an appreciable proportion of histones becoming tightly bound to the DNA.[280] Solubilization of chromatin irreversibly altered by acid treatment was so difficult that strong denaturants such as SDS, high pH, or phenol had to be used. As can be seen in Table 1, in this first group of methods the most accepted dissociating agent is SDS.

Following the findings of Carusi and Sinsheimer[281] about the usefulness of SDS in dissociation of DNA from accompanying proteins, Bonner and co-workers[80,134] have used this anionic detergent at a concentration of 1% (or 3.47×10^{-2} *M*) in 50 m*M* TrisCl, pH 8.0 at 37°C, to remove acid-insoluble proteins from the DNA. Since then, SDS has rapidly become a powerful agent available for molecular dispersion of proteins and for the removal of up to 90 to 95% of NHCP[80,96,141] from DNA. The major advantage of this ionic detergent is that it keeps the NHCP solubilized in the supernatant after high-speed ultracentrifugation. In the experiments of Allfrey and associates[40,41,187] the yield of NHCP from dehistonized chromatin amounted 87 ± 2%.

More recently, Sevaljevic and Konstantinovic[50] reexamined the SDS dissociation procedure. These authors confirmed the findings of Elgin and Bonner[80] concerning the yield of NHCP and demonstrated that the lower third of the high-speed supernatant contains almost half of the total proteins and about 10% of the total DNA. Moreover these authors have demonstrated that the proteins and DNA are distributed in the 100,000 × g supernatant unequally, thus influencing the NHCP electrophoretic patterns depending on the region and volume of supernatant used for examination.

2. Dissociation of Whole Chromatin

Other workers[145,146,174] have avoided the many pitfalls associated with acid dehistonization of chromatin by dissociating chromatin in buffer solutions containing various combinations of high salt and urea concentrations (Table 2). These neutral or slightly alkaline solutions containing multimolar NaCl and urea are frequently considered as "mild conditions",[223] as

Table 1
DISSOCIATION MEDIA USED TO SEPARATE NHCP FROM DNA BEFORE ULTRACENTRIFUGATION OF DEHISTONIZED CHROMATINS OR NUCLEI

No.	Dissociation media	Tissues and ref.
	(a) Dehistonized by Acid	
1.	1% SDS-50 m*M* TrisCl, pH 8.0	Rat liver,[80,134,183] kidney,[80] and brain;[184] chicken liver and erythrocytes;[80] duck erythroid cells;[74] sea urchin embryo;[50] pea bud;[80] potato tuber[51]
2	0.1% SDS-10 m*M* TrisCl, pH 8.0	Rat brain, kidney, liver, lung, spleen, thymus, and thyroid[185]
3	1% SDS-0.14 *M* 2-ME-10 m*M* Na-phosphate, pH 7.2	Rat liver[186]
4	0.4 *M* GdnCl-6 *M* urea-0.1% 2-ME-0.1 *M* Na_2HPO_4, pH 7.4	Rat colon normal epithelia and DMH-induced tumors;[40,41,187] human colon normal epithelia and adenocarcinomas,[40,41] and polyps[41]
5	4*M* CsCl-10 m*M* 2-ME-5 m*M* EDTA-20 m*M* lysine, pH 11.5[189] or 11.6[188]	Rat liver;[188] ovariectomized rat uterus, liver, spleen, and calf endometrium[189]
	(b) Dehistonized by Calcium Chloride[190]	
6	4 *M* urea-1 *M* sucrose-20 m*M* lysine, pH 11.6	Slime mold *Physarum polycephalum;*[191] brain of guinea pig, rabbit, and rat[192]
	(c) Dehistonized by Salt-Urea, pH 5.0[193] or 6.0[194-196]	
7	2.5 *M* NaCl-5 *M* urea-10 m*M* TrisCl, pH 8.0	Calf thymus;[196] rat liver,[193,196-199] prostate and testis,[199] pancreas normal[199,200] and regenerating;[200] hepatomas; Novikoff,[196-198] 30D,[196] 3′-MDAB-induced,[197] and Walker carcinosarcoma[196]
8	2% SDS-50 m*M* Na-phosphate, pH 7.0	Chicken erythrocytes and reticulocytes[48]

compared to the extremes of pH used in acid extractions. It should be emphasized, however, that these conditions are denaturing, although it is possible to remove the denaturants and recover some native proteins.

In studies on chromosomal protein dissociation from DNA, Bekhor et al.[145] reported that only 69% of the total chromosomal proteins were dissociated in 2 *M* NaCl, among them all the histones and all the chromosomal RNA, but only a portion of the NHCP. A considerable spectrum of NHCP was found to remain attached to the DNA under these conditions. This parallels the results of Ohlenbusch et al.[282] and Fambrough and Bonner[283] on the dissociation of calf thymus and pea bud nucleohistones, respectively.

Chaudhuri[148] has examined different concentrations of NaCl (2, 3, and 4 *M*) and urea (5 and 7 *M*) either alone or in combination in an attempt to achieve maximal dissociation of chromatin proteins from DNA and also to keep the extraction medium as mild as possible. Urea alone did not remove any significant amount of protein but 3 *M* NaCl-5 *M* urea removed 72 to 80% of chromatin proteins. Solution of 3 *M* NaCl-7 *M* urea-1 m*M* 2-ME-10 m*M* TrisCl, pH 8.3, however, has dissociated 91 to 94% of these proteins. Ultracentrifugation has removed more than 99% of DNA from the protein solution.

Solubilization of chromatin following the Levy et al. procedure[169] and its modifications (Table 2, No. 15) also leads to dissociation of 90 to 95% of chromatin proteins from DNA. Similar values, 90 and 93%, respectively, have been obtained in the systems of Arnold and Young[221] (Table 2, No. 14) and Yoshida et al.[68] (Table 3c). Monahan and Hall[93] (Table 2, No. 8) have reported even higher protein recovery values (99.9%), attributing them not

Table 2
DISSOCIATION MEDIA USED TO SEPARATE TOTAL CHROMATIN PROTEINS FROM DNA BEFORE ULTRACENTRIFUGATION

No.	Dissociation media	Tissues and ref.
1	1% SDS	Sea urchin sperm[141]
2	3% SDS-0.14 *M* 2-ME	Rabbit heart, liver, and skeletal muscle[97]
3	2 *M* NaCl-5 *M* urea-10 *M* TrisCl, pH 7.5,[201] pH 8.0,[62,145,202] pH 8.2[203] or pH 8.3;[57,146,204-206] also with: 1 m*M* 2-ME[57,62,206] and 1 m*M* EDTA[57]	Chicken erythrocytes and reticulocytes;[201] pea cotyledon;[145] rabbit liver,[62,202,204] submandibular salivary glands,[62] bone marrow, kidney, and thymus;[204] calf thymus;[146] rat liver,[57,203,205] kidney and thymus;[205] hen oviduct[206]
4	2.5 *M* NaCl-10 m*M* TrisCl, pH 8.2	Rat liver and thymus[207,208]
5	3 *M* NaCl-10 m*M* TrisCl, pH 8.0	Beef liver[209]
6	4 *M* NaCl-10 m*M* TrisCl, pH 8.0	Rat liver[210,211]
7	2.5 *M* NaCl-5 *M* urea-10 m*M* TrisCl, pH 7.2	HeLa cells[212]
8	3 *M* NaCl-5 *M* urea-10 m*M* TrisCl, pH 8.0[93] or pH 8.3;[148,213-217] also with: 2 m*M* 2-ME-2 m*M* EDTA[93]	HeLa S3;[148,213,215,216] WI-38 human diploid fibroblasts;[214,216] rat liver[217]
9	3 *M* NaCl-7 *M* urea	Epithelial cheek carcinoma KB cells[93] pig cerebellum and pituitary;[174] rat heart[218]
10	3 *M* NaCl-7 *M* urea in: 10 m*M* Na-acetate, pH 6.0;[36] 1 m*M* 2-ME-10 TrisCl, pH 8.3;[148] 50 m*M* TrisCl, pH 7.0[219] or pH 8.0[36,107]	Novikoff hepatoma;[36] mouse liver and primary hepatocellular carcinoma;[219] baby hamster kidney normal and DMN-transformed[107]
11	3 *M* NaCl-7 *M* urea-0.5 m*M* aprotinin-0.2 *M* TrisCl, pH 8.0 in 30% CsCl	Mouse liver and epidermis[220]
12	60% HCOOH-0.2 *M* NaCl-8 *M* urea	Rat liver[81]
13	27 or 37% GdnCl-0.2% DTT-18 m*M* Na-pyrophosphate, pH 8.0	Sea urchin embryo[163]
14	2.5 *M* GdnCl-0.1 *M* TrisCl, pH 8.0	Rat liver[221]
15	0.4 *M* GdnCl-6 *M* urea-0.1% 2-ME in: 0.1 *M* Na-phosphate, pH 7.0[157,169,186,222] or 10 m*M* TrisCl, pH 8.3[60]	Rabbit livers;[169] guinea pig lymphoid cells;[222] duck erythroid cells;[157] rat liver,[186] Burkitt hybrid cells[60]
16	0.4 *M* GdnCl-6 *M* urea-10 m*M* NaCl-10 m*M* TrisCl, pH 7.2	HeLa S3[223]
17	0.4 *M* GdnCl-6 *M* urea-0.83% 2-ME-5 m*M* Na-phosphate, pH 7.0	Baby hamster kidney normal and DMN-transformed[107]
18	0.4 *M* GdnCl-6 *M* urea-10 m*M* EDTA-10 m*M* TrisCl, pH 6.7	Rat liver[224]
19	4 *M* CsCl-10 m*M* 2-Me-5 m*M* EDTA-20 m*M* lysine, pH 11.5	Ovariectomized rat uterus and calf endometrium[189]
20	3.38 *M* CsCl-4 *M* urea-50 m*M* 2-ME-2 m*M* EDTA-10 m*M* lysine, pH 11.6	Chang's hepatoma;[77,225] rat brain,[226] liver, cerebral cortex, cerebellum, mixture of rat midbrain and brainstem;[77] chicken embryo[77]

only to the salt and urea concentration but also to the presence of 2 m*M* EDTA and 2-ME. In contrast to the Levy et al. procedure[169] which does not separate the RNA from the NHCP fraction, these authors were able to fractionate small samples of chromatin into its four principal components (DNA, RNA, histones, and NHCP).

In light of the above-mentioned solvents that are likely to denature proteins, many early efforts should be mentioned which employ only high NaCl concentrations for the dissociation of chromatin (Table 2), deoxynucleohistone precipitation (Table 3a, b), gel filtration, HAP-chromatography (Table 4a), and PEG-dextran partition (Table 5b). Dissociation of total chromatin at high salt concentrations is one of the most important methods for separation chromatin proteins from DNA under nondenaturing conditions (Tables 2, 3, 4, and 5), with

Table 3
DISSOCIATION MEDIA USED IN THE ISOLATION OF NHCP FROM THE WHOLE CHROMATIN BEFORE DNH OR DNA PRECIPITATION

No.	Dissociation media	Tissues and ref.
	(a) Before DNH Precipitation at 0.14 *M* or 0.15 *M* NaCl	
1	1 *M* NaCl;[176,227-234] also in: 10 m*M* Na-citrate;[225] 10 m*M* TrisCl, pH 8.0;[207,208] 20 m*M* TrisCl, pH 7.5[235] or 1 m*M* DTT-20 m*M* TrisCl, pH 7.7[99]	Rat liver[176,207,208,228] and thymus;[207,208] chicken erythrocytes;[229] calf thymus;[227,230] dog liver;[231,232] osteogenic mouse sarcoma;[233] pig liver and kidney;[234] Chang's hepatoma;[225] Ehrlich ascites tumor;[99] Yaffe L_6 line of myoblasts[235]
2	1 *M* KCl	Calf thymus[236]
3	2 *M* NaCl	Chicken erythrocytes[229]
4	2 *M* NaCl in: 20 m*M* TrisCl, pH 8.0,[237,238] or 50 m*M* TrisCl, pH 8.0[239]	Calf thymus;[237] Ehrlich ascites tumor[238,239]
5	2 *M* NaCl-2 m*M* EDTA-50 m*M* TrisCl, pH 7.0	Sea urchin sperm and embryo[240]
6	2 *M* NaCl-5 *M* urea;[241] also in: 10 m*M* TrisCl, pH 7.5[243,244] or pH 8.0;[242] 50 m*M* TrisCl, pH 7.5[244]	Sea urchin embryo;[241] rat liver[242-244]
	(b) Before DNH Precipitation at 0.4 *M* NaCl	
7	1 *M* NaCl-20 m*M* TrisCl, pH 7.5[245-248] or pH 8.0[249]	Rat liver;[245,246] normal and SV-transformed WI-38 human diploid fibroblasts;[247] chick embryo fibroblasts;[248] calf thymus[249]
8	2 *M* NaCl-0.7 m*M* Na-phosphate, pH 7.0	Rat liver[250]
	(c) Before DNA Precipitation with $LaCl_3$ (13.5 m*M*)	
9	2 *M* NaCl-5 *M* urea-50 m*M*[251] or 0.1 *M*[68] TrisCl, pH 7.9	Calf thymus[68,251]

yields amounting 83.3 ± 11.1% in the system described by Sevall et al.[210,211] (Table 2, No. 6).

In context with chromatin dissociation by 2 *M* NaCl,[282] the suggestion of Sonnenbichler and co-workers[284] should be noted: sedimentation of chromatin proteins with DNA in 2 *M* NaCl is not due to their specific binding to DNA but represents primarily a co-sedimentation phenomenon, caused by the aggregation of proteins with the highest tendency to aggregate, i.e., predominantly the arginine-rich histones, and, to a lesser extent, the slightly lysine-rich histones and some nonhistone proteins. Centrifugation of chromatin proteins in the absence of DNA yielded essentially the same amount of protein in the sediment as that found after centrifugation of chromatin. When chromatin was centrifuged in 2 *M* NaCl plus 5 *M* urea, only very small amounts of protein were pelleted, if any at all. This was interpreted to indicate that 5 *M* urea prevented aggregation. However, according to these authors,[284] the prevention of aggregate formation was not the only effect of urea on the sedimentation behavior of proteins or protein-DNA mixtures. The addition of 5 *M* urea to 2 *M* NaCl increased the viscosity of the solution by a factor of 1.47 and the density increased from 1.077 (2 *M* NaCl) to 1.151. The increase in viscosity influenced the sedimentation behavior of proteins and DNA equally, whereas the increase in density had more effect on the sedimentation of proteins than on that of DNA, thus favoring a better separation of proteins from DNA.

Additional to these considerations may be the report of Bidney and Reeck[285] on the effectiveness of NaCl, at concentrations of 0.15 to 3 *M* to dissociate NHCP of an established

Table 4
DISSOCIATION MEDIA USED TO SEPARATE TOTAL CHROMATIN PROTEINS FROM DNA BY CHROMATOGRAPHY

No.	Dissociation media	Tissues and ref.
	(a) Before Gel Filtration	
1	1 % SDS (Bio-Gel® A-150)	Sea urchin sperm[141]
2	3 *M* NaCl;[157-161,174] also in: 10 m*M* TrisCl, pH 7.9[252] or pH 8.0[30] (Bio-Gel® A-50 m)	Pig cerebellum and pituitary;[174] chick embryo brain;[161] duck erythroid cells;[157] rat liver[30,252]
3	3 *M* NaCl-5 *M* urea in: 10 m*M* TrisCl, pH 8.0[253] or 5 m*M* 2-ME-10 m*M* TrisCl, pH 8.3[254] (Bio-Gel® A-50)	Pea embryonic axis bud and cotyledon;[253] seeds of conifers[254]
4	27 or 37% GdnCl-0.5 mg% DTT-18 m*M* sodium pyrophosphate, pH 8.0 (Sepharose® 4B)	Sea urchin embryo[163]
5	4 *M* GdnCl-5 m*M* DTT-5 m*M* TrisCl, pH 8.0 (Bio-Gel® A 5m)	Rat Walker 256 carcinosarcoma[159]
	(b) Before HAP-Chromatography	
6	2 *M* NaCl-5 *M* urea in: 1 m*M* sodium phosphate, pH 6.8[103,110] or pH 8.0;[244] 2 m*M* TrisCl-1 m*M* sodium phosphate, pH 6.8[161] or 0.5 m*M* aprotinin-2 m*M* TrisCl-1 m*M* sodium phosphate, pH 6.8[218]	Brain, kidney and liver of calf[103] and mouse;[103,162] calf thymus and duck erythrocytes;[103] Friend 707 cells;[162] mouse spleen, [103] liver and epidermis;[218] rat liver,[110,244] brain, kidney and Landschültz ascites cell[110]
7	2 *M* NaCl-8% glycerol-5 m*M* sodium phosphate, pH 6.8	Rat leukemia cells DBLA-6[100] and liver[59,256]
8	2.5 *M* NaCl in: 1 m*M* sodium phosphate, pH 7.0[207,208] or 10 m*M* TrisCl, pH 7.0[255]	Rat liver and thymus[207,208,255]
9	3 *M* KCl-10 m*M* potassium phosphate, pH 6.8	Calf thymus[236]

line of rat hepatoma cells labeled in vivo with [^{3}H]tryptophan (proteins), [Me-^{3}H]thymidine (DNA) and [5-^{3}H]uridine (RNA). The percentage of released nonhistones increased with rising NaCl concentrations up to 0.75 *M*; 1 and 3 *M* NaCl were not significantly more effective. A maximum of only 50% of nonhistone protein was recovered free of DNA. The release of nonhistones (amount and electrophoretic patterns) from sheared and unsheared chromatin was similar. Higher recoveries of NaCl-dissociated nonhistones were obtained with sucrose gradient centrifugation than with centrifugation in the absence of sucrose. This may be due to the fact that sucrose is known to minimize protein-protein interactions.[286] These authors[285] concluded that sodium chloride is an attractive dissociating agent, which unlike many other salts, has little effect even at high concentrations on the thermal stability of proteins. It is judged to have only little effect on hydrophobic protein interactions and therefore is unlikely to denature proteins. Two major conclusions may be drawn from their results: (1) since 0.75 *M* NaCl seems to be just as effective as any higher NaCl concentrations in releasing nonhistones, and because DNA can be much more easily separated from the dissociated proteins by centrifugation in 0.75 *M* than at higher concentrations, the 0.75 *M* NaCl is preferable to higher NaCl concentrations for dissociating most nonhistones and (2) since the release of nonhistones from sheared and unsheared chromatins is similar and because it is easier to separate the DNA of unsheared chromatin from dissociated proteins, unsheared chromatin preparations should be used.

Table 5
DISSOCIATION MEDIA USED TO SEPARATE NHCP OR CHROMATIN FROM DNA BY TWO-PHASE PARTITION SYSTEMS

No.	Dissociation media	Tissues and ref.
	(a) Two-Phase Water-Phenol System	
1	Concerning dehistonized chromatin or nuclear residue after extraction of the 0.14 *M* NaCl-soluble proteins, histones, and lipids[48,175,186,257-272] Phenol saturated with: 0.14 *M* 2-ME-10 m*M* EDTA-0.1 *M* TrisCl, pH 8.2;[270-272] pH 8.4[48,175,186,257-269] or pH 9.4,[259] i.e., with buffer for suspending chromatin or nuclei	Rat brain and spleen,[258] kidney,[175,258] liver,[175,186,257,258,260,262,273] and embryo;[263] adrenalectomized rat heart, kidney and liver;[266] pig lymph nodes;[267] Syrian hamster cardiomyopathic skeletal muscle;[265] Syrian hamster liver and Kirkman-Robbins hepatoma;[269] goose liver, mature erythrocytes, and regenerating erythroid cells;[259] duck adult erythrocytes and embryo erythroid cells;[261] normal and leukemic human leukocytes;[264] chicken erythroblasts, reticulocytes,[268] erythrocytes,[48,268] and thrombocytes;[48] *Physarum polycephalum*;[270,271] and *Acanthamoeba castellani*[272]
2	Concerning nuclear residue after extraction of 0.14 *M* NaCl-soluble proteins[273] Phenol saturated with: 4 *M* urea-10 m*M* sodium phosphate, pH 7.0-6 m*M* NaF-0.1 m*M* PMSF Nuclear residue suspended in: 1% SDS-0.15 *M* 2-ME-10 m*M* sodium phosphate, pH 7.0	
	(b) Two-Phase PEG-Dextran System	
3	5 *M* NaCl-15 m*M* 2-ME-10 m*M* TrisCl, pH 8.1	Wheat germ[274]
4	2 *M* NaCl-8% glycerol-1% DMSO-5 m*M* sodium phosphate, pH 7.9-1 m*M* PMSF	Leukemia DBLA-6;[100] rat liver[59,256]

C. DNA Removal

1. Common Procedures for DNA Removal from Dehistonized and Whole Chromatin

a. Ultracentrifugation

From dissociated dehistonized (Table 1) or whole chromatin (Table 2), DNA is most frequently removed by prolonged high speed sedimentation, e.g., for 48 hr at 85,000 to 88,000 × g,[174,214] 100,000 × g,[169] 125,000 × g,[77,225,226] 150,000 × g,[215,216] or 180,000 × g;[255] for 44 hr at 100,000 × g;[107] for 43 hr at 120,000 × g;[220] for 36 hr at 200,000 × g;[187] for 24 hr at 85,000 × g,[141] 165,000 × g,[203] 200,000 × g,[93,185] 269,000 × g;[205] for 20 hr at 215,000 × g;[210,211,287] for 15 hr at 300,000 × g;[57] for 12 hr at 408,000 × g[212] or even for 10 hr at 485,000 × g and less than 10 hr at 503,000 × g.[223]

Separation of DNA from NHCP by equilibrium density centrifugation (Table 1, No. 5) belongs to the earliest methods.[188,189] Teng and Hamilton[189] have employed 4 *M* CsCl in lysine buffer, pH 11.5, to separate the NHCP from acid-dehistonized chromatin, or total chromosomal proteins from whole chromatin (Table 2, No. 19). In the experiments of Fujitani and Holoubek[77,225,226] (Table 2, No. 20) only 3 to 5% of the proteins sedimented with DNA as a firm pellet.

The disadvantage of DNA separation by ultracentrifugation of chromatin dissociated at high salt and urea concentrations lies in the fact that it takes a long time to pellet the DNA. Moreover, this must be followed by prolonged dialysis to remove salt and urea before the chromatin proteins can be concentrated. The absence of urea in the NaCl solvent or the use of GdnCl decreases the time and speed of ultracentrifugation for pelleting nucleic acids.[5]

b. SDS-PAGE in the Presence of DNA

As was already mentioned, most of the methods of DNA and NHCP separation require several days or weeks of preparative work. For this reason, SDS-electrophoresis of dehistonized or whole chromatin without prior removal of DNA was introduced almost simultaneously by three groups of investigators.[108,183,277]

In the experiments of Smith and Chae,[183] dehistonized or whole chromatin was dissociated in buffered SDS-2-ME and dialyzed against this solution at 37°C for 16 to 18 hr before SDS-PAGE.[140] It was demonstrated that in the presence of DNA, which did not enter into the gel, the electrophoretic patterns obtained by this technique were the same as the patterns of chromatin proteins from which the DNA was removed, e.g., by prolonged high-speed centrifugation. This method is simple, rapid, yields good resolution of the chromatin proteins, and requires only small amounts of chromatin for analysis. It has found many applications with some modifications concerning mainly the composition of chromatin dissociation buffers.[46,56,65,66,72,73,115,278,279]

2. Additional Procedures for DNA Removal from Dehistonized Chromatin

a. DNase Digestion

Wilson and Spelsberg[301] introduced DNase treatment for the DNA removal from NHCP. By three simple steps, i.e., chromatin dehistonization with acid (0.2 *M* HCl), short incubation with DNase I (25 μg DNase per 1 mg chromatin) and acid (0.4 *M* $HClO_4$) precipitation, the bulk of NHCP could be quickly and quantitatively separated from DNA, with more than 95% yield.

b. Two-Phase Partition System: Water-Phenol

Removal of DNA can be also achieved in a two-phase partition system. The first example of this system was the isolation of phenol-soluble nonhistone proteins by removing the DNA in aqueous phase. This procedure, introduced by Steele and Busch,[288] became the subject of numerous modifications (Table 5a).

c. Analytical Chemical Separation of DNA

For studies where the objective is limited to analytical determinations, chemical composition, specific radioactivity, etc., the nucleic acids in acid-dehistonized chromatin can be hydrolyzed by digestion with 5 to 10% trichloroacetic acid (TCA)[5,111] or 0.5 *M* perchloric acid (at 90°C for 15 min)[5] and the insoluble residue, representing the NHCP, is solubilized in alkali (e.g., 0.2 *M* NaOH)[111] or various solubilizers used in liquid scintillation counting. The proteins thus obtained are completely denatured and cannot be used for most biochemical investigations.[5] Using this method, the NHCP have been reported to vary quantitatively and metabolically during development and chromatin fractionation.[290,291] Chanda and Cherian[292] have shown mercury-binding proteins in HCl- and TCA-insoluble residue of chromatin.

3. Additional Procedures for DNA Removal from Whole Chromatin

In the second group of NHCP isolation methods (Figure 3), the following techniques are used for DNA removal: (a) gel filtration, (b) precipitation of DNA with histones at low ionic strength, (c) precipitation of DNA with lanthanum chloride, (d) chromatography on hydroxylapatite, and (e) a two-phase partition system: polyethylene glycol-dextran.

a. Gel Filtration

Since the first molecular weight sieving of NHCP, performed by Shirey and Huang[141] (Table 4, No. 1) several modifications of this technique have been introduced (Table 4a). According to Shaw and Huang,[174] this procedure is capable of removing 90 to 96% of chromatin proteins from the DNA. A disadvantage of the gel filtration method lies in the fact that shearing of chromatin is required for the DNA to enter the resin.[30,174] This results in the formation of a small quantity of aggregated chromatin (approximately 5 to 10%) which needs to be centrifuged out of solution before exclusion chromatography.[174] Application of unsheared chromatin to the BioGel® column results in no separation of chromosomal components.[30] In the experiments of Graziano and Huang,[161] the sheared, salt-dissociated chromatin (Table 4, No. 2) was fractionated into a nucleic acid pool containing an average of 66.6% of the applied DNA and 13.2% of the applied protein, and into a protein pool containing an average of 72.1% of the applied protein and of 5.8% of the applied DNA. Similarly, the protein pool resulting from exclusion chromatography performed by Van den Broek et al.[30] (Table 4, No. 3) contained 84.3% of the total protein content of sheared chromatin.

Chromatin dissociated into its macromolecular components in GdnCl was also separated by gel filtration (Table 4, No. 4 and 5). In the experiments of Tuan et al.[159] about 90 to 95% of chromatin DNA was recovered in the first peak, together with 30 to 40% of NHCP and approximately 50% of chromatin RNA; 60 to 70% of NHCP and all the histones were recovered in the second and third peaks. Hence, it is obvious that the cross-contamination of chromosomal proteins by nucleic acids and nucleic acid by proteins, the low ratio of sample/gel-bed required for satisfactory resolution of macromolecules by gel filtration, and the slow flow rates severely limit the use of molecular weight sieving for large-scale isolation of NHCP.[5]

b. Removal of DNA by Precipitation of the Deoxyribonucleohistone

It has long been known that the solubility of deoxyribonucleohistone (DNH) is minimal at 0.14 *M* NaCl and increases above and below this value.[129] Patel and Wang[227] and Wang[176] were first to exploit this phenomenon for the removal of DNA. They found that nonhistone proteins are released from the DNP complex when it is solubilized in 1 *M* NaCl, at which ionic concentration the histones are not completely dissociated from DNA.[225,282,293] Rapid lowering of the salt concentration to 0.14 *M* NaCl results in precipitation of reconstituted DNH leaving a significant amount of the soluble NHCP in supernatant. Generally, Wang's technique yields only a part of the total NHCP pool, contaminated with nucleic acids. However, it is frequently used by many investigators (Table 3a). According to Hacha and Fredericq[236] the use of a denaturing agent, after dissociation of nucleohistone in 1 *M* KCl and reconstitution by dialysis in 0.15 *M* KCl, resulted in a supernatant containing 10 to 13% proteins (partly acidic) and 4 to 10% DNA.

The separation of NHCP by precipitation of DNH at 0.4 *M* NaCl has been employed by Langan[245] for the isolation of phosphoproteins. This principal idea of partial precipitation of DNH after its rapid reconstitution by dilution to 0.4 *M* NaCl from solution in concentrated NaCl, and dehistonization by Bio-Rex® 70 resin of the nonhistone proteins remaining in supernatant, has been utilized by Kamiyama and Wang[246] and subsequently by other investigators (Table 3b) for NHCP isolation.

c. Precipitation of DNA with $LaCl_3$

Yoshida and Shimura[251] have precipitated nucleic acids of the dissociated chromatin (Table 3c) by addition of $LaCl_3$ at concentration 13.5 m*M*. They have shown that the selectivity of nucleic acid precipitation depends on the presence of both concentrated NaCl and urea, $LaCl_3$ concentration, and pH. Under optimal conditions (500 μg/mℓ DNA, pH 7.9, and higher than 8 m*M* concentration of $LaCl_3$) no protein was precipitated and the supernatant had a typical protein absorption spectrum (A_{275} to A_{260} ratio 1.66). This procedure has the advantage of rapid separation of nucleic acids from protein at very high yields (about 93%). Recently, this method was applied by Yoshida et al.[160] to chromatin depleted of 0.35 *M* NaCl-soluble proteins.

d. Chromatography on Hydroxylapatite

MacGillivray et al.[103,295] and others[294] have introduced one-step separation of three principal classes of chromatin macromolecules (histones, NHCP, and nucleic acids) by column chromatography on hydroxylapatite (HAP). In this approach, the dissociated chromatin (Table 4b) is applied to HAP columns; the nucleic acids and NHCP are retained. Differential elution of various macromolecules is effected by stepwise increase of phosphate buffer concentration in the elution solvent. At 1 m*M* phosphate, histones, nonadsorbed to HAP, appear in the break-through peak. The bulk of NHCP is eluted by 50 m*M* phosphate buffer; additional, smaller amounts of nonhistones are eluted along with RNA at higher phosphate concentrations (0.1 and 0.2 *M*). Virtually all the DNA applied to the column is eluted by 0.5 *M* phosphate buffer. Depending on the method of chromatin preparation, 41 to 100% of total chromatin proteins can be recovered by this technique. While not completely eliminating interactions between histones and NHCP and also between NHCP and nucleic acids, this approach has the advantage of separating all the chromatin macromolecules in a single operation and for this reason it is widely used by many investigators (Table 4b).

e. Polyethylene Glycol-Dextran Two-Phase Partition System

Aqueous solution containing substantial concentrations of two water-soluble polymers will often separate into two distinct phases, each rich in one of the two polymeric components, yet both containing 80 to 95% of water. Such phase separation system may be used for gentle fractionation of biological materials, since its highly aqueous nature allows retention of biological activity.[296] By only a few manual extractions it enables large-scale fractionation of nucleic acids on the basis of their degree of double-helical structure, as well as a rapid separation of DNA from RNA and gentle removal of proteins from either nucleic acid.

Following the experiment of Turner and Hancock,[297] Simon and Becker[274] developed a nondenaturing procedure for the separation of salt-dissociated chromosomal proteins from nucleic acids, mainly DNA. Chromatin is solubilized in high salt (5 *M* NaCl) and mixed with 0.35 volume of 30% PEG, followed by the addition of 0.3 volume of 20% dextran. These empirically determined conditions of PEG, dextran, and NaCl concentrations lead, after stirring (30 min) and low-speed centrifugation (2000 × g, 10 min) to the recovery of most of the chromatin proteins (85 and 65% of the labeled histones and NHCP, respectively) in the upper (PEG-rich) phase, yet contaminated with less than 1% of the initially present DNA. This separation technique not only avoids acid extraction and denaturing conditions, but also does not require time-consuming centrifugation or chromatographic steps. It is rapid, gentle, and provides a complementary alternative to other procedures[176,245] with which it compares favorably, especially with respect to efficiency of nucleic acid removal, time requirement, and applicability to large-scale isolation. In addition, it has been reported that the polymers may stabilize enzyme activity.[298]

The PEG-dextran two-phase partition system has been also used by Kikuchi and Sato[100] for the separation of chromosomal proteins from DNA in nuclei sonicated after their dissociation (Table 5b). In the presence of 0.5 volume of 20% dextran and 0.5 volume 30% PEG in the above buffer, pH 7.9, nuclear proteins were effectively partitionated into the upper (PEG-rich phase) before two-step chromatography on HAP. More recently, the method of Kikuchi and Sato[100] has been applied by Kilianska et al.[59,256] to the removal of the nucleic acids before four-step HAP chromatography of nuclear[59] and chromosomal proteins[256] isolated from the whole chromatin and chromatin devoid of the 0.35 *M* NaCl-soluble fraction.

D. Separation of NHCP from Histones

The removal of histones which results in dehistonized chromatin in the first group of NHCP isolation method (Figure 3) is most often achieved by dilute mineral acid, e.g., HCl at concentrations: 0.1 *M*,[188] 0.2 *M*,[134] 0.25 *M*[74,189] or 0.4 *M*;[51] or H_2SO_4 at similar concentrations: 0.1 *M*,[66,114] 0.125 *M*,[80,108] 0.2 *M*[46,73,115,185,186,299,300] or even 0.4 *M*.[93] Readily soluble histones are rapidly separated by low-speed centrifugation from the insoluble material containing nucleic acids and NHCP. However, exposure of chromatin to dilute mineral acids denatures most of its components, thus limiting the usefulness of this relatively simple method.[5] Indeed, examination (by immunochemistry, circular dichroism, and the ability of the DNA to serve as a template in the in vitro DNA-dependent RNA synthesis), of the effects of histone removal by acid has demonstrated that this procedure induces marked conformational alterations in the proteins and DNA, and damages the NHCP antigenic sites.[299] Therefore, other methods are recommended for removing histones, e.g., high salt or salt and urea.[193-196] This holds especially true for studies on biological interactions of NHCP, regardless of the fact that the dissociation methods result in quantitative and qualitative loss of NHCP when compared with the acid removal of histones.[301]

Many studies seeking milder conditions for histone removal have been undertaken. Hnilica and co-workers[194,195] have shown that histones can be released (along with some of the NHCP and RNA) from chromatin with 2 *M* NaCl-5 *M* urea-50 m*M* Na acetate, pH 6.0. Selectivity of histone extraction depends critically upon the pH, since at higher pH all macromolecular components of chromatin begin to dissociate.[195] In a modification of this method, histones have been also extracted by 2.5 *M* NaCl-5 *M* urea-50 m*M* Na succinate, pH 5.0[193,197] or 2 *M* NaCl-5 *M* urea-10 m*M* sodium phosphate buffer, pH 6.8.[196,302] This last kind of chromatin dehistonization enhanced its immunospecificity[303] and for this reason has been used in preparing NHCP-DNA complexes collected after ultracentrifugation (e.g., at 100,000 × g for 36 hr) and resuspended in 2 m*M* TrisCl buffer, pH 7.5, for immunization.[302-304] In the experiments of Adolph et al.,[305] dextran sulfate and heparin were used to dehistonize the chromosomes rather than 2 *M* NaCl. The use of 1 *M* $CaCl_2$ for histone removal[190] prior to solubilization of the residual NHCP from slime mold *Physarum polycephalum*[191] and from the brain tissue of three species (rabbit, guinea pig, and rat)[192] by the modified Benjamin and Gelhorn[188] method should be also mentioned.

In the report of Stein et al.,[215] histones were selectively dissociated from chromatin by the method of Smart and Bonner[306] using sodium deoxycholate. The order of histone extraction with this ionic detergent (0.005 to 0.1 *M*) was: slightly lysine-rich, arginine-rich, and last, the very lysine-rich. Sodium deoxycholate was ten times as effective as sodium perchlorate, which was about twice as effective as NaCl or GdnCl.

A different method of NHCP separation from histones was used by Fujitani and Holoubek.[226] The majority of chromosomal proteins (95 to 97%) in the supernatant after pelleting the DNA by ultracentrifugation, was precipitated by addition of TCA to 10% final concentration. The precipitate, dried with ethanol and ether after the removal of histones with 0.2 *M* HCl, represented total NHCP, which were analyzed by PAGE. A similar extraction of histones has been applied by Teng and Hamilton[189] to total chromatin proteins separated from DNA by equilibrium density centrifugation.

Milder conditions for the separation of NHCP from histones can be achieved by: (1) ion-exchange chromatography, including chromatography on hydroxylapatite, (2) nucleohistone separation, and (3) electrophoretic procedures (Figure 3).

1. Cation and Anion Exchange Chromatography

Following the separation of total chromatin proteins from DNA, the NHCP can be separated from histones by chromatography on a variety of cation and anion exchange resins. Since histones and nonhistones form insoluble complexes at low ionic strength[176,228] their chromatographic separation is usually performed in the presence of urea (5 *M*). The elution is generally achieved with a salt gradient. The order of acidic and basic protein elution depends mostly on the choice of resin. To elute nonhistone proteins first in the flow-through fraction or at low ionic conditions, cation exchange resins have been frequently employed (Table

Table 6
CATION-EXCHANGE RESINS USED FOR THE SEPARATION OF NHCP FROM HISTONES

No.	Resin	Running and NHCP eluting buffer with ref.
1	SP-Sephadex® C-25	7 *M* urea-0.23 *M* NaCl-10 m*M* soldium acetate, pH 5.2[161,201,252]
2	SE-Sephadex® C-25	25% formic acid-10 *M* urea[81]
3		5 *M* urea-50 m*M* NaCl-10 m*M* sodium acetate, pH 5.0; concentration gradient of NaCl to 0.6 *M*[221]
4	CM-Sephadex® C-50	0.37 *M*[163] or 0.4 *M*[159] GdnCl-6 *M* urea-10 m*M* sodium pyrophosphate, pH 7.8
5	CM-cellulose	5 *M* urea in: 10 m*M* TrisCl, pH 7.9[251] or pH 8.3[147]
6	Amberlite® CG-50	0.5 *M* NaCl in: 10 m*M* sodium phosphate, pH 7.2;[207,208] 50 m*M* sodium phosphate, pH 7.2[255] or pH 5.7[203]
7	Bio-Rex® 70	0.35 *M*[186,222] or 0.4 *M*[157,169,210,217] GdnCl-6 *M* urea-0.1% 2-ME-0.1 *M* sodium phosphate, pH 7.0
8		0.3 *M* GdnCl-5 *M* urea-6 m*M* 2-ME-0.1 *M* sodium phosphate, pH 7.0[54,275]
9		0.4 *M* GdnCl-6 *M* urea-20 m*M* sodium phosphate, pH 6.1[212,223]
10		0.35 *M* NaCl-5 *M* urea-10 m*M* Tris-acetate, pH 7.2[253]
11		50 m*M* NaCl-5-*M* urea-0.5 m*M* $MgCl_2$-1 m*M* 2-ME-1 m*M* EDTA-10 m*M* TrisCl, pH 8.3[205]
12		50 m*M* NaCl-15 m*M* 2-ME-50 m*M* sodium phosphate, pH 7.0[274]
13		0.4 *M* NaCl in: 10 m*M* TrisCl, pH 7.0,[30,38,209] 20 m*M* TrisCl, pH 7.0[58,293] or 15 m*M* sodium borate, pH 7.5[250]

Table 7
ANION-EXCHANGE RESINS USED FOR THE SEPARATION OF NHCP FROM HISTONES

No.	Resin	Loading buffer with ref.	Salt in NHCP eluting buffer
1	DEAE-cellulose	1 m*M* EDTA-1 m*M* 2-ME-15 m*M* TrisCl, pH 8.0[246]	0.15 *M* and 0.3*M* NaCl
2		5 *M* urea-10 m*M* TrisCl, pH 8.0[253]	0.4 *M* NaCl
3	DE 52 DEAE-cellulose	5 *M* urea-2 m*M* EDTA-2 m*M* 2-ME-10 m*M* TrisCl, pH 8.0[93]	Salt gradient: 0.0 *M* to 1 *M* NaCl
4	QAE-Sephadex® A-50	5 *M*[147,204] or 8 *M*[221] urea-10 m*M* TrisCl, pH 8.3	2 *M* NaCl
5	A-50	5 *M* urea-10 m*M* TrisCl, pH 8.3,[148,213,214] also with 1 m*M* 2-ME[148,218]	3 *M* NaCl
6	QAE-Sephadex® A-25	8 *M* urea-50 m*M* sodium phosphate, pH 7.6[276]	Salt gradient: 0.1 *M* to 3.0 *M* NaCl

6). Conversely, the NHCP can be retained by anion exchange resins (Table 7) at low ionic conditions and then eluted by increasing ionic strength (after washing the column with the starting solvent to remove unadsorbed histones). Trace amounts of nucleic acids that may contaminate the proteins can be also separated by manipulating the ionic strength of elution solvents.[5]

Because of the high affinity of the basic amino groups of lysine and arginine for negatively charged resins, Graziano and Huang[161] have employed a strong sulfophenyl ion-exchanger, i.e., SP-Sephadex® C-25, not only for the separation of histones from other chromatin proteins but also for fractionation of NHCP. Following application of the dissociated chromatin protein pool to the equilibrated column (Table 6, No. 1), the proteins were eluted with four 7-*M*-urea-10-m*M* Na acetate buffers, pH 5.2, containing 0.23, 0.3, 0.4, and 0.8 *M* NaCl, respectively. At least 90% of the total NHCP of chromatin were eluted from the column at 0.23 *M* NaCl. The remaining 10% or less of NHCP were partially recovered (together with histones) at 0.3 *M* and 0.4 *M* NaCl, and could not be separated by rechromatography. The successful separation of these proteins depended upon the protein complement of chromatin being always in solution, never concentrated, and never lyophilized before application to the ion-exchange column. Lyophilization of chromatin proteins before application to the SP-Sephadex® column caused them to aggregate making their subsequent solubilization extremely difficult, if not impossible.[161]

Carboxymethyl Sephadex® C-50 chromatography, initially reported by Hill et al.[163] was utilized in the experiments of Tuan et al.[159] with certain advantages, particularly regarding the recovery. More than 95% of the NHCP were recovered in the first peak (Table 6, No. 3). The temperature of the sample was carefully controlled at 4°C at all times (instead of 20 to 25°C as in the procedure of Hill et al.[163]) which was essential for successful separation. The recovered proteins, after gel filtration (Sepharose® 4B[163] or Bio-Gel® A 5m[159]), can be alkylated with iodoacetate[163] or iodoacetamide[159] thus minimizing the possibility of degradation during further manipulations. Unlike the histones, the NHCP recovered from CM-Sephadex® were not pure in that they contained a small amount of RNA (approximately 3% by weight), some DNA (1%), and considerable amounts of carbohydrates (32%).[159] The nucleic acid contamination could be eliminated by DEAE-cellulose chromatography.[169] According to Levy et al.,[169] elution with equilibrating buffer provided the NHCP at 0.3 *M* NaCl and nucleic acid at 0.6 *M* NaCl. However, in experiment of Tuan et al.,[159] under slightly different conditions, the nucleic acids were eluted between 0.2 *M* and 0.32 *M* NaCl from DEAE-cellulose together with some nonhistones, as revealed by gel electrophoresis.

Arnold and Young[221] have separated NHCP from histones using QAE-Sephadex® batch procedure (after Gilmour and Paul[146]) or chromatography on SE-Sephadex® C-50. Recovery of protein from the batch operation was approximately 90% and from the column usually about 70%. Although the column method was more laborious than the bulk QAE-Sephadex® procedure, this technique gave a more complete separation of the two classes of proteins. The application of SE-Sephadex® C-25 chromatography was reported by Elgin and Bonner[81] (Table 6, No. 2).

Experiments of Umansky et al.[203] on the separation of NHCP from histones, in which three types of resins were used (DEAE- and CM-cellulose, and also Amberlite® CG-50) have shown that NHCP were most successfully recovered (85 to 90% of total NHCP) in the unadsorbed fraction when total chromatin proteins were chromatographed on Amberlite® CG-50 (Table 6, No. 6). Nonhistone proteins obtained by the technique of Umansky et al.[203] were well-soluble at low and high ionic strength. Yoshida and Shimura[251] reported that even in 5 *M* urea (Table 6, No. 5) some of the acidic proteins aggregated to histones and were retained by CM-cellulose.

Five types of resins, three anion exchangers (QAE-Sephadex® A-50, DEAE-Sephadex® A-50, and DEAE-cellulose), and two cation exchangers (Bio-Rex® 70 and CM-cellulose) were tried by Chaudhuri[148] for the separation of [^{3}H] tryptophan-labeled HeLa S_3 chromatin proteins. Following the batch procedure of Gilmour and Paul[204] to remove histones from NHCP, Chaudhuri[148] modified this technique using a QAE-Sephadex® A-50 column to separate the NHCP from histones and then to fractionate them on the same column. In the presence of buffered urea, pH 8.3 (Table 7, No. 5) histones and 10% of NHCP were not retained by the anion exchange column while the rest of NHCP remained tightly bound. The total recovery of NHCP was 62%; 52% was eluted from the QAE-Sephadex® with 5 *M* urea buffer in the presence of 3 *M* NaCl. To improve the recovery, the NHCP were first separated from histones. However, when chromatographed on cation exchange resins it was found that almost 60% of the tryptophan-labeled proteins remained bound to the column, in contrast to the Levy et al.[169] report about complete separation of nonhistone proteins from histones using Bio-Rex® 70 columns (Table 6, No. 7). It should be noted that Richter and

Sekeris[205] also used QAE-Sephadex® A-25 column to fractionate NHCP after an additional step, i.e., elimination of even traces of histones by chromatography on Bio-Rex® 70.

Among the four cation exchangers (CM-cellulose, CM- and SP-Sephadex® and Bio-Rex® 70) which were investigated by Van den Broek et al.,[30] the best results were obtained on the Bio-Rex® 70 column with samples dissolved in buffered 0.4 *M* NaCl (Table 6, No. 13), a procedure similar to that of Langan.[245] Histones, which were retained by this weak cation exchanger, were eluted with buffered 1 *M* NaCl. Several attempts were made by Van den Broek et al.[30] to purify the NHCP from a solution of total chromatin proteins in 5 *M* or 7 *M* urea-10 *M* TrisCl, pH 8.0, using anion exchange resins (DEAE-cellulose, DEAE-Sephadex®, QAE-Sephadex®, AE-cellulose, or Bio-Rex® 5). In all cases a relatively small proportion of the nonhistones was retained, less than 10% of the total protein applied to the resin. The retained protein included substantial amounts of histone, especially in the case of QAE-Sephadex®.

Simon and Becker[274] (Table 6, No. 12) found that dialysis of the chromatin protein-containing PEG-rich phase in the presence of Bio-Rex® 70 to lower the salt concentration (from 5.0 to 0.05 *M* NaCl) resulted in the adsorption of both histones and NHCP to the resin, from which they then could be differentially eluted with a NaCl gradient. Nonhistones eluted from the resin between 0.05 and 0.5 *M* NaCl, while histones came as a sharp peak at salt concentrations exceeding 0.55 *M* NaCl. Careful electrophoretic analysis revealed no detectable cross-contamination between nonhistones and histones eluted from Bio-Rex® 70 in this way. Limited fractionation of the nonhistones was also achieved by this elution protocol.

2. *Separation of NHCP from Histones by Preparative Gel Electrophoresis*

First separation of the chromatin protein constituents by preparative PAGE in Canalco equipment was performed by Shirey and Huang.[141] The total chromatin proteins, released from DNA by SDS and freed of excess SDS by urea dialysis-$BaCl_2$ precipitation procedure,[141] were separated from each other according to their electrophoretic mobilities.[307] Shaw and Huang[174] reported the resolution of NHCP and histone using the urea-polyacrylamide gel system, pH 7.2, of Panyim and Chalkley.[29] After electroporesis of the entire set of gels one gel was stained and used as a template to cut the NHCP region from the other gels which had been quickly frozen. The NHCP containing pieces of gel were minced, homogenized in 10 *M* urea, and after centrifugation and dialysis of the supernatant against distilled water, lyophilized. These two techniques of NHCP separation from histones by preparative PAGE have not become as popular as all the other methods.

Fujitani and Holoubek[77,225] separated NHCP from histones on an analytical scale by a modification of the electrophoretic system described by Fambrough and Bonner.[283] The solution of chromosomal proteins in lysine-urea buffer, pH 11.6, containing 2-ME (50 m*M*) and EDTA (2 m*M*) was acidified to pH 3.0 and electrophoresed in 0.35 *M* β-alanine buffer, pH 4.5.

III. FRACTIONATION OF NHCP ON ANALYTICAL SCALE

Attempts to isolate and fractionate NHCP and to understand their functional and structural roles in chromatin and nucleus revealed their exceptional complexity. The results of various manipulations must be monitored by methods that allow rapid and reproducible examination of individual proteins, often present in minute amounts in highly heterogeneous mixtures.[309] Among the available techniques, zone electrophoresis, depending on a molecular filtration effect, represents a principal way of NHCP analytical-scale fractionation and has contributed considerably to the rapid advancement of biochemistry and biology of this complex class of proteins.

Generally, electrophoresis in acrylamide polymers, which can be prepared in wider range of gel concentration than starch gels, provides an elegant and reproducible method for rapid quantitative and qualitative characterization of nuclear proteins.[310] There are two principal types of zone electrophoresis in polyacrylamide gels: one using differences in electric mobility at a constant pH (PAGE) and the other employing a pH gradient (IFPA). The electrophoresis at a constant pH may be performed at a constant gel concentration or at a gradient gel concentrations, either continuous[311] or discontinuous.[307,312] In the method using continuous gradient gel concentration,[311] protein molecules are driven through a gel of progressively decreasing pore size toward a "pore limit",[313] where they are arranged according to their molecular size.

The "molecular sieve" effect[307,314,315] of polyacrylamide gel is closely related to its structure, i.e., to the concentration of the polymer, which may be varied over wide limits. The "average pore size" appears to be roughly 2, 5, and 15 nm at polyacrylamide concentrations of 20, 7.5, and 3%, respectively.[307] Electrophoretic separation is improved by selecting a suitable concentration of acrylamide to give a pore size optimal to the dimensions of the molecules being separated.[307] However, what is optimal for one protein need not be so for another. Hence, in complex mixtures of components such as total NHCP extract, high resolution of one part of the protein spectrum may only be attained at the expense of another.[315]

Generally PAGE and IFPA techniques may serve three different functions: physical characterization by exploiting differences in the major properties of macromolecules, namely size and net charge, "macromolecular mapping", and isolation of native macromolecular species.[316] This section will be concerned only with the first two aspects of NHCP examination.

A. One-Dimensional Polyacrylamide Gel Electrophoresis

Among many methods of PAGE, some are used more often. These are electrophoresis at a constant pH in denaturing conditions (SDS-PAGE)[139,312] which fractionates proteins on the basis of differences in molecular weight, and electrophoresis at pH gradient in urea-polyacrylamide gel systems (IFPA), in which the separation also reflects the isoelectric point of the proteins.

1. SDS-Polyacrylamide Gel Electrophoresis

Unidimensional gel electrophoresis of NHCP in the presence of SDS is the most popular tool for monitoring and initial characterization of the isolated NHCP or their fractions. It is most often carried out in 5 to 10% polyacrylamide gels (rods or slabs) in two general procedures, i.e., in phosphate-buffered gels (technique of Shapiro et al.[139] or Weber and Osborn[140] and their modifications) or in Tris-glycine buffered gels (the discontinuous system of Laemmli[312] and its modifications). In Laemmli's system, the gel consists of two or three sections, i.e., a large-pore anticonvection gel containing the protein sample, a large-pore spacer gel in which electrophoretic concentration takes place, and a small-pore gel in which electrophoretic separation is accomplished.[317] It should be noted that the double gel technique can also be used in the phosphate-buffered system[139,140] with stacking (2.5%[183,318] or 3%[48,101] acrylamide) gel on top of the separating gel.

Sodium dodecyl sulfate is one of the most powerful agents available for molecular dispersion and electrophoretical separation of proteins.[319-322] It binds to the protein and disperses high molecular weight aggregates to fairly homogeneous and negatively charged SDS-protein-complexes, eliminating virtually all protein-protein interactions. However, elimination of the positive charges on NHCP by SDS makes impossible the electrophoretic separation of differentially acetylated or phosphorylated NHCP.

Following calibration of the sodium phosphate system with SDS-proteins of known molecular weight larger than 10,000 daltons, the techniques of Shapiro et al.[139] or Weber and

Osborn[140] may be used for the rapid and simple estimation of the molecular weights of monomeric proteins or of subunits of oligomeric proteins. As SDS minimizes the native charge differences, all proteins migrate as anions and separate electrophoretically on the basis of molecular size. If the relative distance of migration (abscissa) is plotted against the logarithm of molecular weight (ordinate) a straight line can be fitted from molecular weight 15,500 to 165,000.[139] The study of Weber and Osborn[140] on 40 different proteins (globular and rod-shaped in native state), showing that their molecular weights can be estimated with an accuracy of better than $\pm 10\%$ for polypeptide chains between 15,000 and 100,000 mol wt, bears testimony to the usefulness of this technique described originally by Shapiro et al.[139] There are some deviations from this principle when the SDS-Tris-glycine discontinuous method of Laemmli[312] is used. However, regardless of this disadvantage, this method offers superior resolution of NHCP bands of molecular weight greater than 18,000. Proteins that appeared as single bands by the SDS-phosphate gels are frequently resolved into 3 or 4 bands by the SDS-Tris-glycine gels.[79]

By electrophoretic criteria the nonhistone proteins of eukaryotic chromatin are heterogeneous. However, the extent of their specificity is somewhat controversial (for review see References 3 and 5). More than a decade ago, Elgin et. al. observed that NHCP electrophoretic patterns from different rat tissues were strikingly similar and concluded that the NHCP exhibit only limited heterogeneity and specificity.[80,81,185] At least several dozen distinct NHCP polypeptides (12 to 18 major and at least 20 to 40 minor components[185]) can usually be detected in a single tissue, although the exact number varies with the resolving power of the electrophoretic system employed. The SDS-phosphate[139] and SDS-Tris/glycine[312] polyacrylamide gel systems have enabled these investigators to resolve in rat liver about 21 (14 major and 7 minor)[80,81,185] and 45 bands,[185] respectively. According to the electrophoretic analysis of the middle molecular weight NHCP (40,000 to 100,000) it was possible[185] to categorize the tissues examined in three groups: (1) liver, lung, spleen, and kidney, (2) thyroid and thymus, and (3) brain; i.e., in this region some electrophoretic tissue specificity was observed, especially in brain, characterized by a unique electrophoretic pattern with a preponderance of high molecular NHCP. The limited heterogeneity of NHCP has also been supported by the observations of other investigators.[50,78,94,103,225,323.]

However, this limited heterogeneity may be caused among others, by incomplete dissociation of NHCP by SDS or by freezing of the animal tissues prior to the isolation of nuclei or chromatin. Platz and Kleinsmith[324] have shown that NHCP isolated from frozen tissues have a marked decrease in heterogeneity and loss of high molecular weight components. It should be emphasized that the concept of limited heterogeneity refers only to the most prominent species of NHCP and does not apply to the extensive heterogeneity which may exist in protein species present in small amounts. The data of Platz and Kleinsmith[324] indicate that steps involving selective extraction and concentration of NHPC can reveal their greater heterogeneity. It should be pointed out that essentially all the conclusions concerning NHCP heterogeneity depend on SDS-acrylamide gel electrophoresis. This technique of separation has at least three inherent limitations[80,324] which could result in an underestimation of actual heterogeneity of NHCP:

1. This system separates polypeptide chains solely on the basis of size (molecular weight). If two proteins existed with the same amino acid backbone but with differing side chain modifications they would still band together in this system. This is especially important when studying enzymatically modified proteins which may vary greatly in the number of their modifications.
2. The detergent treatment in the SDS-electrophoresis dissociates any subunit structure which might be present in the native proteins. If subunit structures do exist then the association of subunits in varying combinations would allow for additional heterogeneity.

3. A third drawback in demonstrating heterogeneity by SDS-electrophoresis is that components present in relatively small amounts would most likely go undetected. Minor bands are not visualized because of the required total protein load limit and the limit of dye sensitivity. On the other hand, all proteins regardless of their isoelectric point, do migrate on these gels.

Although one-dimensional PAGE has serious limitations, several investigators have been able to improve upon the technique and show that at least some tissue-specific NHPC can be found in brain tissues,[325] normal liver and hepatoma,[58] winter wheat embryos,[46] jack pine seedlings,[254] and others. It should also be mentioned that the detection of specific NHCP can be greatly enhanced by immunological staining with appropriate antisera.[326,327]

2. Isoelectric Focusing Analysis of NHCP

Isoelectric precipitation as well as isoelectric fractionation by electrical transport have been, for a long time, among the standard procedures of protein separation.[328] The latter is based on sending direct current through a system of electrolytes, whose pH increases gradually from anode to cathode. If the pH gradient can be kept reasonably stable for the duration of the experiment, proteins or other ampholytes will be repelled by both electrodes and focus in a region where the local pH is identical with the isoelectric point of the ampholyte.[329]

In the original procedure[330,331] the pH gradient was produced in a specially designed column by applying voltage to a mixture of carrier ampholytes stabilized in a sucrose density gradient. The result of the fractionation was determined by analyzing fractions collected from the column. In 1968 this procedure was simplified[332] for rapid analysis of small amounts of proteins using the principle of isoelectric focusing in tubes of polyacrylamide gel containing a protein sample and carrier ampholytes. A number of samples may be analyzed simultaneously in several hours, compared with several days for a single fractionation by the original procedure. Ease of this operation, substantial savings in the expensive ampholytes used to form the gradient, and increased speed of protein separation are the most significant advantages of this system.

The technique of isoelectric focusing has proved to be an invaluable adjunct to traditional electrophoretic procedures, differing from the latter in that various proteins in the sample move under the applied voltage to their isoelectric points in a preestablished pH gradient.[333] This is important, especially where the desired protein must be separated from others of similar size. Isoelectric focusing relies on the generation of reproducible linear pH gradients formed by complex mixtures of amphoteric substances prepared by copolymerization of polyamines with suitable reactants to introduce acidic groups. Although a broad range of ampholytes covering pHs between 3 and 10 is quite useful, better resolution of protein components can be accomplished through the use of narrow-range ampholytes.[334] Until a few years ago the choice of pH ranges for IFPA was limited by the commercially available pI ranges of synthetic mixtures of carrier ampholytes of the Ampholine® type (e.g., Ampholine®, Servalyte®, Pharmalyte®, etc.). Today, there is much greater freedom in the choice of carrier constituents to suit the chemical predilection of the molecule of interest, the desired pH range, carrier constituent molarity, etc.[316]

Analytical gel electrofocusing can be carried out in tubes, in horizontal flat beds, or vertical gel slabs.[316] In any case, a gel is chosen with a pore size which offers low resistance to the passage of proteins and thus acts mainly as an anticonvectant medium. One can usually apply the sample at the desired gel surface in the same manner as for PAGE, i.e., after addition of a dense medium, by layering under the electrolyte, or in the case of sample volumes exceeding 200 $\mu\ell$ by overlayering the electrolyte.[316] The resolved proteins can be visualized as white precipitation bands by immersing the gels in trichloroacetic acid (5%). The gels, washed repeatedly in this solution to remove the ampholytes, can be then stained, e.g., in amido-black or Coomassie® blue and destained by washing in acetic acid (7%). The pH gradient in gels is determined by cutting 2-mm sections from an unstained gel run simultaneously with stained gels, soaking each section in 2 mℓ water for several hours and measuring the pH of the extracts. The pH gradient is reproducible for gels run simultaneously.[332]

First separations of NHPC by isoelectric focusing were performed by Gronow and Griffiths[335] and by Elgin and Bonner,[81] using modifications of the existing procedures,[336,337] in 5% and 3.7% polyacrylamide gels in 8 *M* and 10 *M* urea, respectively, and using 10 m*M* phosphoric acid as anolyte and 20 m*M* sodium hydroxide as catholyte.[336,337] Since that time, isoelectric focusing in polyacrylamide gel has been extensively used by many workers for the separation and analysis of NHCP during one-dimensional and the first run of two-dimensional electrophoresis. It revealed a great number (about 20 to 30)[104] protein components having a pI spectrum between pH 3.7 and 9.0, with the majority of bands in the range of pH from 5.0 to 7.0.[81,104,221,338] Blocking SH groups by *N*-ethylmaleimide prior to comparative analysis of NHCP from four rat tissues in experiments of Gronow and Thackrah[102] improved the resolution in that it prevented protein polymerization by disulfide formation.

B. Two-Dimensional Electrophoresis of NHCP

One-dimensional PAGE methods could detect only partial heterogeneity of NHCP since each separation is based on a single parameter, e.g., molecular weight[80,259] or isoelectric point.[81,335] As the introduction of two-dimensional PAGE proved of great value in improving the resolution of several complex mixtures, e.g., plasma[339] or ribosomal proteins,[340] this technique was rapidly applied to the examination of the NHCP. First two-dimensional electrophoretic analyses of NHCP were performed by Barret and Gould[104] and Busch and associates.[323] Since then, two two-dimensional systems using isoelectric focusing[104] or acid-urea[323] electrophoresis in cylindrical gel for the first dimension and a SDS-containing slab gel in the second dimension, have been employed by many investigators to study NHCP heterogeneity.

The system of Busch and co-workers[323,341,342] used for comparison of nuclear proteins by a combination of acid-urea and SDS gel electrophoresis revealed 98 protein components in the normal rat liver and 111 components in the Novikoff hepatoma ascites cell 0.2 *M* H_2SO_4-extracts[323] (among them, 11 qualitative and 5 quantitative differences). However, this system, in which the acrylamide concentration in the first and second dimension can be as high as 6 and 8%,[343] respectively, distributed the protein spots in a diagonal fashion rather than across the entire surface of the gel, thus suggesting that separation according to differences in molecular size had to occur in both the first (at acid pH) and second (in SDS) dimensions. To optimize resolution, each dimension should separate proteins according to their independent parameters. For this reason, Busch and Takami[52] have modified their two-dimensional gel system,[344] following the method of Barret and Gould[104] which employs isoelectric focusing and SDS-electrophoresis in the first and second dimension, respectively. Since these two parameters are unrelated it is possible to obtain an almost uniform distribution of protein spots across the two-dimensional gel. This technique provided a highly reproducible separation of NHCP from normal rat liver and Novikoff hepatoma[52] on the basis of isoelectric points over the pI range of 4.5 to 8.5 in the cylindrical gels and of molecular weights on the SDS-slab gels containing uniform concentration of buffered acrylamide. The combination of this analytical procedure with successive extraction of nuclei of normal rat liver and Novikoff hepatoma using five buffered solutions considerably improved the resolution and analysis of NHCP. Each fraction contained 108 to 200 spots. In the Novikoff hepatoma nuclei, 18 polypeptides from a total number of 483 different polypeptides were not found in the normal rat liver, and 12 polypeptides from a total number of 427 protein spots were not found in the Novikoff hepatoma nuclei.[52]

As was already mentioned, Barret and Gould[104] adapted the successful technique of Martini and Gould for separation of ribosomal proteins,[340] using isoelectric focusing in 4% acrylamide according to Gronow and Griffiths[335] and SDS-electrophoresis in uniform 5% acrylamide slab gel and applied it to the analysis of NHPC extracted from nuclear or chromatin preparations by the method of Gronow,[345] i.e, with buffered 8 *M* urea (pH 7.6). Two-dimensional electrophoretic NHCP patterns from three tissues studied were reproducible and each protein could be identified by its coordinates, i.e., molecular weight and isoelectric point. The major proteins in the molecular weight range 30,000 to 70,000 were studied in detail. There were 18 such proteins in rat liver, 11 in chicken liver, and 13 in chicken reticulocyte. There was a much closer resemblance between the patterns of the two liver tissues than that of the reticulocytes. Loeb and Creuzet,[346] using a one-dimensional electrophoretic system, came to a similar conclusion.

Even better resolution of the NHCP was obtained by MacGillivray and Rickwood[338] who used two-dimensional PAGE and employed a combination of modified isoelectric focusing[335] of the reduced protein on SDS electrophoresis according to MacGillivray et al.[103] in a flat-bed apparatus and a slab gel prepared from small-pore and large-pore gel solutions. The authors analyzed [^{32}P]-labeled NHCP fractions obtained by HAP chromatography of salt-urea dissociated chromatin. Electrophoretic comparison of the components of HAP2 and HAP3 protein fractions showed that many of the phosphorylated and nonphosphorylated proteins were common to mouse liver, kidney, and brain chromatins. Only a few protein species were found to be tissue specific.

At present, the most powerful procedure for the resolution of NHCP by two-dimensional electrophoresis involves the application of O'Farrel's technique.[347] This high-resolution method which has resolved 1100 different components from *Escherichia coli*[347] and up to 1600 components of total cell protein from a rat hepatoma cell line,[76] and which is potentially able to resolve some 7000 proteins, has opened a new era in the analysis and detection of proteins in complex biological systems. It has initiated the idea of "protein explosion"[348] and "molecular anatomy".[349,350] Indeed, using this technique the detection of proteins by both stain and autoradiography is very sensitive. Staining with Coomassie® blue or with the recently developed silver staining procedure can detect less than 0.01 μg of protein, and calibration curves for autoradiography show that a spot containing as little as one disintegration per minute of either ^{14}C or ^{35}S can be quantified after 20-day exposure (a protein which constitutes 10^{-4} to 10^{-5}% of total can be detected by autoradiography).[347]

O'Farrel's two-dimensional gel procedure, combined with the S1 nuclease method of Peterson and McConkey[351] for preparation of DNA-free NHCP samples before electrophoresis, revealed approximately 470 components in total HeLa NHCP with nearly identical gel patterns for nonhistones of chromatins prepared from nuclei isolated by nonaqueous (in glycerol) and aqueous techniques (Figure 4). Few additional proteins were found when the pH gradient was expanded from 5.9 to 7.5 to pH 5.0 to 7.8. The prominent NHCP had molecular weights between 40,000 and 100,000 but there were also many proteins with molecular weights between 10,000 and 40,000. The complexity of NHCP patterns was comparable to that of the cytoplasm (532 proteins detected) and greater than that of nucleoplasm (300 proteins detected) on autoradiograms of gels exposed for 32 days. Most of the NHCP were rare (less than 10,000 copies per haploid genome) and were not detectable in the cytoplasm. According to the authors,[351] these two protein populations were strikingly different. Approximately 64 abundant proteins were found in chromatin, nucleoplasm, and cytoplasm (one appears to be actin) which may point to a dynamic relationship between the cytoplasm and nucleus.

Recently[98] two-dimensional PAGE according to O'Farrel[347] and one-dimensional SDS-PAGE have been employed to compare NHCP from: (1) three human cell lines, i.e., HeLa, Raji, and phytohemagglutinin-stimulated peripheral blood lymphocytes from normal adult, (2) three Chinese hamster cell lines, i.e., CHO-K1, cultured embryo fibroblasts, and T78-

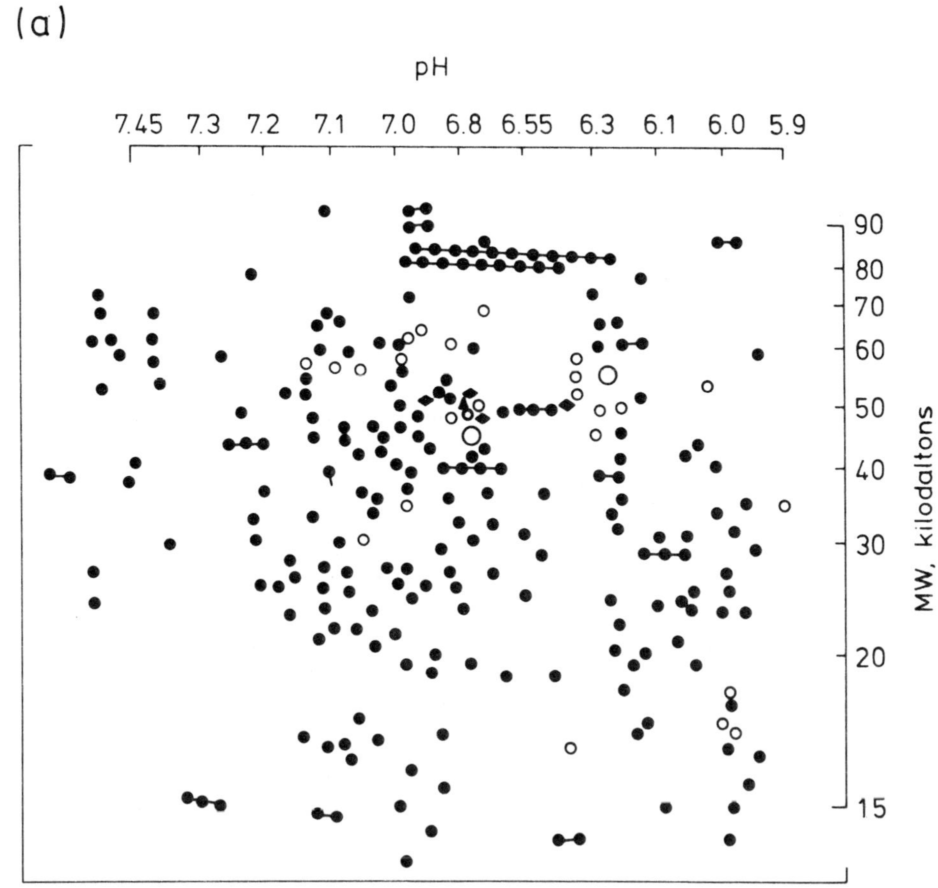

A

FIGURE 4. (a) Schematic diagram comparing two-dimensional electrophoretic resolution[347] of nonhistone proteins of chromatins obtained from nuclei isolated by glycerol and aqueous methods from HeLa S3 cells grown for 3 days in medium containing 5 μCi of [^{35}S] methionine per mℓ. Chromatin pellet was resuspended in 10 *M* urea-0.1% SDS-1 m*M* TrisCl, pH 7.4, at 25°C, at a concentration of 20 A_{260}/mℓ; 0.1 volume of 10 *M* urea-1% SDS-25 m*M* $ZnCl_2$-0.3 *M* lysine, pH 3.8 was added and DNA was digested with 70 units of S1 nuclease per A_{260} for 5 min at 45°C. Protein from approximately 1 A_{260} of chromatin in 10 *M* urea-0.1% SDS-5% 2-ME-2% NP40-0.1 *M* TrisCl, pH 7.4/2% ampholines (pH 3 to 10) was loaded on each gel. Since histones have pI above the pH range of the ampholines used, and lower methionine content than NHCP, the autoradiograms concern the "glycerol" and "aqueous" NHCP superimposed to identify proteins that could be detected in both types of chromatins. Black circles represent proteins found in both types of chromatin, black rhomboids represent proteins found in the glycerol but not in the aqueous chromatin, and the black triangle, a single protein found in the aqueous but not in the glycerol chromatin. The open circles correspond to the spots that appeared on autoradiograms when glycerol nuclei were mixed with [^{35}S] methionine-labeled aqueous cytoplasm and are assumed to be cytoplasmic proteins bound to chromatin. (b) A diagrammatic comparison of the NHCP, nucleoplasmic (NU), and cytoplasmic (CY) proteins. NHCP, proteins found only in the chromatin; NHCP-NU, proteins found in the NHCP and nucleoplasm only in NHCP-CY, proteins found in the NHCP and cytoplasm only; NHCP-NU-CY, proteins found in all three cell fractions. (From data of Peterson, J. L. and McConkey, E. H., *J. Biol. Chem.*, 251, 548, 1976. With permission.)

60 (chemically transformed fibroblasts), and (3) seven types of human-hamster cell hybrids. The gels were stained and fixed according to Fairbanks et al.[352] and processed for fluorography by the method described by Bonner and Laskey.[353] The NHCP were either extracted by the Gronow[345] procedure after RNase digestion, or total chromosomal proteins were

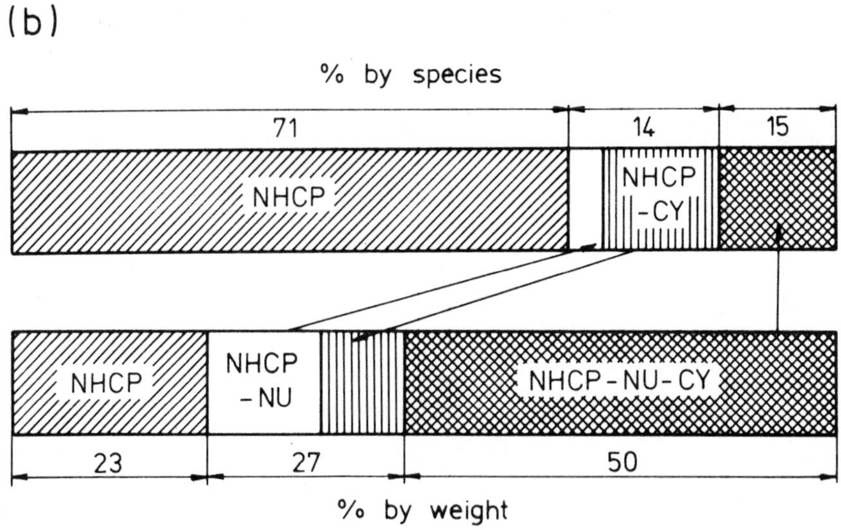

FIGURE 4B

analyzed after or without S1 nuclease treatment.[351] Over 90% of the 230 most prominent NHCP in each type of human cell were indistinguishable electrophoretically from the NHCP of the hamster cells. Of the roughly 10% NHCP which were distinct, 11 may be human specific, since they were found only in each type of human cell examined. The NHCP proteins in seven different stable human-hamster hybrid cell lines, which in total, contained markers for 21 of the 24 human chromosomes were electrophoretically indistinguishable from those of the parental hamster cell lines, with the exception of a single protein of roughly 300,000 daltons synthesized by one hybrid line (HB-5), as well as by each type of human cells investigated.

Until recently, methods for visualizing protein bands after isoelectric focusing and before separation in the SDS-slab gel have always been indirect, i.e., dividing the isoelectric focusing gel into longitudinal parts or calibration with the aid of reference gels, both used for staining. This indirect procedure is avoided in Jackle's[354] modification of the O'Farrel[347] method. After isoelectric focusing, the cylindrical gels are fixed and stained with Coomassie® blue (before SDS-slab electrophoresis). This modification does not alter the protein patterns or their resolution but provides sharper spots.[354,355]

IV. FRACTIONATION OF NHCP ON PREPARATIVE SCALE

Attempts to fractionate NHCP and study their biochemical and biophysical properties have encountered great difficulties, chiefly due to the poor solubility of these proteins in conventional solvents and their considerable complexity. There are four principal approaches to the fractionation strategy of NHCP on a preparative scale, i.e., preparative electrophoresis (PAGE and IFPA), selective precipitation, selective extraction, and chromatography (ion-exchange, including HAP, gel filtration, hydrophobic, and affinity chromatography).

A. Preparative Electrophoresis of NHCP

As in the case of the analytical scale fractionation, the resolution of NHCP by preparative electrophoresis may be performed at a constant pH (PAGE) and at a gradient of pH (IFPA). This is achieved either in special apparatus for preparative disc electrophoresis or more often in an instrument used routinely for one-dimensional (cylindrical or slab gel) and two-dimensional resolution.

1. Preparative Polyacrylamide Gel Electrophoresis of NHCP

Fractionation of NHCP by preparative PAGE at a constant pH may be performed both in nondenaturing (without SDS) and denaturing (in the presence of SDS) conditions. Since SDS-PAGE fractionation deals with the denatured SDS-complexes of proteins or protein subunits, the isolated fractions are suitable either for chemical analysis only or the SDS must be removed. In the latter case, the method of Weber and Kuter[144] using chromatography on Dowex® AG1-X2 resin is especially effective. It is noteworthy that Elgin and coworkers[356,357] were able to obtain antibodies in rabbits immunized with *Drosophila* chromatin or total nonhistone chromosomal proteins dissolved in SDS (0.1%). These antibodies reacted with NHCP in 0.1% SDS and were directed against several proteins as determined by indirect immunofluorescent staining of antibody-antigen complexes in SDS polyacrylamide gels.[358] The presence of low concentration of SDS did not seem to affect either the production of antibodies or their general specificity and the selectivity of the distribution patterns of individual NHCP on *Drosophila* polytene chromosomes.[357,359]

Quite useful for preparative purposes under nondenaturing conditions has become the so-called acid-urea PAGE, usually performed in 0.9 *M* acetic acid-2.5 *M* urea at low pH (e.g., 2.7) originally developed for the analysis of histones in 15% acrylamide.[29,360] It affords separation on the basis of charge, molecular weight, and shape.[361] This procedure has been applied by Busch and associates to the separation of NHCP in 6%[42,300] or 10%[53,362] acrylamide slab gels in 0.9 *M* acetic acid-4.5 *M*[53,300] or 10 *M*[42] urea or in 90 m*M* TrisCl-2 m*M* EDTA-80 m*M* borate, pH 8.1.[362] With the use of gel elution device[363] several electrophoretically homogeneous nuclear proteins, e.g., A24,[53] NAg-1,[362] C-14,[42] and BA[300] were thus obtained. Quite recently two-dimensional electrophoresis, i.e., acid-urea-PAGE and SDS-PAGE was adapted for the isolation of small amounts of NHCP nucleolar proteins purified to homogeneity.[364]

2. Fractionation of NHCP by Preparative Isoelectric Focusing

Isoelectric focusing fractionation of NHCP on a preparative scale may be performed in special apparatus[325,365,366] or as one-dimensional separation either in cylindrical, or more often, slab gels, vertical[359] or horizontal.[367] Pollow and Kruger fractionated the NHCP of cow udder[365] or neuronal and glial tissues[325] after so-called first and second dissociation into several IFPA fractions using Valmet's zone convection electrofocusing apparatus. This enabled them to obtain purified proteins in milligram quantities. The apparatus chambers, holding about 50 mℓ, contained 88 and 62 mg of protein after the first and second dissociation,[365] respectively, in 7 *M* urea-2 m*M* DTT-3%[365] or 4%[325] ampholines, pH 3.5 to 10, with 50 m*M* sodium bicarbonate, pH 10, as catolyte, and 1% acetic acid, pH 2.8, as anolyte; both in 12% sucrose. Preparative isoelectric focusing of NHCP extracted after first dissociation resulted in 7 fractions from cow udder[365] and 15 fractions from neuronal tissue[325] in the pH range of 3.5 to 9.7 as well as 15 fractions from glial tissue in the pH range of 3.7 to 10.3. The second dissociation conditions provided 7 fractions in the pH range of 4.1 to 9.5 from cow,[365] 10 fractions from neuronal (pH range 3.9 to 10.5) and 10 fractions from glial (pH range 3.7 to 9.9) tissues.[325] All IFPA fractions were dialyzed and lyophilized before further chemical analysis and microgel electrophoresis.

Because the NHCP aggregate during focusing in vertical pH gradients stabilized by sucrose, MacGillivray[367] adopted the horizontal thin-layer isoelectric focusing technique of Radola[368] for analysis of NHCP fractions separated by HAP chromatography. They were able to separate 5 mg of NHCP into five subfractions in a thin layer of Sephadex® G-50 superfine which was divided into five sections based on the distribution of [^{32}P]-radioactivity in the pH gradient. The proteins were extracted by incubation of the Sephadex® sections in several volumes of 6 *M* GdnCl-10 m*M* DTT-0.3 *M* TrisCl, pH 8.3 at 37°C for 3 hr followed by removal of Sephadex® by filtration or centrifugation, and analyzed by two-dimensional analytical gel electrophoresis (IFPA and SDS-PAGE).

Sequential preparative vertical slab gel isoelectric focusing and SDS polyacrylamide slab gel electrophoresis were employed by Silver and Elgin[359] for the isolation of three *Drosophila* embryo NHCP with isoelectric points at pH 5.22, 5.35, and 5.86 and molecular weights of 21,000, 54,000, and 26,000, respectively, to prepare antisera and study the distribution of these proteins in polytene chromosomes.[358] The preparative IFPA was performed with NHCP isolated under mild conditions (8 *M* urea, pH 7.6)[335,345] using slab gel of 5% acrylamide in 8 *M* urea-1 m*M* DTT-1.6% ampholines, pH 5 to 8-0.8% ampholines, pH 4 to 6. Major protein bands observed on stained IFPA slab gels were excised, cut into long strips and, after lyophilization, either stored at −20°C or separated on 10% acrylamide SDS slab gels according to Laemmli.[312] A well-separated protein band at pI 5.22 produced a single band on the SDS slab gel. A doublet at pI 5.35 and a band at pI 5.86 were separated into one major and four minor bands, and into three major and many minor bands, respectively. From 10 mg of total NHCP it was possible to obtain 10 to 50 μg of a single polypeptide characterized by its isoelectric point and molecular weight.[359]

It is of interest that proteins with pI higher or lower than the limits of the established pH gradient will migrate into the electrode solutions. Gronow and Thackrah described such a phenomenon when separating NHCP from rat liver nuclei.[366] The cathode solution (20 m*M* NaOH) and outside the pH range 4.0 to 8.6 gradient contained a low molecular weight protein (approximately 10,000) with an acidic amino acid composition. It bound five to six times its own weight of basic ampholine carrier ampholytes to give a complex with a pI of 8.82. The complex could be dissociated by dialysis against 1 *M* NaCl.

B. Selective Extraction

This fractionation method takes advantage of the differential solubility or DNA binding properties of NHCP. Proteins are sequentially solubilized from DNA by treatment of chromatin with increasing amounts of salt or urea (or both) at various pH. Urea is used to facilitate extraction of hydrogen-bonded molecules and to maintain the extracted proteins in soluble state. As a rule, each dissociation step is followed by high-speed centrifugation to pellet the non-dissociated protein-DNA complex and to yield the solubilized NHCP in the supernatant. On basis of their solubility, the NHCP can be divided into three major groups, i.e., weakly, tightly, and very tightly bound.

The weakly bound NHCP may be defined as those proteins which are solubilized from the chromatin complex by treatment with low concentrations (up to 0.35 *M*) of NaCl, KCl, or with urea (5 *M*) in the absence of salt.

The tightly bound NHCP represent those proteins which remain associated with chromatin in 0.35 *M* NaCl. These proteins can be dissociated from the DNA by treatment with NaCl ($\geq$ 2 *M*) and urea ($\geq$ 5 *M*) or GdnCl.

The very tightly bound NHCP represent so-called residual nonhistone proteins and SDS or other detergents are necessary for their dissociation.

1. Some Schemes of NHCP Selective Extraction

Several groups of investigators have elaborated schemes of selective extraction of chromosomal proteins. Some of these schemes include chromatographic or electrophoretic techniques for separation or purification of nonhistone fractions.

The laboratory of Boffa and Allfrey especially concerned themselves with selective extraction of the NHCP nuclei isolated from colonic epithelial tumors induced in rats by the administration of 1,2-dimethyhydrazine (DMH).[40] Four fractions of nonhistone proteins with differential extractability were obtained from nuclei thoroughly prewashed with 0.14 *M* NaCl-10 m*M* $NaHSO_3$, and 50 m*M* TrisCl, pH 8.0-0.1 m*M* PMSF to remove nucleoplasm. There were NHCP soluble in (1) 0.35 *M* NaCl, (2) 0.6 *M* NaCl-TrisCl, pH 8.0, (3) 1 *M* NaCl-TrisCl, pH 8.0, and finally (4) 0.4 *M* GdnCl-6 *M* urea-35 m*M* NaCl (all solutions made 0.1 m*M* PMSF). Electrophoretic analysis of proteins in the lyophilized extracts revealed two prominent specific NHCP bands with molecular weights of approximately 44,000 (TNP_1) and 62,000 (TNP_2). The TNP_1 was extractable in 0.35 *M* NaCl, while TNP_2 required high salt concentrations (2 *M* NaCl) or the use of denaturants such as urea and GdnCl for its displacement.

These two nonhistone proteins were found to accumulate during DMH carcinogenesis,[369] and were also found in human colonic adenocarcinoma as clearly visible protein bands in the total NHCP population extracted with 0.4 *M* GdnCl-6 *M* urea-0.1% 2-ME-0.1 *M* sodium phosphate, pH 7.4-0.1 m*M* PMSF.[41,369] These two proteins were not uniformly distributed throughout five nuclear subsets, derived from rat colonic epithelial tumors, and obtained by buoyant density centrifugation in discontinuous sucrose gradient ranging from 2.30 to 2.70 *M*.[40] They were absent from normal rat and human colonic epithelial cells[40,41] and also from human nonmalignant tumors.[41] The synthesis of TNP_1 and TNP_2 was enhanced at very early times in carcinogenesis, long before any morphological signs of malignancy.[187,369]

Another procedure of NHCP fractionation on the basis of their sequential extractability from chromatin has been proposed by Bekhor and co-workers.[62] In the initial approaches, purified unsheared chromatin was sequentially extracted with 0.0, 0.5, 1.0, 1.5, and 2.0 *M* NaCl in 5 *M* urea-10 m*M* TrisCl, pH 8.0-2.85 m*M* PMSF. The five solubilized NHCP fractions were recovered from high-speed supernatants, dialyzed against 10 m*M* TrisCl, pH 8.0-2.85 m*M* PMSF, and lyophilized. Under these conditions it was found that 94% of chromosomal proteins were dissociated from DNA in the first three supernatants, i.e., in 5 *M* urea and up to 1.0 *M* NaCl, and 6% dissociated between 1.0 and 2.0 *M* NaCl in 5 *M* urea. This technique was used to characterize the NHCP from four rabbit tissues (brain, heart, liver, and submandibular glands).[62,370] The isolated fractions were extremely heterogeneous and exhibited significant differences in their molecular weight profiles. Many of the components were common to various tissues, yet some distinct qualitative differences were found. It was concluded[370] that each organ may be classified by the NHCP composition as resolved by disc gel electrophoresis.

The principal objective of this group[54,275,371,372] became the characterization of three NHCP fractions designated M_0, M_1, and M_3 (i.e., obtained by extraction with 0 *M*, 1 *M*, and 3 *M* NaCl in 5 *M* urea, in 10 m*M* TrisCl buffer) pH 8.0-1 m*M* PMSF-0.2 m*M* DTT. An observation that the DNA-tightly bound NHCP fraction M_3 exhibited significant accumulation of transcriptionally active genes stimulated its further characterization. It contained about 4% of total chromatin protein (chicken liver and reticulocyte),[373] was highly heterogeneous,[374] and bound to DNA by hydrophobic interactions[375] which were not abolished by high ionic strength up to 3.0 *M* NaCl.[147,375] The template activity of M_3-DNA was of the same magnitude as that of free DNA,[375] and as was already mentioned, the M_3-DNA complex was significantly enriched in transcriptionally active DNA sequence. The M_3 proteins were presumed to contain gene regulatory proteins.

Busch and co-workers[342] used 0.2 *M* H_2SO_4 for the removal of histones. NHCP were then solubilized by the Wilson-Spelsberg[301] procedure employing DNase I treatment of dehistonized chromatin. The acid-insoluble NHCP of DNase-digested dehistonized chromatin were precipitated by making the DNase digestion solution (cooled to 4°C) 0.4 *M* with respect to $HC10_4$ and collected by centrifugation. The protein pellet was solubilized in electrophoresis sample buffer (0.9 *M* acetic acid-10 *M* urea-1% 2-ME) and centrifuged (30,000[342] or 40,000,[376] 30 min). The supernatant designated as "Chromatin Fraction II" represented 40.0% (rat liver) or 45.4% (hepatoma), and the pellet, i.e., ''Chromatin Fraction III'' constituted 0.4% (rat liver) or 0.6% (hepatoma) of total nuclear protein. Approximately 98% of the protein remaining in chromatin after the acid extraction was soluble in the two-dimensional (acid/urea-PAGE) sample buffer. The NHCP of the ''Chromatin Fraction II'' from six different tissues, i.e., from normal rat liver and kidney, the regenerating and the thioacetamide-treated rat liver, as well as from Novikoff hepatoma and Walker 256 carcinosarcoma, were subjected to comparative studies by two-dimensional PAGE[376] (acid-urea-PAGE and SDS-PAGE) that revealed many similarities as wel as some differences between these tissues.

Sequential extraction with three 10 m*M* Tris-Cl buffers, pH 8.0 containing (1) 0.35 *M* NaCl, (2) 0.6 *M* NaCl, and (3) 3 *M* NaCl-7 *M* urea of nuclei first washed with 75 m*M* NaCl-25 m*M* EDTA and also with 10 m*M* TrisCl buffer, pH 8.0, is another example of selective extraction procedures used by Takami and Busch.[52] These three classes of chromosomal proteins were isolated from the nuclei of normal rat liver, Novikoff hepatoma,[52] regenerating and fetal rat liver, slow-growing Morris hepatoma 9618A, fast-growing Morris hepatoma 3924A,[377] and analyzed by two-dimensional PAGE. Many of the protein spots were common to all of these tissues. Some proteins, e.g., with molecular weights of 98,000 and 150,000 and pI 8.4 and 8.2, respectively, were found in the fetal liver and three studied hepatomas; they most likely represented oncofetal proteins. The overall electrophoretic patterns for the fast-growing tumors were similar to those of fetal liver.

In an attempt to avoid the multiple lengthy extractions needed to achieve fractionation of nuclear proteins, Gronow[345,378] introduced a procedure which solubilized the bulk of NHCP (70 to 80%) under mild conditions (8 *M* urea-50 m*M* sodium phosphate, pH 7.6) from isolated nuclei or chromatin, together with some RNA and 80% of the nuclear thiol. The histones could then be extracted with dilute acid leaving residual proteins and DNA.[335] In most preparations *N*-ethylmaleimide (1 m*M*) was added to block all thiol groups present.[102] No histone or DNA was removed by this method if the nuclear isolation procedure was rapid and the temperature was not allowed to rise sufficiently to allow the activation of endonucleases.[45] Because of its advantages, this technique was employed by other investigators[98,359,379] and a similar solvent, i.e., 8 *M* urea-10 m*M* TrisCl-10 m*M* Mg-acetate-1 m*M* EDTA-1 m*M* 2-ME was used by Olson and Busch[364] to extract a group of nucleolar proteins. Recently it was modified by Tsitilou et al.[184]

A somewhat different fractionation scheme developed by Hnilica and co-workers[196,197,380] utilizes differences not only in salt concentrations but also in the composition and pH of buffers in the extraction steps. It is based on extraction of chromatin by three buffer solutions containing urea (5 *M*) and in two later stages high salt concentration (2.5 *M* NaCl). After three centrifugations (two high-speed and extensive) this procedure yields three types of chromatin proteins, i.e., UP (buffered urea-soluble proteins), HP (histones), and NP (NHCP tightly bound to DNA), in three supernatants, and DNA complexed with non-extracted proteins in the final pellet. In other words, the histones are removed at pH 5.0 and the NP proteins are extracted from the pellet with 2.5 *M* NaCl-5 *M* urea-50 m*M* TrisCl, pH 8.0. They represent about 3 to 5% of total chromatin protein content, and the final pellet is DNA with a small amount of associated protein, i.e., very tightly bound to DNA. This fractionation procedure[380] has recently been modified by Yaneva et al.,[381] mainly by first removing the loosely bound NHCP with 0.35 *M* NaCl, and applied to electrophoretic study of tightly bound NHCP from three rat tissues (liver, spleen, and kidney). The general appearance of the densitometric scans showed three major low molecular weight (14,000 to 22,000) components, similar in liver and spleen but rather different in kidney. Additionally, about 20 minor bands of higher molecular weight were also present. Each tissue was characterized by its own pattern of protein bands and contained at least one unique band, i.e., molecular weight 30,500, 27,000, and 37,500 in liver, spleen, and kidney, respectively. Some quantitative differences between the common protein bands were also observed.

Because of the time-consuming centrifugations and relatively large volumes, the buffered urea-salt extraction procedure was modified by Fujitani et al.[43,380] by employing HAP column chromatography instead of centrifugation to separate histones and the NP proteins from DNA. The DNA-bound proteins (NP fraction) were shown to be tissue specific by electrophoretic and immunological criteria.[4,380,382] They exhibited affinity for homologous DNA, with a significant preference for native DNA.[196,198] They were responsible for the tissue-specific binding of steroid hormone complexes with cytoplasmic receptors.[199] This fraction, further resolved on HAP (50 m*M* sodium phosphate buffer eluate), became the source for isolation, purification, and characterization of nuclear antigens p39, p49, and p56 from Novikoff hepatoma chromatin.[327]

Krüger and Pollow[365] described an NHCP fractionation scheme which includes: (1) differential dissociation of chromatin in buffers containing 0.35 *M* GdnCl in the first, and 3 *M* NaCl in the second dissociation step, (2) cation exchange chromatography on Bio-Rex® 70, (3) preparative fractionation of NHCP nonadsorbed on Bio-Rex® 70 by isoelectric focusing in a Valmet's zone convection electrofocusing apparatus, and (4) continuous linear polyacrylamide pore microgradient SDS-gel electrophoresis.

Another scheme of NHCP fractionation (into four groups of "acidic" proteins AP_1, AP_2, AP_3, and AP_4) was introduced by Spelsberg et al.[383,384] It demonstrated alterations in the chromatin of chick oviduct during estrogen-mediated development and differentiation. The AP_1 fraction, containing all the histones, and the AP_2 fraction were dissociated from chromatin with 2.0 *M* NaCl-5 *M* urea-sodium phosphate buffer, pH 6.0. Ultracentrifugation sedimented the AP_2 and a DNA-AP_3-AP_4 complex, termed nucleoacidic protein (NAP). The pellet was solubilized in 2 m*M* TrisCl buffer, pH 7.5, and centrifuged to separate the insoluble AP_2 proteins from NAP which was then further fractionated using the same 2 *M* NaCl-5 *M* urea solvent, but at pH 8.5 . The dissociated NAP was centrifuged to sediment the DNA bound at AP_4 and leaving the AP_3 in the supernatant. Histones could be separated from AP_1 by dialysis against water and acidification. Nonhistone fractions AP_2 and AP_3 represented the major portion of the total NHCP and the fraction AP_2 was largely responsible for the quantitative changes observed in the total nonhistones of chromatin during estrogen-mediated oviduct development. The AP_4 represented about 10% of total oviduct NHCP and was only removed from DNA by phenol or protease treatment. This procedure was employed by Gevorkian[385] in binding studies of hydrocortisone-receptor complex with various groups of rat liver NHCP. It should also be mentioned that this fractionation scheme is being employed for purification of steroid hormone nonhistone acceptor proteins by Spelsberg and co-workers.[386] This topic is discussed in detail in another chapter of this series.

Stein and co-workers[387] used a combination of the procedures of Langan[245] and Kish and Kleinsmith[388] to fractionate chromosomal proteins. This method employed 1 *M* NaCl-50 m*M* TrisCl, pH 7.5 extraction of HeLa S3 cell chromatin, and reconstitution of the nucleohistone at 0.4 *M* NaCl. Centrifugation at 80,000 × g (1 hr) provided two groups of chromatin proteins, i.e., "80 × g supernatant proteins" and "80,000 × g pellet proteins". The first were freed of histones by Bio-Rex® 70 chromatography and separated on calcium phosphate gel into nonbinding proteins, and phosphoproteins which did bind to this gel. The second group of NHCP was dissociated in 3 *M* NaCl-5 *M* urea-50 m*M* TrisCl, pH 7.2, and separated from DNA by high-speed centrifugation (250,000 × g, 24 hr). This fractionation procedure separated NHCP into three electrophoretically distinguishable classes with reproducible patterns.

Several modifications of the salt-urea extraction procedures were employed by Wang and co-workers for the fractionation of NCHP.[38,239] In more recent experiments,[389] rat prostate chromatin was fractionated by sequential extractions into 0.35 *M* NaCl-soluble, 2 *M* NaCl-soluble and insoluble fractions. It was found that beside specific androgen-receptor binding-site located in the 2 *M* NaCl-insoluble residual fraction,[199,390] the rat prostate chromatin contained a second androgen receptor located in 0.35 *M* NaCl-soluble fraction. According to the authors, the androgen-binding protein component in the 0.35 *M* NaCl-soluble fraction was associated with actively transcribing elements.[390]

A modified procedure of Wang and associates[38] was used to selectively extract NHCP from euchromatin and heterochromatin,[293] isolated by controlled mechanical shearing of rat liver chromatin and subsequent glycerol gradient centrifugation.[391] It consisted of two extractions with 0.35 *M* and 2 *M* NaCl in TrisCl buffers, pH 7.5 and 8.0, respectively, followed by Bio-Rex® 70 chromatography. Three classes of NHCP, i.e., loosely (0.35 *M* NaCl-soluble), tightly (2.0 *M* NaCl-soluble), and 2 *M* NaCl-residual NHCP were obtained. This last group pelleted with DNA in 2.0 *M* NaCl and represented NHCP very tightly bound to DNA. Quantitative as well as qualitative differences in NHCP protein distribution between the two chromatin fractions were described by the authors.[293] Somewhat similar observations were reported by Rakowicz-Szulczynska and Horst[392] who analyzed NHCP from spleen chromatin of normal and immunized mice, and also by Blüthmann and Illimensee.[393]

A sequential fractionation procedure has been recently developed for the preparation of representative fractions of NHCP in substantial quantity (50 to 400 mg from 1 kg of pig thymus).[160] It represents a combination of methods based on the three intrinsic properties of NHCP, i.e., their binding in chromatin, molecular weight, and their isoelectric points. This procedure includes: (1) dividing by salt extraction the whole NHCP pool into loosely (0.35 *M* NaCl-soluble), and tightly chromatin bound NHCP (soluble in 2 *M* NaCl-5 *M* urea-50 m*M* Tris, pH 7.9), (2) separation of the loosely bound proteins into 2% TCA-precipitable and 10% TCA-precipitable fractions, (3) fractionation by gel filtration on SP-Sephadex® and Sephadex® G-100 of loosely bound NHCP into 11 subfractions, and of tightly bound NHCP into 6 subfractions, 2 of which were histones. All the subfractions were compared by SDS-PAGE, IFPA, and amino acid analysis. The loosely bound NHCP in chromatin were acidic to neutral in contrast to the tightly bound components which were basic. In addition, the high molecular weight NHCP were acidic to neutral, while the lower molecular weight components were mainly basic. The loosely bound NHCP, extracted from the chromatin with 0.35 *M* NaCl, accounted for about 17% of the total chromosomal protein and approximately 54% of total chromatin NHCP. The relative protein contents of the 2% TCA-precipitable and 10% TCA-precipitable fractions were about 78 and 22%, respectively, of the unfractionated 0.35 *M* NaCl-extract. The major component of the 10% TCA-precipitable fraction (component J) was similar to the HMG-1 or HMG-2 proteins from calf thymus[178] (molecular weight, N-terminal amino acid, and amino acid composition).

2. *Some NHCP of Selective Solubility*

Among the reports on NHCP isolation and characterization are many which concern the NHCP soluble in only one solvent. These are (1) acid-soluble, (2) phenol-soluble, (3) alkali-soluble, and (4) 0.35 *M* NaCl-soluble NHCP, as well as (5) NHCP very tightly bound to DNA.

a. Acid-Soluble NHCP

The observation that some of the NHCP are acid-soluble is not new. Grogan et al.[394] reported that 27% of the nucleolar proteins are extractable with 0.25 *M* HCl. The presence of basic proteins among the NHCP is also known from the experiments of Wang and Johns,[228] who used DEAE-cellulose chromatography to analyze three rat liver NHCP fractions. In all three, there was a "break-through" protein peak unadsorbed on DEAE-cellulose, which in one of the three fractions was basic and had a higher basic than acidic amino acid content. This "break-through" peak represented about 2% of the total NHCP.

The experiments of Hasuma et al.[395] have shown that basic NHCP which were coextracted with histones in 0.25 *M* HCl from rat liver nuclei were preferentially phosphorylated by the nuclear protein kinase NII. When the [^{32}P] phosphorylated proteins soluble in 0.25 *M* HCl were analyzed by Bio-Rex® 70 column chromatography, two [^{32}P] peaks were recognized; one in the flow-through (5% of the applied proteins), the second in the wash-out (40% GdnCl) fractions. These two peaks represented 50% and 24%, respectively, of the total recovered radioactivity. To check whether the histones H3 and H4 (also present in the 40% GdnCl eluate) were phosphorylated, the [^{32}P]-labeled whole histone preparation was subjected to two-dimensional PAGE. Autoradiograms revealed that [^{32}P]-labeled proteins were localized exclusively in the region of nonhistone proteins of 32,000 to 72,000 mol wt range. The basic properties of these proteins were confirmed by isoelectric focusing. As much as 71 to 72% of the [^{32}P] radioactivity was distributed in the pH region higher than 70. Since the proteins with pI greater than 9.0 must have migrated out of the gel during isoelectric focusing, the actual percentage of the phosphorylated basic proteins may have been even higher than the observed value. The authors suggested that some of the basic proteins which are substrates for protein kinase NII interact directly with DNA or acidic NHCP in chromatin, and that introducing negative charges to these basic proteins by their phosphorylation may alter their conformation and interaction with DNA and other chromatin components. It should be emphasized that the nuclear protein kinase NII was found to phosphorylate some HMG proteins (HMG-17 and probably HMG-14) to a considerable extent. Hence, this kinase may modulate chromatin activity by phosphorylating a particular set of nuclear proteins.[395]

Another class of acid-soluble NHCP is the recently reported H1 histone satellite acid-soluble NHCP.[396] These NHCP represent the low-mobility protein group (LMG) soluble in 2% TCA, and extractable together with histones H1 by 5% perchloric acid.[178] The H1 satellite NHCP constitute less than 1 to 2% of the total protein eluted from P-60 Bio-Gel® during purification of the 5% perchloric acid extract. It should be mentioned that for a single polyacrylamide gel in the case of liver chromatin, material was needed from 50 mice, and for spleen chromatin, from 200 mice.

b. Phenol-Soluble NHCP

The NHCP resulting from the water-phenol two-phase partition procedure used for DNA removal belong to a special class known for its extensive phosphorylation. These nuclear phosphoproteins (for review see Reference 181) represent between 5 and 10% of total dry weight of the cell nuclei and account for roughly 15% (calf thymus) to 50% (rat liver) of the total NHCP fraction.[397,398] Since these proteins are discussed in detail elsewhere in this book, their review will be excluded from this chapter.

c. Alkali-Soluble NHCP

The dissociation of nonhistone proteins from chromatin at alkaline pH, e.g., at pH 11.5[189] or pH 11.6[77,188,192,225,226] is perhaps the oldest method for their extraction. In the experiments of Murphy and Bonner,[399] chromatin suspended in 10 m*M* $NaHSO_3$-10 m*M* TrisCl, pH 8.0, was titrated to the desired pH, over a range between pH 8 to 11. It was possible to enrich for a rather specific subset of NHCP by first titrating the chromatin to pH 10, removing the dissociated material and then titrating the chromatin again to pH 11. Russev et al.[94] showed that two major groups of NHCP with different stability toward alkaline dissociation, i.e., alkali-labile and alkali-resistant, exist in rat liver chromatin. The first group NHCP, alkali-labile was easily dissociated in the pH interval 10.5 to 11.8 and comprised about 70% of the total NHCP. This was interpreted to mean that these proteins are complexed with chromatin by means of ionic interactions only. The remaining 30% exhibited a strong resistance toward alkaline dissociation. They remained bound to chromatin even at pH 13, which indicated that they are held by nonionic forces. SDS-PAGE[312] without prior DNA removal revealed that the majority of the high molecular weight NHCP, along with a few characteristic fractions between 40,000 and 60,000 mol wt, remained in the alkali-resistant group. [^{14}C] Leucine pulse-labeling experiments showed that in normal and regenerating rat liver the specific activity of the alkali-labile NHCP was two to three times higher than that of the alkali-resistant NHCP, which, in turn, had the same specific activity as the histones. It was concluded that alkali-resistant nonhistone proteins and the histones have simlar metabolic rates, suggesting that this fraction of NHCP remains metabolically conserved.

It shoud be noted that reports from Gronow et al.[308,400] point to the instability of chromatins in alkaline solutions and of their NHCP extracted with buffered 8 *M* urea, pH 7.5. These NHCP were shown to be degraded in 0.1 *M* or 1 *M* NaOH solutions, both at 4 and 18°C. Treatment in 1 *M* NaOH at 18°C for 24 hr was sufficient to break down approximately 77% of these proteins to low molecular weight peptides.[400] Electrophoretical analysis (IFPA[335] and SDS-PAGE[140]) demonstrated the destruction of proteins with pI values above 5.5, correlating with rapid loss of high molecular weight bands in SDS electrophoresis.[140] These results indicate that exposure of NHCP to alkaline pH conditions should be avoided.[400]

d. The 0.35-M NaCl-Soluble NHCP

The use of 0.35 *M* NaCl as a nondenaturing medium for the extraction of weakly bound NHCP revealed that a highly heterogeneous protein mixture results from this procedure. Although, according to Fujitani and Holoubek,[225] the gel electrophoretic patterns of NHCP loosely and tightly bound to DNA were similar, the 0.35 *M* NaCl-soluble fractions contained many unique proteins. The loosely bound NHCP of Ehrlich ascites tumor chromatin were found to contain a subfraction of highly phosphorylated proteins. This subfraction was isolated from 0.35 *M* NaCl-tumor chromatin extract as either a phenol-soluble protein following the procedure of Teng et al.,[258] or with the use of DNA-affinity chromatography on heterologous *(E. coli)* and homologous (Ehrlich ascites) DNA-cellulose columns. It consisted of relatively low molecular weight proteins (mostly 36,000 and less) and contained 0.9% alkali-labile phosphorus. This phosphoprotein fraction was enhanced 25- to 30-fold in phosphorus content as compared with the total 0.35 *M* NaCl-soluble chromatin protein extract. It was able to stimulate DNA-dependent RNA synthesis in vitro by binding selectively to the unique DNA sequences and activating their transcriptional initiation.[38]

In an extensive comparative program, Busch and associates[401] isolated the 0.35 *M* NaCl- and phenol-soluble phosphoproteins from normal and thio-acetamide-treated rat liver (nondividing tissues) and from regenerating rat liver and Novikoff hepatoma (dividing tissues). Four [^{32}P]-labeled protein spots in two-dimensional PAGE (CU′, C13p, C21p, and CMp) were found only in the Novikoff hepatoma, and two spots (B6 and B10) were found in liver but not in the tumor. The 0.35 *M* NaCl-soluble proteins from chromatin of Novikoff hepatoma became the source for purification of NCHP C14.[42] This protein was isolated in high purity as shown by its migration as a single spot on two-dimensional PAGE and identification of its single aminoterminal amino acid (lysine). It was partially characterized by its relatively high molecular weight (approximately 70,000), high acidic-to-basic amino acid ratio (1.8), its ability to stimulate DNA transcription system by 30% (homologous or *E. coli* polymerase), and its high affinity for DNA.

The 0.35 *M* NaCl-soluble nonhistone proteins isolated from nuclei preextracted with 0.14 *M* NaCl-0.1 m*M* PMSF may be also considered as 0.35 *M* NaCl-soluble NHCP.[402] It was shown that they were distinctive for several Morris hepatomas (7777, 7288CTC, 9618A_2, 5123C, and 7787) and livers of tumor-bearing rats. Their IFPA revealed an increase in fast-growing hepatoma 7777 of proteins that have isoelectric points below pH 6.0 and a decrease in two proteins with pI 8.2 and 8.5. Similar changes were observed in the more slowly growing hepatomas 5123C and 7787. The two basic proteins were isolated by preparative IFPA and characterized by their relatively low molecular weights: 26,000 for pI 8.2, and two bands with 25,000 and 30,000 mol wt for pI 8.5 protein and with higher acidic-to-basic amino acid ratios (1.93 to 2.57) than were expected from their isoelectric points. The amino acid compositions of these two proteins distinguished them from histones by their low lysine, arginine, and histidine content, and from most NHCP by their high glycine content. The virtual absence of the pI 8.2 and pI 8.5 proteins from neoplastic tissue raised questions regarding their function. Should such proteins exert an inhibitory effect on cell growth then their continued absence may account for the uncontrolled growth of tumor cells.[402]

The NHCP loosely bound to DNA, extracted by 0.35 *M* NaCl from young and old mouse and rat liver chromatin[403] and also from mouse liver, hepatoma, and spleen chromatin[69] were fractionated into three groups: water-soluble, 0.14 *M* NaCl-soluble, and 0.35 *M* NaCl-soluble. It was found by SDS-PAGE that the aging of liver cells was associated with increase in number and quantities of high molecular weight NHCP of the water-soluble group. The pattern of water-soluble, 0.35 *M* NaCl chromatin extract was also found to be tissue specific and to change during malignant growth.[69]

e. NCHP "Very" and "Most" Tightly Bound to DNA

Besides NHCP loosely (buffered urea-soluble or 0.35 *M* NaCl-soluble) and tightly bound to DNA, that need high concentrations of salt or salt and urea to be dissociated from chromatin, there exists a special class of NHCP, called residual NHCP. These proteins remain bound to DNA at very high ionic strength, e.g., in 3.0 *M* NaCl[157] or 2.5 *M* NaCl-5.0 *M* urea system.[196,212,401,404] These residual NHCP can be further divided into proteins that are very tightly and most tightly bound to DNA. The first group may be dissociated by SDS-solutions (after heating) and the second group represents proteins which may be covalently bound to the DNA. The amount of very tightly bound NHCP depends to some extent on tissue, but mainly on the composition, ionic strength, and pH of solution systems used for their dissociation. For example, they constitute less than 3% of total duck erythroid cell chromatin proteins after dissociation in 3 *M* NaCl;[157] 8% of total HeLa chromatin proteins[212] (20% of total NHCP) after dissociation with 2.5 *M* NaCl-5 *M* urea, pH 7.2, but less than 1% of total chromatin protein (about 2% of total NHCP) in the same solvent at pH 8.0;[401] 3 to 5% of rat brain chromosomal proteins after ultracentrifugation of chromatin dissolved in 3.38 CsCl-4 *M* urea-50 m*M* 2-ME-2 m*M* EDTA-10 m*M* lysine, pH 11.6.[226]

Pederson and Bhorjee[212] attempted to characterize the tightly bound proteins which pelleted with DNA at very high ionic strength (when over 92% of the total chromatin protein was released). They could be solubilized in 1% SDS and possessed physical and metabolic properties distinct from other chromatin proteins. Their distribution in chromatin was not random, with over 96% being confined to genetically inactive regions. Assuming the uniform distribution of the tightly bound proteins in chromatin (except for their exclusion from template-active regions), there would be 5.6 molecules of these proteins per 10^6 daltons of DNA, i.e., approximately one molecule of the tightly bound protein per 275 DNA base pairs. Although the precise function of these proteins was not elucidated, they represent the first particular class of NHCP which are distributed asymmetrically between transcriptionally active and inactive chromatin regions.

It was shown more recently that certain proteins cannot be removed by phenol, protease, alkali treatment, or by any combination thereof, from highly purified animal cell DNA (Ehrlich ascites tumor cells, rat liver, calf thymus, sea urchin sperm)[405] or from DNA of plant cells *(Sinapis alba,* hypocotyls, *Pisum sativum* epicotyls, *Cymbidium pumillum* protocorms, and *Rhoeo discolor* leaves).[406] The presence of these proteins was demonstrated by radioiodination of purified DNA samples. After degradation of DNA the iodinated material was analyzed by SDS-PAGE followed by autoradiography. The characteristic band patterns revealed the presence of proteins in the 54,000 to 68,000 mol wt range. Since these proteins could not be removed from DNA except by its degradation,[405] it was concluded that these proteins were covalently bound to DNA. The biological significance of the alkali-stable, DNA-associated proteins is unknown. They represent 1 to 2% of the DNA mass and they possibly play a role in maintaining the stability of the superhelical loops and three-dimensional structure of chromatin. They could also be involved in nondeoxynucleotide linking between adjacent DNA strands.[407] It is noteworthy that similar proteins were also found in nuclear ghosts,[408,409] histone-depleted chromosomes,[410,411] and in salt-treated polytene chromosomes.[412]

C. Fractionation of NHCP by Chromatography

1. Ion-Exchange Chromatography

A variety of techniques have been employed for chromatographic NHCP separation, e.g., chromatography on DEAE-cellulose,[169,206,228,246,413,414] QAE-Sephadex®,[148,205,216,218,415-417] SE-Sephadex®,[86] SP-Sephadex®,[68,161] Bio-Rex®,[206] etc.

a. DEAE-Cellulose Chromatography

This separation technique was first used by Wang and Johns[228] to further fractionate four fractions obtained by selective precipitation of chromatin. Chromatographic profiles of these fractions were different, and heterogeneity of the main protein peaks was studied by starch gel electrophoresis. Neither direct ionic exchange chromatography on DEAE-cellulose nor starch gel electrophoresis was successful in demonstrating a definite fractionation pattern of these multicomponent proteins.

A more successful method was introduced by Kamiyama and Wang.[246] Following the Langan method[245] for phosphoprotein isolation, nuclei, thoroughly preextracted with 0.15 *M* NaCl-3.3 *M* $CaCl_2$-10 m*M* TrisCl, pH 7.2, were homogenized in 1 *M* NaCl-20 m*M* TrisCl, pH 7.5. After removing nucleohistones by precipitation from buffered 1 *M* NaCl-extract by diluting it to a final concentration of 0.4 *M* NaCl, Bio-Rex® 70 was added to the supernatant, with stirring. The NHCP not adsorbed to Bio-Rex® 70, i.e., histone-free, were applied to the DEAE-cellulose column, which was eluted successively with three NaCl solutions (0.05, 0.15, and 0.3 *M*) in 1 m*M* EDTA-1 m*M* 2-ME-15 m*M* TrisCl buffer, pH 8.0. The elution with 0.3 *M* NaCl provided two fractions which were further studied in respect of their effects on chromatin transcription. In further experiments,[413,414] NHCP fractions of different protein kinase activity were eluted from DEAE-cellulose with various NaCl concentrations (0.15, 0.3, and 0.5 *M*[413] or 0.4 and 0.6 *M*[414]) in 50 m*M* TrisCl buffer, pH 8.0.

Chromatography on DEAE-cellulose, in the presence of 3 *M* urea with either a linear (from 0 to 0.3 *M*) or complex NaCl gradient (from 0 to 0.25 *M* NaCl), was used by Levy et al.[169] to resolve NHCP (separated from histones on Bio-Rex® 70 and from RNA on DE-52 columns) into several distinct fractions which were analyzed by SDS-PAGE. The first major peaks contained primarily the higher molecular weight NHCP. The second group contained lower molecular weight NHCP. Complete separation of any particular protein was not achieved.

b. QAE-Sephadex® Chromatography

Many investigators adopted the QAE-Sephadex® batch technique of Paul and Gilmour,[146,204,418] not only for the separation of total chromatin proteins into histones and nonhistones but also for NHCP fractionation.[148,205,216,218,417] Richter and Sekeris[205] fractionated a nonhistone pool on QAE-Sephadex® A-25. Elution profiles of the NHCP from three tissues (rat liver, kidney, and thymus) were very reproducible in their complexity. According to the authors, this separation procedure can serve as a preliminary step in a systematic classification of the NHCP.

Chaudhuri[148] used a single QAE-Sephadex® A-50 column for the separation of histones from nonhistones and fractionation of the [^{3}H] tryptophan-labeled NHCP into six major groups by their displacement with 0.05, 0.15, 0.35, 0.6, 1.0, and 3.0 *M* NaCl. Most of the radioactivity bound to the column was present in 0.15 and 0.35 *M* NaCl-eluted fractions

(13.8 and 26.0% respectively, from 51.8% of the total NaCl eluate). SDS-PAGE showed that each of the six major fractions was still heterogeneous. The protein distribution was similar to that observed in DEAE chromatography,[171] i.e., the high molecular weight proteins were eluted at lower ionic strength than the low molecular weight proteins. The QAE-Sephadex® chromatography of NHCP was also successfullly applied by MacGillivray and co-workers[338,415,416] to further resolve fractions (HAP2 and HAP3) separated on hydroxylapatite.

c. SE-Sephadex® C-25 Chromatography

Elgin and Bonner[81] developed a multistep NHCP fractionation and purification procedure for chromatography on SE-Sephadex® C-25, using 25% HCOOH-10 *M* urea with a gradient of NaCl from 0.2 to 0.8 *M* (pH 2.5) as eluant. The proteins eluted in order of increasing basicity. Initial fractionation of chromosomal proteins on this ion-exchange resin, followed by secondary fractionation by gel filtration (Sephadex® G-100 or G-75) or phosphocellulose chromatography yielded four NHCP fractions. Two were single proteins (θ and ϵ, and two others contained 4 to 6 and 4 to 7 different polypeptides in SDS-phosphate[139,140] and isoelectric focusing[337,419] gel electrophoresis. Proteins in these fractions had acidic-to basic amino acid ratios from 1.2 to 2.7 and isoelectric points from less than 3.7 to 8.0

d. SP-Sephadex® C-25 Chromatography

SP-Sephadex® C-25 was employed by Graziano and Huang,[161] not only for the separation of NHCP from histones but also for the fractionation of NHCP, providing a major nonhistone pool (90% of total NHCP) in 0.23 *M* NaCl-eluate. In experiments of Yoshida et al.[68] calf thymus chromosomal proteins (after DNA precipitation with $LaCl_3$) were first subjected to gel filtration on a Sephadex® G-15 column, previously equilibrated with eluting standard solution (5 *M* urea-20 m*M* acetic acid-3 m*M* 2-ME) containing 0.25 *M* NaCl. Total chromosomal proteins which eluted in void volume were further fractionated by sequential SP-Sephadex® C-25 column chromatography. In the first step the proteins were eluted stepwise with seven concentrations of NaCl (0.25, 0.30, 0.35, 0.40, 0.45, 0.50, and 1.0 *M*). Each fraction was then subjected to a second-step chromatography where the eluting solution was 5 *M* urea-3 m*M* 2-ME-20 m*M* Na-acetate buffer, pH 5.5, with an increasing linear NaCl concentration gradient. Under the first-step conditions total chromosomal proteins were separated into 13 fractions. Total recovery of proteins from the column was 80%. All of these 13 protein fractions were separated by the second-step SP-Sephadex® C-25 chromatography into several subfractions. All protein subfractions obtained by the second-step chromatography were examined for homogeneity and molecular size by SDS-PAGE;[140] 20 subfractions were found to be electrophoretically homogeneous. These purified proteins accounted for about 18% of the total chromosomal proteins. Their molecular weights ranged from 7,400 to 19,000; 11 subfractions of these purified NHCP had acidic-to-basic amino acid rations above 1.0 and the 9 remaining subfractions had ratios below 1.0, corresponding to nonhistone proteins of basic character. It should be emphasized that care was taken to prevent proteolytic degradation of the chromosomal proteins during their fractionation by using effective concentrations (1 and 2.5 m*M*) of *p*-chloromercuribenzoate.

e. Bio-Rex® 70 Chromatography

An interesting technique for fractionation of hen oviduct NHCP has been devised by O'Malley et al.[206] It represents the first application of this cation-exchange resin to fractionation of NHCP. Chromatin (from approximately 90 g oviduct) was dissociated in 2 *M* NaCl-5 *M* urea-0.1% 2-ME-10 m*M* TrisCl buffer, pH 8.3. After desalting by gel filtration (Sephadex® G-25), the chromatin proteins were separated into three distinct fractions by stepwise elution: (I) 0.1 *M* NaCl, (II) 0.25 *M* NaCl, and (III) 15% GdnCl. Fractions I and II contained only nonhistones and Fraction III contained histones with a small amount of

NHCP. Further fractionation of Fraction I NHCP was achieved by ammonium sulfate precipitation and DEAE-cellulose chromatography (after its prior separation on DEAE-cellulose column from RNA which was reported to reduce the resolution of NHCP on this weak anion-exchanger[169]). Fraction II was resolved by phosphocellulose chromatography and gel filtration on Bio-Gel® A-15 m. The histones and nonhistones present in Fraction III were separated by gel filtration on Bio-Gel® A-0.5 m. All fractionation steps were monitored by SDS-Tris-glycine[312] and acid-urea[360] PAGE. The integrity of the histones was maintained during this fractionation procedure indicating that proteolytic degradation was unlikely. The nonhistones were separated into distinct fractions with enrichment of some species not apparent prior to fractionation of the chromatin proteins. This technique reproducibly yielded histone-free NHCP which were soluble in buffers lacking urea.

2. Hydroxylapatite (HAP) Chromatography

This relatively simple one-column procedure not only separates the chromosomal proteins from DNA and the NHCP from histones in high yields, but also provides an initial fractionation of these proteins according to their degree of phosphorylation.[162]

Hydroxylapatite chromatography of proteins was originally developed by Tiselius et al.[420] This method was further improved by Bernardi and Cook and used in the fractionation of phosphoproteins[421] and nucleic acids.[422,423] Bernardi and Kawasaki[424] have undertaken precise studies on the mechanism of interactions of proteins with HAP using a number of water-soluble homo- and heteropolypeptides. The adsorption of polypeptides and proteins to HAP is believed to take place because of the interactions between their carboxyl groups and the calcium sites of the HAP crystals. Increasing the molarity of the eluting phosphate buffer progressively reduces these interactions to zero, at which point desorption occurs. Decrease of the interactions appears to be due to a specific competition for the Ca sites of HAP between the phosphate ions of the eluting buffer and the carboxyl groups of polypeptides.

Studies on the chromatographic behavior of native and and denatured proteins (chromatography in the presence or absence of 8 *M* urea)[424] showed that the latter were much less or not at all retained by HAP columns equilibrated with 1 m*M* potassium phosphate buffer. It appears that the disruption of secondary and tertiary structures of proteins caused a diminution of their interactions with HAP. This phenomenon could be explained by the fact that the random coil conformation of the denatured protein will have a decreased number of carboxyl groups available for interaction per unit of protein surface in contact with HAP and therefore will require lesser molarity for its elution. Elution in these experiments was performed with potassium phosphate buffer, pH 6.8. Sodium phosphate buffer, pH 6.8, may be also used at room temperature or even at 4°C, provided that its molarity is not higher than 0.3 *M*.[422] Calcium-complexing agents like citrate and EDTA should be absent from the samples of nucleic acids or proteins to be chromatographed, however, chloroform, isoamyl alcohol, phenol, formaldehyde, and urea may be present. Regenerated HAP was not used in any experiments described by Bernardi.[421-425]

Fractionation procedures used to separate histones from nucleoprotein solutions dissociated in concentrated NaCl solutions[425] followed the observations of Bernardi and Kawasaki[424] and Bernardi[422,423] that the basic histones are not retained by HAP and that the retained components are eluted in the order of: acidic protein, RNA, and DNA. For routine preparation of NHCP relatively free of RNA, elution by a discontinuous system of phosphate buffers in 2 *M* NaCl-5 *M* urea was found to be optimal.[103] Such separations of histones and NHCP are only achieved on columns run in buffers containing high concentrations of sodium (or potassium) ions. These ions effectively block the phosphate sites which otherwise bind basic molecules such as histones.[110,162]

Introduction of concentrated GdnCl solutions, which are efficient solvents of NHCP,[162] increased their recovery from HAP to virtually quantitative yields[162,416] (instead of the 60 to 70% of adsorbed NHCP which could be recovered with sodium phosphate concentrations up to 200 m*M*[103,110]). But in contrast to columns run in 2 *M* NaCl-5 *M* urea, the proteins were not eluted in order of their phosphorylation. An additional disadvantage of this solvent system was that the NHCP fractions were highly contaminated with nucleic acid material. While solutions of GdnCl are extremely effective in dissociating chromatin, the guanidine ions, unlike sodium ions, are unable to completely block the negatively charged phosphate binding sites of HAP.[162] This phenomenon may be caused by steric factors affecting the interactions of guanidine ions with hydroxylapatite. On the other hand, the presence of some NHCP in the histone fraction and increased nucleic acid contamination of the NHCP fractions suggest that GdnCl reduces the affinity of nonhistone proteins and nucleic acids for the calcium sites of HAP.[162]

In the procedure of MacGillivray and co-workers,[162,426,427] chromatin was dissociated in 2 *M* NaCl-5 *M* urea-1 m*M* sodium phosphate buffer, pH 6.8, with addition of 2 m*M* TrisCl to protect the proteins against cyanate degradation,[155,156] and 0.1 m*M* PMSF as protease inhibitor.[32] Its chromatography on a HAP column provided five distinct fractions which were eluted with increasing concentrations of sodium phosphate (50 to 200 m*M*) and 2 *M* GdnCl (HAP4 and HAP5). The HAP1 fraction (over 70% of total chromatin protein) not retained in the loading (i.e., dissociating) buffer, contained histones and no nucleic acid material. However, contamination of this fraction with up to 15% of NHCP was suggested by [^{3}H]-tryptophan labeling. The HAP2 fraction, eluted with 50 m*M* sodium phosphate and containing about 2% RNA, represented just over 40% of the total NHCP (about 12% of total chromatin protein). These proteins were much more phosphorylated than those of the HAP1 fraction. Proteins of the HAP3 fraction, eluted with 200 m*M* sodium phosphate, were two to three times more phosphorylated than the HAP2 proteins. HAP3 contained relatively large amounts of RNA (up to 50% of the total protein) and, occasionally, small amounts of DNA.

Introduction of 2 *M* GdnCl instead of 2 *M* NaCl-5 *M* urea provided the fraction HAP4, which was phosphorylated to a greater extent than the HAP2 proteins but less than HAP3 proteins, and contained about half as much protein as nucleic acid, mainly RNA. Increasing the phosphate concentration in the last solution to 500 m*M* provided fraction HAP5, containing DNA as the major constituent along with RNA and a small amount of protein (about 1% of total chromatin protein). It should be mentioned that in the first version of this procedure,[103] (i.e., without GdnCl in eluant) the total recovery of chromatin proteins eluted from HAP varied between 45 and 100% depending on the method of preparation of nuclei or chromatin. Because of the limited solubility of the sodium phosphate buffers, the columns had to be run at room temperature. When the HAP columns were run in the cold, substituting potassium phosphate for sodium phosphate in the elution buffers, the proteins in all fractions were identical to those of columns run at room temperature with respect to molecular weight, isoelectric point, and phosphorylation.[162]

It is recommended for the HAP to be freshly prepared from $Na_2HPO_4 \cdot H_2O$ and $CaCl_2$ according to the method of Tiselius et al.[420] and stored at 4°C in 1 m*M* sodium phosphate, pH 6.8, containing a few drops of chloroform.[103] According to MacGillivray,[426] HAP prepared by other procedures or obtained from commercial sources either possesses slow flow rates or gives unsatisfactory separations. However, many laboratories have reported satisfactory results with commercial HAP.

For its simplicity and rapidity the HAP chromatography method of MacGillivray[426] was selected by many investigators and occasionally modified. For example, Appels et al.[428] eluted the NHCP only in two fractions. The first was obtained by washing the column with 200 m*M* sodium phosphate in salt-urea and should be equivalent to the above described HAP2 and HAP3 fractions; the second fraction was then eluted using 0.5 *M* NaCl-5 *M* urea-200 m*M* phosphate-0.2% SDS and is probably similar to the HAP4 fraction. In the experiments of Harlow and Wells,[95] the principal NHCP fraction was eluted with 150 m*M* sodium phosphate, pH 6.8, and the second, representing some residual acidic proteins, with 500 m*M* sodium phosphate, pH 8.0, in 2 *M* NaCl-5 *M* urea.

The HAP chromatography as modified by Bluthmann et al.[105] recovered in high yield and with minimal proteolytic degradation two histone and four NHCP fractions (NH1, NH2, NH3, and NH4) from two bovine tissues (liver and lymphocytes). Batchwise adsorption of chromatin dissolved in 0.5 *M* NaCl-5 *M* urea-1 m*M* sodium phosphate, pH 6.8, was performed in the cold and the suspension was poured into a column run at 4°C. The first NHCP fraction (NH1) was eluted with 0.45 *M* NaCl-5 *M* urea-50 m*M* sodium phosphate, pH 6.8, and was routinely rechromatographed to remove contaminating histones. Three additional nonhistone fractions (NH2, NH3, and NH4) were eluted from the column stepwise with 50 m*M*, 100 m*M*, and 175 m*M* potassium phosphate, pH 7.5, in 2 *M* KC1-5 *M* urea. The recovery of histones was 90% or higher for both investigated tissues. The same was true for the lymphocyte NHCP eluates in the fractions NH1 to NH4, whereas chromatography of liver chromatin gave a yield of NHCP only about 70%. The histones and the first two NHCP fractions were virtually free of nucleic acids. Most of the chromatin RNA was eluted in fraction NH4. The four nonhistone fractions exhibited reproducible fraction- and tissue-specific patterns on SDS polyacrylamide gels.[105] Stimulation of bovine lymphocytes with phytohemagglutinin resulted in quantitative as well as qualitative changes in the electrophoretic patterns of four NHCP fractions obtained by HAP [429]

The fraction NH4 from lymphocyte chromatin was shown to contain a prominent band of 30,000 mol wt, comprising 40% of the NH4 HAP fraction, which is characteristic for this tissue and not found in liver chromatin.[105] Affinity chromatography of the NH4 HAP fraction from lymphocytes on a single-stranded DNA-agarose column containing either *E. coli* DNA or lymphocyte DNA resulted in the isolation of a homogeneous NHCP, designated NH30000.[430,431] Equilibrium competition experiments showed that protein NH30000 exhibited higher affinity for lymphocyte DNA as compared to *E. coli* DNA.[430] It interacted preferentially with single-stranded DNA and with AT-rich and repetitive DNA sequences.[431]

Mild conditions for NHCP fractionation were used in the HAP chromatography performed by Kilianska et al.[59,256] It concerned the whole chromatin and chromatin devoid of 0.35 *M* NaCl-soluble fraction, which constituted about 30% of rat liver NHCP.[256] These two kinds of chromatin were dissociated in buffered 2 *M* NaCl, pH 7.9, containing 8% glycerol, 1% DMSO (to produce a relaxation of chromatin structure by destabilization of protein-DNA interactions[432]) and 1 m*M* PMSF, and were released from DNA by a two-phase PEG-dextran partition system before being applied to the HAP column. Chromatography was carried out at 4°C. After elution of histones in the loading buffer (2 *M* NaCl-8% glycerol-5 m*M* sodium phosphate, pH 6.8-0.1 m*M* PMSF), three distinct NHCP fractions NHCP1, NHCP2, and NHCP3 were eluted with increasing phosphate concentrations: 50 m*M* (Na^+ form), 100 m*M*, and 200 m*M* (K^+ form), respectively. Electrophoretic patterns of these three fractions obtained from the two kinds of chromatin revealed heterogeneity, which was somewhat lesser in chromatin depleted of 0.35 *M* NaCl-soluble proteins. The three NHCP fractions represented, in whole chromatin, about 66, 26, and 8% of NHCP recovered from HAP.[256] They were acidic in nature (acidic-to-basic residues ratios were 1.3, 1.6, and 1.3 in NHCP1, NHCP2, and NHCP3, respectively). They were completely devoid of DNA, but not free of RNA or carbohydrates, as was demonstrated by PAS staining.

Bloom and Anderson[433] developed a HAP dissociation method which permits the fractionation of chromosomal proteins according to their chromatin binding properties. This technique is based on the ability of HAP to bind native chromatin in solutions which do not dissociate chromosomal proteins from DNA (e.g., 10 m*M* sodium phosphate, pH 7.0). Chromosomal proteins can be then selectively dissociated from the immobilized chromatin (immobilized DNA or other chromatin components still bound to DNA) by stepwise elution with NaCl, urea or GdnCl, or a combination of the two reagents. This procedure, easily adaptable to large-scale experiments, represents a simple one-step general method for the fractionation of chromosomal proteins which results in their near-quantitative recovery, devoid of nucleic acids. The authors have shown that the specific dissociation patterns result from the differential chromatin-protein binding properties and not from protein-HAP interactions. Their technique makes possible the isolation of three major histone groups (very lysine-rich, moderately lysine-rich, and arginine-rich) in a single operation as well as of the loosely, tightly, and very tightly bound NHCP. The histone H1 was completely dissociated in 0.5 to 0.75 *M* NaCl. The moderately lysine-rich species (H2A and H2B) were eluted with 0.75 to 1.25 *M* NaCl, whereas the histones H3 and H4 were dissociated only at the higher salt concentrations (1.25 to 2 *M* NaCl). Analysis of the NHCP was more difficult because of the marked heterogeneity of these proteins. They were classified into three groups according to their patterns of dissociation from immobilized chromatin. All dissociating solutions were prepared in 80 m*M* sodium phosphate buffer, pH 7.0. A few of the NHCP appeared to be completely dissociated from chromatin with 0.25 *M* NaCl. The second group of NHCP was dissociated only after addition of urea (5 *M*) or GdnCl (4 *M*). The third group was eluted by 4 *M* GdnCl-5*M* urea. The dissociation of NHCP from chromatin was also altered by urea (diminution of nonelectrostatic interactions). More than 80% of the NHCP were dissociated from immobilized chromatin by 5 *M* urea in the absence of NaCl, an observation confirmed by many previous reports using selective extraction methods. Proteins dissociated in 5 *M* urea contained all detectable polypeptides eluted by 0.25 *M* NaCl plus some of the proteins which were eluted in 2 *M* NaCl-5 *M* urea. The very tightly bound NHCP, dissociated by treatment with GdnCl, exhibited similar banding patterns, regardless of whether the columns were initially treated with 0.25 *M* NaCl or 5 *M* urea.

3. Hydrophobic Affinity Chromatography of NHCP

Hydrophobic affinity chromatography,[434] or hydrophobic interaction chromatography,[435] or hydrophobic chromatography,[436] was conceived as a consequence of nonspecific adsorption[436] seen in affinity chromatography systems and attributed to the characteristics of the matrix or support and the nature of ligand and spacer-arms introduced to bridge the ligand to the matrix backbone (for review see Reference 437). Hjerten[435] has demonstrated the usefulness of hydrophobic interaction chromatography for the fractionation of biological materials, described some general aspects of the method,[435] and prepared some neutral adsorbents by coupling aliphatic and aromatic alcohols, instead of amines, to agarose gels.[438] During the last decade, hydrocarbon groups[439,440] and mixed hydrocarbon and ionic groups,[434,436,441] bound to agarose or Sepharose®, have been used to fractionate a variety of proteins. Contrary to histones which bind primarily through ionic forces to negatively charged DNA, some NHCP[442,443] may interact with DNA by hydrophobic interactions, including intercalation of tyrosine and tryptophan residues between the bases of DNA. This phenomenon is partially responsible for the limited solubility of DNA-free NHCP.

First attempts to fractionate the NHCP by hydrophobic chromatography were made by Kikuchi and Sato.[100] During studies[252] on the interaction of daunomycin with rat liver chromatin these authors noticed that daunomycin binds not only to DNA but also to nonhistone proteins. It was shown, by equilibrium dialysis, that NHCP were by far more potent in binding daunomycin than other proteins tested (histones, phosvitin, casein, catalase, and fibrinogen). The amount of daunomycin bound to NHCP increased with increasing concentration of protein until a maximum was reached, which was most likely due to aggregation of NHCP at high concentrations.[249] In an attempt to prevent this aggregation and to increase the daunomycin binding various glycerol concentrations were added to the dialysis buffer.[100] It was found that 8% glycerol was inhibitory. This pointed to hydrophobic interactions. Because of the marked interactions between daunomycin and DNA,[444] the NHCP were tested for possible DNA contamination.[100] DNase treatment had no significant effect on the daunomycin binding indicating that DNA contamination, if any, was negligible. (DNase digested DNA did not react with the drug.[444])

Experiments of daunomycin binding to NHCP were utilized by these authors[100] to fractionate the NHCP from rat leukemia cells (DBLA-6) on daunomycin-CH-Sepharose® 4B. Daunomycin was efficiently coupled through its amino group to the carboxyl group of CH-Sepharose® 4B by the use of *N*-(3-dimethylaminopropyl) -*N*-ethylcarbodiimide. The NHCP were adsorbed to the column using high ionic strength buffer (4 *M* NaCl-5 m*M* sodium phosphate, pH 6.8) and eluted by lowering the ionic strength and by increasing the concentration of glycerol, i.e., by 1 *M* NaCl-20% glycerol-10 m*M* TrisCl buffer, pH 7.5. This method proved to be useful for fractionation of NHCP which aggregate without detergent or at low ionic strength.[169,445]

Comings et al.[158] have fractionated mouse liver NHCP by hydrophobic chromatography on six various columns. A 2 *M* NaCl nuclear extract NHCP fraction eluted with 50 m*M* sodium phosphate buffer, pH 6.8, from the HAP column was employed in the experiments. Two column-loading techniques, termed as the 0.35 *M* NaCl technique and the $(NH_4)_2SO_4$ technique, were used. In the second one, which was more satisfactory, the NHCP were mixed with hydrophobic agarose, dialyzed against 40% $(NH_4)_2SO_4$, eluted initially with this solution, followed with 0.35, 1.0, and 5.0 *M* LiBr and finally 1% SDS. Good fractionation was obtained with ω-amino decyl-agarose, butyl amino-agarose, butyl-agarose, and CPAD-Sepharose®. A pair of relatively pure 15,000 to 20,000 mol wt polypeptides, termed the K proteins, was found in the 40% $(NH_4)_2SO_4$ wash.

4. Bioaffinity Chromatography of NHCP

In recent years there has been considerable interest in the development of numerous techniques utilizing the separation of macromolecules through their specific interactions (binding) with matrix-bound ligands. According to Cuatrecasas and Anfinsen,[446] affinity chromatography combines "procedures used to purify one of the components of a system consisting of two or more species whose reversible interactions reflect affinity with a high degree of biological specificity." In this kind of chromatography "specific biological properties may be exploited for purification as much in the future as the physicochemical properties have been exploited in the past." The basic principle in the affinity chromatography of NHCP is to immobilize one of the components of the interacting system (e.g., the DNA or histone as a ligand) to an insoluble, porous support (water-insoluble polymers, i.e., beaded derivatives of dextran, agarose, or synthetic polyacrylamide gels) which then can be used to adsorb selectively, in a chromatographic procedure, the component (e.g., NHCP) with which it can selectively interact.

Depending on the ligand, two main kinds of affinity chromatography have been used to fractionate NHCP — DNA affinity chromatography and histone affinity chromatography. In both, the nonhistones to be fractionated are passed through a column containing water-insoluble polymer to which a specific chromatin compound (DNA or histone) has been covalently attached. All NHCP without substantial affinity for the ligand will pass directly through the column whereas the ones which recognize some nucleotide or amino acid sequences will be retarded in proportion to its affinity constant.[447] Elution can be achieved by any one of a number of procedures which result in the dissociation of the complex, e,g., by changing such parameters as salt concentration or pH, or by addition of a competitive substance. A special case of affinity chromatography is the application of Cibacron® Blue F3GA-agarose.

a. DNA Affinity Chromatography of NHCP

Chromatography on affinity adsorbents containing polynucleotide ligands dates back to 1968 when Litman[448] reported the successful purification of micrococcal DNA polymerase on a column containing DNA immobilized onto cellulose by irradiation with ultraviolet light. Since then affinity chromatography using immobilized nucleic acids has established itself as a useful technique which has received numerous applications since the initial experiment of Kleinsmith et al.[449] For review see articles by Allfrey and associates.[450-452]

b. Histone Affinity Chromatography of NHCP

The presence of both histone and nonhistone proteins in chromatin raises the possibility that these two types of proteins interact with each other. This question was first addressed by Wang and Johns[228] who found that histones and NHCP precipitate in specific manner when mixed together. The idea that interactions between histone and nonhistone proteins may be biologically meaningful was strengthened by Kaplowitz et al.[453] who showed that such interactions greatly enhanced the rate of NHCP phosphorylation, with histone H1 stimulating phosphorylation the most and H2B the least. More recently the problem of histone-nonhistone interactions was investigated by histone affinity chromatography, a technique that avoids the precipitation of histone-NHCP complexes.

Perhaps the best results of NHCP fractionation on immobilized histones were obtained when, instead of conventional top-loading of the column, [^{32}P]-labeled NHCP were bound to histone (H1 or H2B)-Sepharose® using gradient dialysis,[209] i.e., mixing of NHCP with histone-Sepharose® in the presence of 2 *M* NaCl-4 *M* urea and then gradually lowering the salt concentration. The radioactive NHCP were eluted in several distinct peaks as ionic strength was increased (from 0.15 to 2.0 *M* NaCl) with the elution profiles differing for columns containing histones H1 and H2B. Virtually no radioactivity was bound to the unreacted Sepharose® run under the same conditions. Likewise, only a little [^{32}P]-NHCP was bound to the column of cytochrome c Sepharose®, even though cytochrome c is also a small basic protein.

The histone H1 column bound less total radioactivity (39%) than the H2B histone column (65%). The largest peak of [^{32}P]-NHCP was eluted from the H1 column at a lower salt concentration (0.3 *M* NaCl) than from the H2B column (0.5 *M* NaCl). A given peak, rechromatographed on the same histone column, was eluted at the same NaCl concentration which was originally required for its elution, but at a different NaCl concentration when rechromatographed on another type of histone column. SDS-PAGE and autoradiography of the various NHCP fractions eluted from the histone-Sepharose® columns revealed significant differences in molecular and metabolic patterns of the individual NHCP.

Histone affinity chromatography using total histones or purified H2A, H2B, or H4 fractions immobilized on Affi-Gel® 10 was used by Conner and Comings[454] to examine the cytoplasm and nucleoplasm of mouse liver for histone-binding proteins. Proteins from the nuclear washes showed greater binding to the histone-columns than proteins from cytoplasm. Proteins bound to columns constructed with individual histones were similar to those bound to columns of whole histone, polylysine, and DEAE. A 25,000 mol wt polypeptide (J2), found only in nuclear washes (0.15 *M*-NaCl nuclear extract), was a prominent histone-binding protein. It had an acidic-to-base amino acid ratio of 1.58 and it also bound to hydrophobic columns. The cited results suggest the presence of interactions of some specificity between the NHCP and histones, a finding of potential significance for chromatin structure and function. The polypeptide J2 was recently purified using hydrophobic chromatography.[454]

c. Cibacron® Blue Affinity Chromatography

Special case of NHCP affinity chromatography is their fractionation on Cibacron® Blue F3GA-agarose.[99] In contrast to Blue Dextran-agarose which is specific only for proteins containing the so-called "dinucleotide fold", Cibacron® Blue F3GA-agarose is less specific and may have affinity for all proteins with a nucleotide-binding site.[455,456] Recently Kristensen and Holtlund[99] fractionated Ehrlich ascites tumor NHCP on Cibacron® Blue F3GA-agarose.

The comparison of chromatographic behavior of nonhistones and histones on this matrix revealed their different affinity to Cibacron® Blue and suggested some correlations between the strength of the protein-DNA and the protein-Cibacron® Blue interactions. This seems to indicate that the dye interacts with sites on proteins which normally bind nucleic acids.

V. CONCLUSIONS

As follows from our discussion of chromosomal nonhistone protein research, there are no simple ways, the HMG proteins and some enzymes exempting, which would allow a rapid and clean isolation of the individual members of this highly complex family. Nevertheless, by combining various fractionation techniques, several nonhistone proteins were characterized. Their representative, although incomplete, list is presented in Table 8. Of the methods available for identification of the individual protein species, even in the presence of numerous other proteins, immunology is gaining popularity and will become, without doubt, one of the most important techniques aiding studies on the biochemistry and biology of this unique protein class.

ACKNOWLEDGMENTS

Supported by the Polish National Cancer Program PR-6 (LKS) and by the U.S. Department of Health and Human Services, National Cancer Institute Grants CA 18389, CA 26412, and CA 27338 (LSH). The authors are grateful to Miss Monika Radwan for her able technical help and to Ms. Doris Harris for her excellent assistance in preparing this chapter for publication.

Table 8
FRACTIONATION STEPS USED IN THE ISOLATION OF SOME INDIVIDUAL NHCP EXCLUDING HMG PROTEINS

No.	Protein, some properties and ref.	Source of material	Purification procedures
		(a) Reports of Busch and Associates	
1	A24(27,000 mol wt; A/B[a] = 0.93)[53]	Rat liver nuclei or chromatin	Extract: 0.2 *M* H_2SO_4 or 3*M* NaCl-7*M* urea Purification: (1) Sephadex® G-200 in: 1 *M* urea-1% 2-ME-50 m*M* HCl; ethanol precipitation (2) Preparative slab gel acid-urea-PAGE
2	NAg-1 (26,000 mol wt; A/B: 1 to 1.4, 15 mol gal)[363]	Novikoff hepatoma; Walker 256 carcinosarcoma chromatin	Extract: 0.6 *M* NaCl-10 m*M* TrisCl, pH 8.0-0.1 m*M* PMSF Purification: (1) Sephadex® G-150 (2) Preparative slab PAGE in: 90 m*M* TrisCl-2 m*M* EDTA-80 m*M* sodium borate, pH 8.1
3	C-14 (70,000 mol wt; A/B = 1.8)[42]	Novikoff hepatoma chromatin	Extract: 0.35 *M* NaCl-10 m*M* TrisCl, pH 8.0-0.1 m*M* PMSF (buffer A) Purification: (1) Ammonium sulfate 30—60% pptn (2) Sephadex® G-75 in buffer A (3) Preparative slab gel acid-urea-PAGE
4	BA (31,000 mol wt; A/B = 0.9)[300]	Rat liver chromatin	Extract: Chromatin Fraction II after DNase treatment; solubilized in: 8 *M* urea-50 m*M* sodium phosphate, pH 7.6 Purification: (1) Preparative slab gel acid-urea-PAGE (2) Sephadex® G-100 chromatography
		(b) Reports of Wang and Associates	
5	Transcription stimulator (10,000—170,000 mol wt; P[b] = 0.9%)[38,112]	Ehrlich ascites tumor chromatin	Extract: 0.35 *M* NaCl-20 m*M* TrisCl, pH 7.5 Purification: (1) Bio-Rex® 70 chromatography; DNA affinity chromatography in 50 m*M* NaCl-1 m*M* EDTA-10 m*M* TrisCl, pH 7.4 on: *E. coli*, DNA-cellulose, and ascites DNA-cellulose (2) Phenol procedure
6	1-NHP (11,000 mol wt; A/B = 1.4; P = 2.7)[238,239]	As in No. 5	Extract: 2 *M* NaCl-20 m*M* TrisCl, pH 8.0; DNH pptn at 0.14 *M* NaCl Purification: (1) Phenol procedure
7	I-NHP (30,800 mol wt)[237]	Calf thymus chromatin	As in No. 6
		(c) Some Other Reports	
8	NH 30000 (30,000 mol wt; A/B = 1.39)[430]	Bovine lymphocytes chromatin	(1) HAP chromatography; NH4 eluted with 2 *M* NaCl-5 *M* urea-175 m*M* potassium phosphate, pH 7.5 (2) Affinity chromatography on *E. coli* DNA-agarose and lymphocyte DNA-agarose

Table 8 (continued)
FRACTIONATION STEPS USED IN THE ISOLATION OF SOME INDIVIDUAL NHCP EXCLUDING HMG PROTEINS

No.	Protein, some properties and ref.	Source of material	Purification procedures
		(a) Reports of Busch and Associates	
9	Component 10 (11,000 mol wt; pI 4.5)[416]	Mouse liver nuclei nucleoplasm-free	(1) HAP chromatography; HAP2 eluted with 2 *M* NaCl-5 *M* urea-50 m*M* sodium phosphate, pH 6.8 (2) QAE-Sephadex® A-25 in: 5 *M* urea-0.5 m*M* $MgCl_2$-1 m*M* DTT-1 m*M* EDTA-10 m*M* TrisCl, pH 8.3 or thin-layer isoelectro-focusing in Sephadex® G-75 superfine (3) Sepharose® 6B in: 6 *M* GdnCl-1 m*M* DTT-50 m*M* sodium acetate, pH 6.4
10	Nuclear antigens p39, p49 and p56[43,327]	Novikoff hepatoma chromatin	(1) HAP chromatography of UC extract; HAP2 eluted with 2 *M* KCl-5 *M* urea-2 m*M* TrisCl-50 m*M* potassium phosphate, pH 7.0-0.1 m*M* PMSF (2) Preparative slab gel SDS-PAGE[312] (3) Dowex® AG1x2 for SDS removal[144]
11	Nuclear antigen (90,000 mol wt; A/B = 1.35)[48]	Chicken erythroid chromatin	(1) Sequential extraction: (a) 5 *M* urea-200 m*M* potassium phosphate, pH 7.8; (b) 2.5 *M* NaCl-5 *M* urea-sodium succinate, pH 5.0; (c) 2% SDS-50 m*M* sodium phosphate, pH 7.0 (2) Preparative rod gel SS-PAGE[140] (3) Dowex® AG1x2 for SDS removal[144]
12	Phosphorprotein B_2 (68,000 mol wt; pI 6.2—8.8)[458]	Rat liver nuclei	Extract: phenol procedure Purification: (1) CM-cellulose CM 23 in: 8 *M* urea-20 m*M* glycine-0.05% 2-ME-20 m*M* TrisCl, pH 9.2-0.1 m*M* PMSF; rechromatography at pH 10 (2) Preparative rod gel SDS-PAGE
13	RNA polymerase II stimulator (66,000 mol wt; pI 8.2—9.0)[276]	Rat liver nuclei	Extract: 8 *M* urea-50 m*M* sodium phosphate, pH 7.6 Purification 177-fold: (1) QAE-Sephadex® A-25, linear gradient 0.1—3, *M* NaCl (2) Preparative thin-layer flat bed isoelectrofocusing in Sephadex® G-75 superfine
14	Specific substrate of cAMP-independent kinase (13,000 mol wt; pI 9.0)[459,460]	Calf thymus nuclei	Extract: 0.35 *M* NaCl Purification: (1) Sephadex® G-75 in: 0.35 *M* NaCl-5 *M* urea-20 m*M* sodium borate, pH 9.0 (2) CM-cellulose in: 5 *M* urea-20 m*M* sodium borate, pH 9.0 (3) HAP chromatography of F_2: linear gradient of 10 to 200 m*M* sodium phosphate, pH 7.5 substrate activity in 80—130 m*M* phosphate

[a] A/B = acidic-to-basic amino acid ratio.
[b] P = alkali labile phosphorus

REFERENCES

1. **Spelsberg, T. C., Wilhelm, J. A., and Hnilica, L. S.,** Nuclear proteins in genetic restriction. II. The nonhistone proteins in chromatin, *Sub-Cell Biochem.*, 1, 107, 1972.
2. **Hnilica, L. S.,** *The Structure and Biological Functions of Histones,* CRC Press, Boca Raton, Fla., 1972.
3. **Elgin, S. C. R. and Weintraub, H.,** Chromosomal proteins and chromatin structure, *Annu. Rev. Biochem.*, 44, 725, 1975.
4. **Chiu, J. F. and Hnilica, L. S.,** Nuclear nonhistone proteins: chemistry and function, in *Chromatin and Chromosomal Structure,* Li, H. J. and Eckhardt, R. A., Eds., Academic Press, New York, 1977, 193.
5. **Patel, G. L.,** Isolation of the nuclear acidic proteins, their fractionation, and some general characteristics, in *Acidic Proteins of the Nucleus,* Cammeron, I. L. and Jeter, J. R., Jr., Eds., Academic Press, New York, 1974, 29.
6. **Birnie, G. D.,** Isolation of nuclei from animal cells in culture, in *Methods in Cell Biology,* Vol. 17, Stein, G. S.,Stein, J., and Kleinsmith, L. J., Eds., Academic Press, New York, 1978, 13.
7. **Garrard, W. T. and Hancock, R.,** Preparation of chromatin from animal tissues and cultured cells, in *Methods in Cell Biology,* Vol. 17, Stein, G. S., Stein, J., and Kleinsmith, L. J., Eds., Academic Press, New York, 1978, 27.
8. **Stellwagen, R. H., Reid, B. R., and Cole, R. D.,** Degradation of histone during the manipulation of isolated nuclei and deoxyribonucleoprotein, *Biochim. Biophys. Acta,* 155, 581, 1968.
9. **Davies, P., Rita, G. A., Krakauer, K., and Weissman, G.,** Characterization of a neutral protease from lysosomes of rabbit polymorphonuclear leucocytes, *Biochem. J.*, 123, 559, 1971.
10. **Davies, P., Krakauer, K., and Weissmann, G.,** Calf thymus histone as a substrate for neutral and acid proteases of leucocyte lysosomes and other proteolytic enzymes, *Anal. Biochem.*, 45, 428, 1972.
11. **Destree, O. H. J., D'Adelhart-Toorop, H. A., and Charles, R.,** Cytoplasmic origin of the so-called nuclear neutral histone protease, *Biochim. Biophys. Acta,* 378, 450, 1975.
12. **Heinrich, P. C., Raydt, G., Puschendorf, B., and Jusic, M.,** Subcellular distribution of histone-degrading enzyme activities from rat liver, *Eur. J. Biochem.*, 62, 37, 1976.
13. **Haas, R., Weiss, E., Jusic, M., and Heinrich, P. C.,** Limited proteolysis of the very-lysine-rich histone H1 by a membrane-bound proteinase from rat liver, *FEBS Lett.*, 63, 117, 1976.
14. **Haas, R., Nagasawa, T., and Heinrich, P. C.,** The localization of a proteinase within rat liver mitochondria, *Biochem. Biophys. Res. Commun.*, 74, 1060, 1977.
15. **Kravchenko, L. S., Oksman, A. Ya., Glubovskaya, I. O., and Tereshin, I. M.,** Effect of polyene antibiotics on nuclear membrane-bound and free DNP fractions from dog kidney, *Biokhimyia,* 41, 133, 1976.
16. **Hoffmann, P. and Chalkley, R.,** Procedures for minimizing protease activity during isolation of nuclei chromatin and histone, in *Methods in Cell Biology,* Vol. 17, Stein, G. S., Stein, J., and Kleinsmith, L. J., Eds., Academic Press, New York, 1978, 1.
17. **Carter, D. B., Efird, P. H., and Chae, C. B.,** Chromatin-bound protease and their inhibitors, in *Methods in Cell Biology,* Vol. 19, Stein, G. S., Stein, J., and Kleinsmith, L. J., Eds., Academic Press, New York, 1978, 175.
18. **Furlan, M. and Jericijo, M.,** Protein catabolism in thymus nuclei. II. Binding of histone-splitting nuclear proteases to deoxyribonucleic acid, *Biochim. Biophys. Acta,* 147, 145, 1967.
19. **Kurecki, T. and Toczko, K.,** Purification and partial characterization of protease from calf thymus chromatin, *Acta Biochim. Pol.*, 21, 225, 1974.
20. **Carter, B. D., Efird, P. H., and Chae, C. B.,** Chromatin-bound protease: [^{3}H]diisopropyl fluorophosphate labelling patterns of chromatin, *Biochemistry,* 15, 2603, 1976.
21. **Kurecki, T. and Toczko, K.,** Isolation and partial purification of protease from calf thymus chromatin, *Bull. Acad. Polon. Sci., Ser. Sci. Biol.*, 20, 543, 1972.
22. **Kurecki, T., Kowalska-Loth, B., Toczko, K., and Chmielewska, I.,** Evidence that neutral protease from calf thymus chromatin is a serine type enzyme, *FEBS Lett.*, 53, 313, 1975.
23. **Furlan, M. and Jericijo, M.,** Protein catabolism in thymus nuclei. I. Hydrolysis of nucleoproteins by proteases present in calf-thymus nuclei, *Biochim. Biophys. Acta,* 147, 135, 1967.
24. **Chae, C. B. and Carter, D. B.,** Degradation of chromosomal proteins during dissociation and reconstitution of chromatin, *Biochem. Biophys. Res. Commun.*, 57, 740, 1974.
25. **Stein, G. S., Stein, J. L., and Thompson, J. A.,** Chromosomal proteins in transformed and neoplastic cells: a review, *Cancer Res.*, 38, 1181, 1978.
26. **Lipinska, A. and Klyszejko-Stefanowicz, L.,** The activity of chromatin-bound protease extracted selectively with histone H2B from calf thymus and rat liver, *Int. J. Biochem.*, 11, 299, 1980.
27. **Lipinska, A., Krawczyk, Z., Krajewska, W., Klyszejko-Stefanowicz, L., and Chorazy, M.,** Activity of chromatin-bound protease in histone fractions from rat liver and Morris hepatoma, *Neoplasma,* 27, 409, 1980.

28. **Panyim, S., Jensen, R. H., and Chalkley, R.,** Proteolytic contamination of calf thymus nucleohistone and its inhibition, *Biochim. Biophys. Acta,* 160, 252, 1968.
29. **Panyim, S. and Chalkley, R.,** High resolution acrylamide gel electrophoresis of histones, *Arch. Biochem. Biophys.,* 130, 337, 1969.
30. **Van den Broek, H. W. J., Nooden, L. D., Sevall, J. S., and Bonner, J.,** Isolation, purification and fractionation of nonhistone chromosomal proteins, *Biochemistry,* 12, 229, 1973.
31. **Furlan, M., Jericijo, M., and Suhar, A.,** Purification and properties of a neutral protease from calf thymus nuclei, *Biochim. Biophys. Acta,* 167, 154, 1968.
32. **Nooden, L. D., Van den Broek, H. W. J., and Sevall, J. S.,** Stabilization of histones from rat liver, *FEBS Lett.,* 29, 326, 1973.
33. **Fahrney, D. E. and Gold, A. M.,** Sulfonyl fluorides as inhibitors of esterases. I. Rates of reaction with acetylcholinesterase, α-chymotrypsin and trypsin, *J. Am. Chem. Soc.,* 85, 997, 1963.
34. **Garrels, J. I., Elgin, S. C. R., and Bonner, J.,** A histone protease of rat liver chromatin, *Biochem. Biophys. Res. Commun.,* 46, 545, 1972.
35. **Bartley, J. A. and Chalkley, R.,** Further studies of a thymus nucleohistone-associated protease, *J. Biol. Chem.,* 245, 4286, 1970.
36. **Ballal, N. R., Goldberg, D. A., and Busch, H.,** Dissociation and reconstitution of chromatin without appreciable degradation of the proteins, *Biochem. Biophys. Res. Commun.,* 62, 972, 1975.
37. **Carter, B. D. and Chae, C. B.,** Chromatin-bound protease: degradation of chromosomal proteins under chromatin dissociation conditions, *Biochemistry,* 15, 180, 1976.
38. **Kostraba, N. C., Montagna, R. A., and Wang, T. Y.,** Study of the loosely bound non-histone chromatin proteins, *J. Biol. Chem.,* 250, 1548, 1975.
39. **Paulin, D., Nicolas, I. F., Jacquet, M., Jakob, H., Gros, F., and Jacob, F.,** Comparative protein patterns in chromatins from mouse teratocarcinoma cells, *Exp. Cell Res.,* 102, 169, 1976.
40. **Boffa, L. C. and Allfrey, V. G.,** Characteristic complements of nuclear nonhistone proteins in colonic epithelial tumors, *Cancer Res.,* 36, 2678, 1976.
41. **Boffa, L. C. and Allfrey, V. G.,** Changes in chromosomal proteins in colon cancer, *Cancer,* 40, 2584, 1977.
42. **James, G. T., Yeoman, L. C., Matsui, S., Goldberg, A. H., and Busch, H.,** Isolation and characterization of nonhistone chromosomal protein C-14 which stimulates RNA synthesis, *Biochemistry,* 16, 2384, 1977.
43. **Fujitani, H., Chiu, J. F., and Hnilica, L. S.,** Purification of nuclear antigens in Novikoff hepatoma, *Proc. Natl. Acad. Sci. U.S.A.,* 75, 1943, 1978.
44. **Lesser, B. H. and Comings, D. E.,** Specific interaction between mouse liver non-histone chromosomal proteins and mouse DNA demonstrated by a sequential DNA-proteins binding procedure, *Biochim. Biophys. Acta,* 521, 117, 1978.
45. **Gronow, M., Thackrah, T. M., and Jardine, N. J.,** Isolation of chromatin fragments rich in non-histone protein after endonuclease digestion of rat liver nuclei, *Exp. Cell Res.,* 119, 381, 1979.
46. **Sugita, M., Yoshida, K., and Sasaki, K.,** Germination-induced changes in chromosomal proteins of spring and winter wheat embryos, *Plant Physiol.,* 64, 780, 1979.
47. **Diondjurow L., Ivanova, E., and Tsanev, R.,** Two chromatin fractions with different metabolic properties of non-histone proteins and newly synthesized RNA, *Eur. J. Biochem.,* 97, 133, 1979.
48. **Krajewska, W. M., Briggs, R. C., Chiu, J. F., and Hnilica, L. S.,** Immunologically specific complexes of chromosomal nonhistone proteins with deoxyribonucleic acid in chicken erythroid nuclei, *Biochemistry,* 19, 4667, 1980.
49. **Falkner, G. F. and Biessmann, H.,** Nuclear proteins in *Drosophila melanogaster* cells after heat shock and their binding to homologous DNA, *Nucleic Acids Res.,* 8, 943, 1980.
50. **Sevaljevic, L. and Konstantinovic, M.,** Identification of factors affecting electrophoretic properties of chromatin nonhistone proteins, *Biochimie,* 60, 1193, 1978.
51. **Kahl, G., Schäfer, W., and Wechselberger, M.,** Changes in non-histone chromosomal proteins during the development of potato tubers; their involvement in wound- and hormone-induced processes, *Plant Cell Physiol.,* 20, 1217, 1979.
52. **Takami, M. and Busch, H.,** Two-dimensional gel electrophoretic comparison of proteins of nuclear fraction of normal liver and Novikoff hepatoma, *Cancer Res.,* 39, 507, 1979.
53. **Goldknopf, I. L., Taylor, C. W., Baum, R. M., Yeoman, L. C., Olson, M. O. J., Prestayko, A. W., and Busch, H.,** Isolation and characterization of protein A24, a "histone-like" non-histone chromosomal protein, *J. Biol. Chem.,* 250, 7182, 1975.
54. **Bekhor, I. and Feldman, B.,** Assembly of DNA with histones and nonhistone chromosomal proteins, *Biochemistry,* 15, 4771, 1976.
55. **Hardy, K., Chiu, J. F., Beyer, A., and Hnilica, L. S.,** Immunological properties of fractionated avian erythroid nuclei, *J. Biol. Chem.,* 253, 5825, 1978.

56. **Tsanev, R. and Hadjiolov, D.,** Chromosomal proteins in hepatocarcinogenesis, *Z. Krebsforsch.,* 91, 237, 1978.
57. **Weiss, E. and Heinrich, P. C.,** The effect of inhibitors of transcription and translation on chromosomal proteins, *Mol. Pharmacol.,* 14, 1148, 1978.
58. **Rodriguez, L. V. and Becker, F. F.,** Alterations in chromosomal proteins in C_3H/HeN mice with spontaneous primary hepatocellular carcinomas, *Int. J. Cancer,* 24, 806, 1979.
59. **Kilianska, Z., Sklodowska, M., and Klyszejko-Stefanowicz, L.,** Characterization of nuclear non-histone proteins from rat liver isolated in the polyethylene glycol-dextran system and separated by hydroxylapatite chromatography, *Acta Biochim. Pol.,* 27, 89, 1980.
60. **Lin, J. C. and Pagano, J. S.,** Effect of 5′-iodo-2′-deoxyuridine on physical properties and non-histone chromosomal proteins of chromatin from Burkitt somatic cell hybrids, *Arch. Biochem. Biophys.,* 200, 567, 1980.
61. **Baugnet-Mahieu, L., Baeyens, W., and Maisin, J. R.,** Chromosomal proteins in normal and leukaemic rat lymphoid tissues, in *Biology of the Cancer Cell,* Proc. 5th Meet. Eur. Assoc. Cancer Res. Kugler Publications, Amsterdam, 1980, 415.
62. **Bekhor, I., Lapeyre, J. N., and Kim, J.,** Fractionation of nonhistone chromosomal proteins isolated from rabbit liver and submandibular salivary glands, *Arch. Biochem. Biophys.,* 161, 1, 1974.
63. **Starbuck, W. C., Mauritzen, C. M., Taylor, C. W., Saroja, I. S., and Busch, H.,** A large scale procedure for isolation of the glycine-rich, arginine-rich histone and the arginine-rich, lysine-rich histone in a highly purified form, *J. Biol. Chem.,* 243, 2038, 1968.
64. **Douvas, A. S., Harrington, C. A., and Bonner, J.,** Major non-histone proteins of rat liver chromatin: preliminary identification of myosin, actin, tubulin and tropomyosin, *Proc. Natl. Acad. Sci. U.S.A.,* 72, 3902, 1975.
65. **Fujii, K., Yonemasu, Y., Kitamura, K., and Tsubota, Y.,** A genetic aspect of chromatin protein of human brain tumors, *J. Neurol.,* 223, 23, 1980.
66. **Miyazaki, K., Hagiwara, H., Nagao, Y., Matuo, Y., and Horio, T.,** Tissue-specific distribution of non-histone proteins in nuclei of various tissues of rats and its change with growth of Rhodamine sarcoma, *J. Biochem.,* 84, 135, 1978.
67. **Conner, B. J. and Comings, D. E.,** Nuclear proteins. V. Studies of histone-binding proteins from mouse liver by affinity chromatography, *Biochim. Biophys. Acta,* 532, 122, 1978.
68. **Yoshida, M., Kikuchi, A., and Shimura, K.,** Fractionation, isolation and characterization of nonhistone chromosomal protein from calf thymus, *J. Biochem.,* 77, 1007, 1975.
69. **Medvedev, Zh. A., Medvedeva, M. N., and Robson, L.,** Age-associated changes in the electrophoretic pattern of the high molecular weight nonhistone proteins from mouse liver, hepatoma and spleen chromatin, *Age,* 3, 74, 1980.
70. **Burkholder, G. D. and Weaver, M. G.,** Differential accessibility of DNA in extended and condensed chromatin to pancreatic DNase I, *Exp. Cell Res.,* 92, 18, 1975.
71. **Djondjurov, L. P., Ivanova, E. C., and Tsanev, R. G.,** Metabolic behavior of nonhistone chromosomal proteins in proliferating and resting fibroblasts, *Eur. J. Biochem,.* 77, 545, 1977.
72. **Naha, P. M. and Sorrentino, R.,** Biochemical controls of the G1 phase of mammalian cell cycle. I. Analysis of chromatin proteins in temperature-sensitive variants, *Cell Biol. Int. Rep.,* 4, 365, 1980.
73. **Eastment, C. E., Scott, R. B., Shelton, K. R., and Gorgan, W. McL.,** Characteristics of nuclear proteins during granulocyte development, *Blut,* 41, 119, 1980.
74. **Ruiz-Carrillo, A., Wangh, L. J., Littau, V. C., and Allfrey, V. G.,** Protein composition of avian erythroid cells at different stages of maturation, *J. Biol. Chem.,* 249, 7358, 1974.
75. **Lindell, T. J.,** Inhibition of eukaryotic DNA-dependent RNA polymerase release from isolated nuclei and nucleoli by phenylmethylsulfonylfluoride, *Arch. Biochem. Biophys.,* 171, 268, 1975.
76. **O'Farrell, P. H. and O'Farrel, P. Z.,** Two-dimensional polyacrylamide gel electrophoretic fractionation, in *Methods in Cell Biology,* Vol. 16, Stein, G. S., Stein, J., and Kleinsmith, L. J., Eds., Academic Press, New York, 1977, 407.
77. **Fujitani, H. and Holoubek, V.,** Presence of the same types of nonhistone chromosomal proteins in different tissues, *Experientia,* 30, 474, 1974.
78. **Sevaljevic, L. and Konstantinovic, M.,** Characterization of two major non-histone protein fractions from chromatin of sea urchin embryos, *Int. J. Biochem.,* 7, 281, 1976.
79. **Elgin, S. C. R.,** Methods for isolation and characterization of nonhistone chromosomal proteins, in *Methods in Enzymology,* Vol. 40, O'Malley, B. W. and Hardman, J. G., Eds., Academic Press, New York, 1975, 144.
80. **Elgin, S. C. R. and Bonner, J.,** Limited heterogeneity of the major non-histone chromosomal proteins, *Biochemistry,* 9, 4440, 1970.
81. **Elgin, S. C. R. and Bonner, J.,** Partial fractionation and chemical characterization of the major nonhistone chromosomal proteins, *Biochemistry,* 11, 772, 1972.

82. **Chong, M. T., Garrard, W. T., and Bonner, J.,** Purification and properties of a neutral protease from rat liver chromatin, *Biochemistry,* 13, 5128, 1974.
83. **Ramponi, G., Nassi, P., Liguri, G., Cappugi, G., and Grisolia, S.,** Purification and properties of a histone-specific protease from rat liver chromatin, *FEBS Lett.,* 90, 228, 1978.
84. **Justic, M., Seifert, S., Weiss, E., Haas, R., and Heinrich, P. C.,** Isolation and characterization of membrane-bound proteinase from rat liver, *Arch. Biochem. Biophys.,* 177, 355, 1976.
85. **Hagiwara, H., Miyazaki, K., Matuo, Y., Yamashita, J., and Horio, T.,** Novel protease bound with chromatins in normal and tumorous tissues of rats, *Biochem. Biophys. Res. Commun.,* 94, 988, 1980.
86. **Dounce, A. L. and Lan, T. H.,** Isolation and properties of chicken erythrocyte nuclei, *Science,* 97, 584, 1943.
87. **Hodes, M. E., Palmer, C. G., and Williams, D. S.,** Possible use of synthetic surface-active agents for the preparation of tumor cell nuclei, *Nature (London),* 182, 529, 1958.
88. **Fisher, H. W and Harris, H.,** The isolation of nuclei from animal cells in culture, *Proc. R. Soc. B.,* 156, 521, 1962.
89. **Zalta, J. P., Rozencwajg, R., Carasso, N., and Favard, P.,** Isolement d'une fraction de noyaux cellulaires dont la pureté est controlée avec microscope électronique, *C. R. Acad. Sci. Paris,* 255, 412, 1962.
90. **Hubert, M. T., Favard, P., Carasso, N., Rozencwajg, R., and Zalta, J. P.,** Méthode d'isolement de noyaux cellulaires a partir de foie de rat, *J. Microscopie,* 1, 435, 1962.
91. **Penman, S.,** RNA metabolism in the HeLa cell nuclei, *J. Mol. Biol.,* 17, 117, 1966.
92. **Aaronson, R. P., Wilt, F. H., and Wartiovaara, J.,** Characterization of pulse-labelled nuclear RNA in sea urchin embryos, *Exp. Cell Res.,* 72, 309, 1972.
93. **Monahan, J. J. and Hall, R. H.,** Fractionation of chromatin components, *Can. J. Biochem.,* 51, 709, 1973.
94. **Russev, G., Anachkova, B., and Tsanev, R.,** Fractionation of rat liver-chromatin nonhistone proteins into two groups with different metabolic rates, *Eur. J. Biochem.,* 58, 253, 1975.
95. **Harlow, R. and Wells, J. R. E.,** Preparation of membrane-free chromatin bodies from avian erythroid cells and analysis of chromatin acidic proteins, *Biochemistry,* 14, 2665, 1975.
96. **Candido, E. P. M., Honda, B. M., and Baillie, D. L.,** Low levels of non-histone chromosomal proteins in trout testis chromatin, *FEBS Lett.,* 61, 260, 1976.
97. **Koffer, A. and Brownson, C.,** Fractionation of nuclear proteins from red and white skeletal muscle, heart and liver of rabbit, *Int. J. Biochem.,* 10, 845, 1979.
98. **Lydersen, B. K., Fa-Ten Kao, Pettijohn, D.,** Expression of genes coding for non-histone chromosomal proteins in human-chinese hamster cell hybrids, *J. Biol. Chem.,* 255, 3002, 1980.
99. **Kirstensen, T. and Holtlund, I.,** Chromatography of chromatin proteins on Cibacron Blue F3G-A-agarose, *J. Chromatogr.,* 192, 494, 1980.
100. **Kikuchi, H. and Sato, S.,** Fractionation of nonhistone proteins on a column of daunomycin-CH-Sepharose 4B, *Biochim. Biophys. Acta,* 532, 113, 1978.
101. **Briggs, R. C., Rothstein, H., and Wainwright, N.,** Cell cycle variations in chromosomal proteins of the lens epithelium, *Exp. Cell Res.,* 99, 95, 1976.
102. **Gronow, M. and Thackrah, T.,** The nonhistone nuclear proteins of some rat tissues, *Arch. Biochem. Biophys.,* 158, 377, 1973.
103. **MacGillivray, A. J., Cameron, A., Krauze R. J., Rickwood, D., and Paul, J.,** The non-histone proteins of chromatin. Their isolation and composition in a number of tissues, *Biochim. Biophys. Acta,* 227, 384, 1972.
104. **Barret, T. and Gould, H. J.,** Tissue and species specificity of non-histone chromatin proteins, *Biochim. Biophys. Acta,* 294, 165, 1973.
105. **Blüthmann, H., Mrozek, S., and Gierer, A.,** Non-histone chromosomal proteins. Their isolation and role in determining specificity of transcription *in vitro, Eur. J. Biochem.,* 58, 315, 1975.
106. **Walmsley, M. E. and Davies, H. G.,** Ultrastructural and biochemical observations on interphase nuclei isolated from chicken erythrocyte, *J. Cell Sci.,* 17, 113, 1975.
107. **Forger, J. M., Choie, D. D., and Friedberg, E. C.,** Non-histone chromosomal proteins of chemically transformed neoplastic cells in tissue culture, *Cancer Res.,* 36, 258, 1976.
108. **Stein, G. S. and Borun, T. W.,** The synthesis of acidic chromosomal proteins during the cell cycle of HeLa S-3 cells. I. The accelerated accumulation of acidic residual nuclear protein before the initiation of DNA replication, *J. Cell. Biol.,* 52, 292, 1972.
109. **Stein, G. and Farber, J.,** Role of nonhistone chromosomal proteins in the restriction of mitotic chromatin template activity, *Proc. Natl. Acad. Sci. U.S.A.,* 69, 2918, 1972.
110. **Rickwood, D., Riches, P. G., and MacGillivray, J.,** Studies on the *in vitro* phosphorylation of chromatin non-histone proteins in isolated nuclei, *Biochim. Biophys. Acta,* 293, 162, 1973.
111. **Seale, R. L.,** Conservation of non-histone chromatin proteins during growth in HeLa cells, *Biochem. Biophys. Res. Commun.,* 63, 140, 1975.

112. **Kostraba, N. C., Montagna, R. A., and Wang, T. Y.,** Mode of action of non-histone proteins in the stimulation of transcription from DNA, *Biochem. Biophys. Res. Commun.,* 72, 334, 1976.
113. **Baugnet-Mahieu, L., Goutier, R., and Maisin, J. R.,** Nonhistone chromosomal proteins in normal and erythroblastic spleen of mice, *Arch. Int. Physiol. Biochim.,* 86, 905, 1978.
114. **Vanfleteren, J. B., Neirynck, K., and Huylebroe, D.,** Nematode chromosomal proteins. I. Isolation of chromatin and preliminary characterization of the chromosomal proteins of *Caenorhabditis elegans, Comp. Biochem. Physiol.,* 62B, 349, 1979.
115. **Seale, R. L. and Aronson, A. I.,** Chromatin-associated proteins of the developing sea urchin embryo. I. Kinetics of synthesis and characterization of non-histone proteins, *J. Mol. Biol.,* 75, 633, 1973.
116. **Hymer, W. C. and Kuff, E. L.,** Isolation of nuclei from mammalian tissues through the use of Triton X-100, *J. Histochem.,* 12, 359, 1964.
117. **Blobel, G. and Potter, V. R.,** Nuclei from rat liver: isolation method that combines purity and high yield, *Science,* 154, 1662, 1966.
118. **Tata, J. R., Hamilton, M. J., and Cole, R. D.,** Membrane phospholipids associated with nuclei and chromatin: melting profile template activity and stability of chromatin, *J. Mol. Biol.,* 67, 231, 1972.
119. **Marushige, K. and Bonner, J.,** Template properties of liver chromatin, *J. Mol. Biol,* 15, 160, 1966.
120. **Widnell, C. C. and Tata, J. R.,** A procedure for isolation of enzymically active rat-liver nuclei, *Biochem. J.,* 92, 313, 1964.
121. **Kartenbeck, J., Jarasch, E. D., and Franke, W. W.,** Nuclear membranes from mammalian liver. VI. Glucose-6-phosphatase in rat liver, a cytochemical and biochemical study, *Exp. Cell Res.,* 81, 175, 1973.
122. **Jackson, R.,** On the identity of nuclear membrane and nonhistone nuclear proteins, *Biochemistry,* 15, 5652, 1976.
123. **Jackson, R.,** Polypeptides of the nuclear envelope, *Biochemistry,* 15, 5641, 1976.
124. **Kirkpatrick, F. H., Gordesky, S. E., and Marinetti, G. V.,** Differential solubilization of proteins phospholipids and cholesterol of erythrocyte membrane by detergents, *Biochim. Biophys. Acta,* 345, 154, 1974.
125. **Miller, D. M.,** Total solubilization of erythrocyte membranes by nonionic detergents, *Biochem. Biophys. Res. Commun.,* 40, 716, 1970.
126. **Weiss, E. and Braun, V.,** Actin-like protein of the cytoplasm in the chromatin of *Dictyostelium discoideum, FEBS Lett.,* 108, 233, 1979.
127. **Putnam, F. W.,** The interactions of proteins and synthetic detergents, in *Advances in Protein Chemistry,* Vol. 4, Anson, M. L. and Edsall, J. T., Eds., Academic Press, New York, 1948, 79.
128. **Sreenivasaya, M. and Pirie, N. W.,** The disintegration of tobacco mosaic, virus preparation with sodium dodecyl sulphate, *Biochem. J.,* 32, 1707, 1938.
129. **Mirsky, A. E. and Pollister, A. W.,** Chromosin, a desoxyribose nucleoprotein complex of the cell nucleus, *J. Gen. Physiol.,* 30, 117, 1946.
130. **Marco, A. M. and Butler, G. C.,** The isolation of sodium desoxyribonucleate with sodium dodecyl sulfate, *J. Biol. Chem.,* 190, 165, 1951.
131. **Kay, E. R. M., Simons, N. S., and Dounce, A. L.,** An improved preparation of sodium desoxyribonucleate, *J. Am. Chem. Soc.,* 74, 1724, 1952.
132. **Marmur, J.,** A procedure for the isolation of deoxyribonucleic acid from micro-organisms, *J. Mol. Biol.,* 3, 208, 1961.
133. **Anson, M. L.,** The denaturation of proteins by detergents and bile salts, *Science,* 90, 256, 1939.
134. **Marushige, K., Burtlag, D., and Bonner, J.,** Properties of chromosomal nonhistone protein of rat liver, *Biochemistry,* 7, 3149, 1968.
135. **Reynolds, J. A. and Tanford, C.,** Binding of dodecyl sulfate to proteins at high binding ratios. Possible implications for the state of proteins in biological membranes, *Proc. Natl. Acad. Sci. U.S.A.,* 66, 1002, 1970.
136. **Reynolds, J. A. and Tanford, C.,** The gross conformation of protein-sodium dodecyl sulphate complexes, *J. Biol. Chem.,* 245, 5161, 1970.
137. **Tanford, C.,** Hydrophobic free energy, micelle formation and the association of proteins with amphiphiles, *J. Mol. Biol.,* 67, 59, 1972.
138. **Nelson, C. A.,** The binding of detergents to proteins. I. The maximum amount of dodecyl sulfate bound to proteins and the resistance to binding of several proteins, *J. Biol. Chem.,* 246, 3895, 1971.
139. **Shapiro, A. L., Vinuela, E., and Maizel, J. V., Jr.,** Molecular weight estimation of polypeptide chains by electrophoresis in SDS-polyacrylamide gels, *Biochem. Biophys. Res. Commun.,* 28, 815, 1967.
140. **Weber, K. and Osborn, M.,** The reliability of molecular weight determinations by dodecyl sulphate-polyacrylamide gel electrophoresis, *J. Biol. Chem.,* 244, 4406, 1969.
141. **Shirey, T. and Huang, R. C. C.,** Use of sodium dodecyl sulphate, alone, to separate chromatin proteins from deoxyribonucleoprotein of *Arbacia punctulata* sperm chromatin, *Biochemistry,* 8, 4138, 1969.

142. **Tanford, C.,** Protein denaturation, in *Advances in Protein Chemistry,* Vol. 23, Anson, M. L. and Edsall, J. T., Eds., Academic Press, New York, 1968, 121.
143. **White, F. H., Jr.,** Regeneration of native secondary and tertiary structures by air oxidation of reduced ribonucleases, *J. Biol. Chem.,* 236, 1353, 1961.
144. **Weber, K. and Kuter, D. J.,** Reversible denaturation of enzymes by sodium dodecyl sulfate, *J. Biol. Chem.,* 246, 2504, 1971.
145. **Bekhor, I., Kung, G. M., and Bonner, J.,** Sequence-specific interaction of DNA and chromosomal proteins, *J. Mol. Biol.,* 39, 351, 1969.
146. **Gilmour, R. S. and Paul, J.,** RNA transcribed from reconstituted nucleoproteins is similar to natural RNA, *J. Mol. Biol.,* 40, 137, 1969.
147. **Bekhor, I.,** Reconstitution of chromatin, in *The Cell Nucleus,* Vol. 5, Busch, H., Ed., Academic Press, New York, 1978, 137.
148. **Chaudhuri, S.,** Fractionation of chromatin nonhistone proteins, *Biochim. Biophys. Acta,* 322, 155, 1973.
149. **Stark, G. R., Stein, W. H., and Moore, S.,** Reactions of the cyanate present in aqueous urea with amino acids and proteins, *J. Biol. Chem.,* 235, 3177, 1960.
150. **Dirnhuber, P. and Schütz, F.,** The isomeric transformation of urea into ammonium cyanate in aqueous solutions, *Biochem. J.,* 42, 628, 1948.
151. **Marier, J. R. and Rose, D.,** Determination of cyanate, and a study of its accumulation in aqueous solutions of urea, *Anal. Biochem.,* 7, 304, 1964.
152. **Cole, R. D.,** Ion exchange chromatography of prolactin in urea-containing buffers, *J. Biol. Chem.,* 236, 1369, 1961.
153. **Poulik, M. D.,** Gel electrophoresis in buffers containing urea, in *Methods of Biochemical Analysis,* Vol. 14, Glick, D., Ed., John Wiley & Sons, New York, 1966, 455.
154. **Cejka, J., Vodrazka, Z., and Salak, J.,** Carbamylation of globin in electrophoresis and chromatography in the presence of urea, *Biochim. Biophys. Acta,* 154, 589, 1968.
155. **Hagel, P., Gerding, J. T., Fleggen, W., and Bloemendal, H.,** Cyanate formation in solutions of urea. I. Calculation of cyanate concentrations at different temperature and pH, *Biochim. Biophys. Acta,* 243, 366, 1971.
156. **Herbert, P. N., Shulman, R. S., Levy, R. I., and Fredrickson, D. S.,** Fractionation of the C-apoproteins from human plasma very low density lipoproteins. Artifactual polymorphism from carbamylation in urea-containing solutions, *J. Biol. Chem.,* 248, 4941, 1973.
157. **Sanders, A. L.,** Isolation and characterization of non-histone chromosomal proteins of developing avian erythroid cells, *Biochemistry,* 13, 527, 1974.
158. **Comings, D. E., Miguel, A. G., and Lesser, B. H.,** Nuclear proteins, VI. Fractionation of chromosomal non-histone proteins using hydrophobic chromatography, *Biochim. Biophys. Acta,* 563, 253, 1979.
159. **Tuan, D., Smith, S., Folkman, J., and Merler, E.,** Isolation of the nonhistone proteins of rat Walker carcinoma 256. Their association with tumor angiogenesis, *Biochemistry,* 12, 3159, 1973.
160. **Yoshida, M., Ohtaki, Y., Fujita, T., and Shimura, K.,** Sequential grouping of nonhistone chromosomal protein from pig thymus, *J. Biochem.,* 88, 425, 1980.
161. **Graziano, S. L. and Huang, R. C. C.,** Chromatographic separation of chick brain chromatin proteins using a SP-Sephadex column, *Biochemistry,* 10, 4770, 1971.
162. **Rickwood, D. and MacGillivray, A. J.,** Improved techniques for the fractionation of non-histone proteins of chromatin on hydroxylapatite, *Eur. J. Biochem.,* 51, 593, 1975.
163. **Hill, R. J., Poccia, D. L., and Doty, P.,** Toward a total macromolecules analysis of sea urchin embryos chromatin, *J. Mol. Biol.,* 61, 445, 1971.
164. **Tanford, C., Kawahara, K., and Lapanje, S.,** Proteins in 6 M guanidine hydrochloride. Demonstration of random coil behaviour, *J. Biol. Chem.,* 241, 1921, 1966.
165. **Tanford, C., Kawahara, K., and Lapanje, S.,** Proteins as random coils. I. Intrinsic viscosities and sedimentation coefficients in concentrated guanidine hydrochloride, *J. Am. Chem. Soc.,* 89, 729, 1967.
166. **Tanford, C., Kawahara, K., Lapanje, S., Hooker, T. M., Jr., Zarlengo, M. H., Salahuddin, A., Aune, K. C., and Takagi, T.,** Proteins as random coils. III. Optical rotatory dispersion in 6 M guanidine hydrochloride, *J. Am. Chem. Soc.,* 89, 5023, 1967.
167. **Lapanje, S. and Tanford, C.,** Proteins as random coils. IV. Osmotic pressures, second virial coefficients, and unperturbed dimensions in 6 *M* guanidine hydrochloride, *J. Am. Chem. Soc.,* 89, 5030, 1967.
168. **Lapanje, S.,** Denaturation of globular proteins by guanidine thiocyanate. II. Intrinsic viscosities in 4.5 *M* guanidine thiocyanate solutions, *Biochim. Biophys. Acta,* 243, 357, 1971.
169. **Levy, S., Simpson, R. T., and Sober, H. A.,** Fractionation of chromatin components, *Biochemistry,* 11, 1547, 1972.
170. **Lilienfeld, L.,** Zur Chemie der Leukozyten, *H.S.Z. Physiol. Chem.,* 18, 473, 1894.
171. **Klyszejko, L. and Khouvine, Y.,** Désoxyribonucléoproteides du pancréas de boeuf, *Bull. Soc. Chim. Biol.,* 46, 111, 1960.

172. **Dounce, A. L. and Hilgartner, C. A.**, A study of DNA nucleoprotein gels and the residual protein of isolated cell nuclei. Relationship to chromosomal structure, *Exp. Cell Res.*, 36, 228, 1964.
173. **Leveson, J. E. and Peacocke, A. R.**, Studies on a complex of deoxyribonucleic acid with non-histone protein, *Biochim. Biophys. Acta*, 123, 329, 1966.
174. **Shaw, L. M. J. and Huang, R. C. C.**, A description of two procedures which avoid the use of extreme pH conditions for the resolution of components isolated from chromatin prepared from pig cerebellar and pituitary nuclei, *Biochemistry*, 9, 4530, 1970.
175. **Teng, C. T., Teng, C. S., and Allfrey, V. G.**, Species-specific interactions between nuclear phosphoproteins and DNA, *Biochem. Biophys. Res. Commun.*, 41, 690, 1970.
176. **Wang, T. Y.**, The isolation, properties, and possible functions of chromatin acidic proteins, *J. Biol. Chem.*, 242, 1220, 1967.
177. **Johns, E. W. and Forrester, S.**, Studies on nuclear proteins. The binding of extra acidic proteins to desoxyribonucleoprotein during the preparation of nuclear proteins, *Eur. J. Biochem.*, 8, 547, 1969.
178. **Goodwin, G. H., Sanders, C., and Johns, E. W.**, A new group of chromatin-associated proteins with a high content of acidic and basic amino acids, *Eur. J. Biochem.*, 38, 14, 1973.
179. **Bonner, J., Chalkley, G. R., Dahmus, M., Fambrough, D., Fujimura, F., Huang, R. C., Huberman, J., Jensen, R., Marushige, K., Ohlenbusch, H., Olivera, B., and Widholm, J.**, Isolation and characterization of chromosomal nucleoproteins, in *Methods in Enzymology*, Vol. 12B, Grossman, L. and Moldave, K., Eds., Academic Press, New York, 1968, 3.
180. **Zubay, G. and Dotty, P.**, The isolation and properties of deoxyribonucleoprotein particles containing single nucleic acid molecules, *J. Mol. Biol.*, 1, 1, 1959.
181. **Kleinsmith, L. J.**, Phosphorylation of nonhistone proteins, in *The Cell Nucleus*, Vol. 6, Busch, H., Ed., Academic Press, New York, 1978, 221.
182. **Goodwin, G. H., Walker, J. M., and Johns, E.**, The high mobility group (HMG) nonhistone chromosomal proteins, in *The Cell Nucleus*, Vol. 6, Busch, H., Eds., Academic Press, New York, 1978, 181.
183. **Smith, M. C. and Chae, C. B.**, A rapid method for the electrophoresis of chromatin proteins, *Biochim. Biophys. Acta*, 317, 10, 1973.
184. **Tsitilou, S. G., Cox, D., Mathias, A. P., and Ridge, D.**, The characterization of the non-histone chromosomal proteins of the main classes of nuclei from rat brain fractionated by zonal centrifugation, *Biochem. J.*, 177, 331, 1979.
185. **Wu, F. C., Elgin, S. C. R., and Hood, L. E.**, Nonhistone chromosomal proteins of rat tissues. A comparative study by gel electrophoresis, *Biochemistry*, 12, 2792, 1973.
186. **Suria, D. and Liew, C. C.**, Isolation and analysis of non-histone chromatin proteins from rat liver nuclei by three different metds, *Can. J. Biochem.*, 52, 1143, 1974.
187. **Boffa, L. C., Vidali, G., and Allfrey, V. G.**, Changes in nuclear non-histone protein composition during normal differentiation and carcinogenesis of intestinal epithelial cell, *Exp. Cell Res.*, 98, 396, 1976.
188. **Benjamin, W. and Gellhorn, A.**, Acidic proteins of mammalian nuclei: isolation and characterization, *Proc. Natl. Acad. Sci. U.S.A.*, 59, 262, 1968.
189. **Teng, C. S. and Hamilton, T. H.**, Role of chromatin in estrogen action in the uterus, II. Hormone induced synthesis of nonhistone acidic proteins which restore histone-inhibited DNA-dependent RNA synthesis, *Proc. Natl. Acad. Sci. U.S.A.*, 63, 465, 1969.
190. **Mohberg, J. and Rusch, H. P.**, Isolation of the nuclear histones from the Myxomycete, *Physarum polycephalum, Arch. Biochem. Biophys.*, 134, 577, 1969.
191. **Ali Khan, M. Y. and Shall, S.**, The isolation of acidic nuclear proteins from *Physarum polycephalum, Biochem. J.*, 121, 43P, 1971.
192. **Chatterjee, R. and Narayanaswami, A.**, Studies on brain nuclear acidic proteins from different species, *Indian J. Biochem. Biophys.*, 14, 251, 1977.
193. **Chiu, J. F., Wang, S., Fujitani, H., and Hnilica, L. S.**, DNA binding chromosomal non-histone proteins. Isolation, characterization and tissue specificity, *Biochemistry*, 14, 4552, 1975.
194. **Spelsberg, T. C. and Hnilica, L. S.**, Deoxyribonucleoproteins and the tissue specific restriction of the deoxyribonucleic acid in chromatin, *Biochem. J.*, 120, 435, 1970.
195. **Spelsberg, T. C., Hnilica, L. S., and Ansevin, A. T.**, Proteins of chromatin in specific restriction. III. The macromolecules in specific restriction of the chromatin DNA, *Biochim. Biophys. Acta*, 228, 550, 1971.
196. **Wakabayashi, K., Wang, S., and Hnilica, L. S.**, Immunospecificity of nonhistone proteins in chromatin, *Biochemistry*, 13, 1027, 1974.
197. **Chiu, J. F., Hunt, M., and Hnilica, L. S.**, Tissue-specific DNA-protein complexes during azo dye hepatocarcinogenesis, *Cancer Res.*, 35, 113, 1975.
198. **Wang, S., Chiu, J. F., Klyszejko-Stefanowicz, L., Fujitani, H., Hnilica, L. S., and Ansevin, A. T.**, Tissue-specific chromosomal non-histone protein interactions with DNA, *J. Biol. Chem.*, 251, 1471, 1976.
199. **Klyszejko-Stefanowicz, L., Chiu, J. F., Tsai, Y. H., and Hnilica, L. S.**, Acceptor proteins in rat adnrogenic tissue chromatin, *Proc. Natl. Acad. Sci. U.S.A.*, 73, 1954, 1976.

200. **Klyszejko-Stefanowicz, L. and Krajewska, W. M.,** Comparative studies on pancreas chromatin proteins: species specificity and behaviour during rat pancreas regeneration, *Int. J. Biochem.,* 13, 851, 1981.
201. **Gadski, R. A. and Chae, C. B.,** Mode of chromatin reconstitution. Elements controlling globin gene transcription, *Biochemistry,* 17, 869, 1978.
202. **Bekhor, I.,** Physical studies on the effect of chromosomal RNA on reconstituted nucleohistones, *Arch. Biochem. Biophys.,* 155, 39, 1973.
203. **Umansky, S. R., Tokarskaya, V. I., Zotova, R. N., and Migushina, V. L.,** Isolation and heterogeneity of nonhistone proteins of rat liver chromatin, *Mol. Biol.,* 5, 270, 1971.
204. **Gilmour, R. S. and Paul, J.,** Role of non-histone components in determining organ specificity of rabbit chromatins, *FEBS Lett.,* 9, 242, 1970.
205. **Richter, K. H. and Sekeris, C. E.,** Isolation and partial purification of non-histone chromosomal proteins from rat liver, thymus and kidney, *Arch. Biochem. Biophys.,* 148, 44, 1972.
206. **Parker, M. G., Sheehan, D. M., and O'Malley, B. W.,** Effects of estrogen on gene expression in the chick oviduct. Isolation and fractionation of chromatin non-histone proteins, *Biochim. Biophys. Acta,* 454, 138, 1976.
207. **Umansky, S. R., Kovalev, Yu. I., and Tokarskaya, V. I.,** Specific interaction of chromatin non-histone proteins with DNA, *Biochim. Biophys. Acta,* 383, 242, 1975.
208. **Umansky, S. R., Kovalev, Yu. I., and Picker, E. G.,** Interaction of chromatin non-histone proteins with homologous and heterologous DNA, *Mol. Biol.,* 9, 683, 1975.
209. **McCleary, A. R., Nooden, L. D., and Kleinsmith, L. J.,** Histone affinity chromatography as a tool for fractionating nonhistone chromatin proteins and studying histone-nonhistone protein interactions, *J. Biol. Chem.,* 253, 5199, 1978.
210. **Sevall, J. S., Cockburn, A., Savage, M., and Bonner, J.,** DNA-protein interactions of the rat liver non-histone chromosomal proteins, *Biochemistry,* 14, 782, 1975.
211. **Jagodzinski, L. L., Castro, C. E., Sherrod, P., Lee, D., and Sevall, S.,** Reassociation kinetics of non-histone-bound DNA sites, *J. Biol. Chem.,* 254, 3038, 1979.
212. **Pederson, T. and Bhorjee, J. S.,** A special class of non-histone protein tightly complexed with template-inactive DNA in chromatin, *Biochemistry,* 14, 3238, 1975.
213. **Chaudhuri, S., Stein, G., and Baserga, R,** Binding of chromosomal acidic proteins to DNA and chromatin, *Proc. Soc. Exp. Biol. Med.,* 139, 1363, 1972.
214. **Stein, G., Chaudhuri S., and Baserga, R.,** Gene activation in WI-38 fibroblasts stimulated to proliferate. Role of non-histone chromosomal proteins, *J. Biol. Chem.,* 247, 3918, 1972.
215. **Stein, G. S., Mans, R. J., Gabbay, E. J., Stein, J. L., Davies, J., and Adawadkar, P. D.,** Evidence for fidelity of chromatin reconstruction, *Biochemistry,* 14, 1859, 1975.
216. **Park, W., Jansing, R., Stein, J., and Stein, G.,** Activation of histone gene transcription in quiescent WI-38 cells or mouse liver by a nonhistone chromosomal protein fraction from HeLa S_3 cells, *Biochemistry,* 16, 3713, 1977.
217. **Yamaizumi, M., Uchida, T., Okada, Y., Furusawa, M., and Mitsui, H.,** Rapid transfer of non-histone chromosomal proteins to the nucleus of living cells, *Nature (London),* 273, 782, 1978.
218. **Akhtar, R. A. and Itzhaki, S.,** Studies in vitro of the effects of adenosine 3′,5′-cyclic monophosphate on the phosphorylation of nuclear proteins in isolated rat heart nuclei, *Biochem. J.,* 161, 487, 1977.
219. **Rodriguez, L. V. and Becker, F. F.,** Rat liver chromatin distribution of histone and nonhistone proteins in eu- and heterochromatin, *Arch. Biochem. Biophys.,* 173, 438, 1976.
220. **Winter, H., Alonso, A., and Goerttler, K.,** A sensitive assay system for detection of rare chromosomal proteins with DNA-binding properties, *Anal. Biochem.,* 105, 39, 1980.
221. **Arnold, E. A. and Young, K. E.,** Isolation and partial electrophoretic characterization of total protein from non-sheared rat liver chromatin, *Biochim. Biophys. Acta,* 257, 482, 1972.
222. **Levy, R., Levy, S., Rosenberg, S. A., and Simpson, R. T.,** Selective stimulation of nonhistone chromatin synthesis in lymphoid cells by phytohemagglutinin, *Biochemistry,* 12, 224, 1973.
223. **Bhorjee, J. S. and Pederson, T.,** Rapid preparative-scale purification of chromatin proteins, *Biochim. Biophys. Acta,* 418, 154, 1976.
224. **Gineitis, A. A., Anachkova, B., and Russev, G.,** Failure of preparative flat bed electrofocusing to resolve rat liver chromosomal proteins, *Anal. Biochem.,* 87, 157, 1978.
225. **Fujitani, H. and Holoubek, V.,** Similarity of the 0.35 *M* NaCl soluble nuclear proteins and the nonhistone chromosomal proteins, *Biochem. Biophys. Res. Commun.,* 54, 1300, 1973.
226. **Fujitani, H. and Holoubek, V.,** Nonhistone nuclear proteins of rat brain, *J. Neurochem.,* 23, 1215, 1974.
227. **Patel, G. and Wang, T. Y.,** Chromatography and electrophoresis of nuclear soluble proteins, *Exp. Cell Res.,* 34, 120, 1964.
228. **Wang, T. Y. and Johns, E. W.,** Study of the chromatin acidic proteins of rat liver: heterogeneity and complex formation with histones, *Arch. Biochem. Biophys.,* 124, 176, 1968.

229. **Kostraba, N. C. and Wang, T. Y.,** Tissue variations of acidic nuclear proteins and their biosynthesis during liver regeneration, *Int. J. Biochem.*, 1, 327, 1976.
230. **Wang, T. Y.,** Isolation of a terminal DNA-nucleotidyl transferase from calf thymus non-histone chromatin proteins, *Arch. Biochem. Biophys.*, 127, 235, 1968.
231. **Patel, N. and Holoubek, V.,** RNA associated with nonhistone chromosomal proteins of dog liver, *Biochem. Biophys. Res. Commun.*, 54, 524, 1973,
232. **Patel, N. T. and Holoubek, V.,** Characterization of low molecular weight nonhistone chromosomal protein from dog liver, *FEBS Lett.*, 46, 154, 1974.
233. **Kono, N., Shima, I., and Ohta, G.,** Immunospecificity of non-histone chromosomal proteins in ^{89}Sr-induced osteogenic sarcoma (mouse), *J. Biochem.*, 81, 1549, 1977.
234. **Caiafa, P., Scarpati-Cioffari, M. R., Allegra, P., Ferraro, A., and Turano, C.,** Fractionation of non-histone chromatin proteins from pig liver and kidney by means of immobilized histone H_3, *Mol. Cell. Biochem.*, 30, 101, 1979.
235. **Leibovitch, M. P., Tichonicky, L., and Kruh, J.,** Chromatin protein kinases and phosphoproteins during myoblast growth and differentiation, *Biochem. Biophys. Res. Commun.*, 81, 623, 1978.
236. **Hacha, R. and Fredericq, E.,** Preparation and some properties of strongly acidic proteins from calf-thymus nucleohistone, *Eur. J. Biochem.*, 52, 83, 1975.
237. **Kostraba, N. C., Loor, R. M., and Wang, T. Y.,** Tissue-specificity of the nonhistone protein that inhibits RNA synthesis *in vitro, Biochem. Biophys. Res. Commun.*, 79, 347, 1977.
238. **Kostraba, N. C. and Wang, T. Y.,** Inhibition of transcription in vitro by a non-histone protein isolated from Ehrlich ascites tumor chromatin, *J. Biol. Chem.*, 250, 8938, 1975.
239. **Kostraba, N. C., Newman, R. S., and Wang, T. Y.,** Selective inhibition of transcription by a nonhistone protein isolated from Ehrlich ascites tumor chromatin, *Arch. Biochem. Biophys.*, 179, 100, 1977.
240. **Gineitis, A. A., Nivinskas, H. H., and Vorobev, V. I.,** Non-histone chromatin proteins from sea urchin *Strongylocentrotus droebachiensis* sperm and embryo, *Exp. Cell Res.*, 98, 248, 1976.
241. **Gineitis, A. A., Stankeviciute, J. V., and Vorobev, V. I.,** Chromatin proteins from normal, vegetalized, and animalized sea urchin embryo, *Dev. Biol.*, 52, 181, 1976.
242. **Patel, G. L. and Thomas, T. L.,** Some binding parameters of chromatin acidic proteins with high affinity for deoxyribonucleic acid, *Proc. Natl. Acad. Sci. U.S.A.*, 70, 2524, 1973.
243. **Thomas, T. L. and Patel, G. L.,** DNA unwinding component of the nonhistone chromatin proteins, *Proc. Natl. Acad. Sci. U.S.A.*, 73, 4364, 1976.
244. **Thomas, T. L. and Patel, G. L.,** Optimal conditions and specificity interaction of distinct class of nonhistone chromosomal proteins with DNA, *Biochemistry*, 15, 1481, 1976.
245. **Langan, T.,** A phosphoprotein preparation from liver cell nuclei and its effects on the inhibition of RNA synthesis by histones, in *Regulation of Nucleic Acid and Protein Synthesis,* Koningsberger, V. V. and Bosch, L., Eds., Elsevier, Amsterdam, 1967, 233.
246. **Kamiyama, M. and Wang, T. Y.** Activated transcription from rat liver chromatin by non-histone proteins, *Biochim. Biophys. Acta,* 228, 563, 1971.
247. **Pumo, D. E., Stein, G. S., and Kleinsmith, L. J.,** Stimulated phosphorylation in SV-40 transformed WI-38 human diploid fibroblasts, *Biochim. Biophys. Acta,* 402, 125, 1975.
248. **Kruh, J., Courtois, Y., and Tichonicky, L.,** Phosphorylation and protein kinase activities of chromosomal nonhistone proteins from chick embryo fibroblasts, *Biochimie,* 57, 1323, 1975.
249. **Kleinsmith, L. J. and Allfrey, V. G.,** I. Isolation and characterization of a phosphoprotein fraction from calf thymus nuclei, *Biochim. Biophys. Acta,* 175, 123, 1969.
250. **Borschevskaya, T. A., Dobretsov, G. E., and Petrov, V. A.,** Isolation of rat liver chromatin non-histone protein, *Mol. Biol.*, 9, 676, 1975.
251. **Yoshida, M. and Shimura, K.,** Isolation of nonhistone chromosomal protein from calf thymus, *Biochim. Biophys. Acta,* 263, 690, 1972.
252. **Kikuchi, H. and Sato, S.,** Binding of daunomycin to nonhistone proteins from rat liver, *Biochim. Biophys. Acta,* 434, 509, 1976.
253. **Lin, P. P. C., Wilson, R. F., and Bonner, J.,** Isolation and properties of nonhistone chromosomal proteins from pea chromatin, *Mol. Cell. Biochem.*, 1, 197, 1973.
254. **Pitel, J. A. and Durzan, D. J.,** Chromosomal proteins of conifers. 1. Comparison of histones and nonhistone chromosomal proteins from dry seeds of conifers, *Can. J. Bot.*, 56, 1915, 1978.
255. **Umansky, S. R., Zotova, R. N., and Kovalev, Y. I.,** Comparison of some properties of chromatin nonhistone proteins and nuclear sap proteins, *Eur. J. Biochem.*, 65, 503, 1976.
256. **Kilianska, Z., Szemraj, J., and Klyszejko-Stefanowicz, L.,** An electrophoretic comparison of nonhistone proteins from rat liver total chromatin and chromatin depleted of 0.35 M NaCl soluble proteins, *Int. J. Biochem.*, 13, 941, 1981.
257. **Shelton, K. R. and Allfrey, V. G.,** Selective synthesis of a nuclear acidic protein in liver cells stimulated by cortisol, *Nature (London),* 228, 132, 1970.

258. **Teng, C. S., Teng, C. T., and Allfrey, V. G.,** Studies of nuclear acidic proteins. Evidence for their phosphorylation, tissue specificity, selective binding to deoxyribonucleic acid, and stimulatory effects on transcription, *J. Biol. Chem.*, 246, 3597, 1971.
259. **Shelton, K. R. and Neelin, J. M.,** Nuclear residual proteins from goose erythroid cells and liver, *Biochemistry*, 10, 2342, 1971.
260. **Johnson, E. M. and Allfrey, V. G.,** Differential effects of cyclic adenosine 3′,5′-monophosphate on phosphorylation of rat liver nuclear acidic proteins, *Arch. Biochem. Biophys.*, 152, 786, 1972.
261. **Vidali, G., Boffa, L. C., Littau, V. G., Allfrey, K. M., and Allfrey, V. G.,** Changes in nuclear acidic protein complement of red blood cells during embryonic development, *J. Biol. Chem.*, 248, 4065, 1973.
262. **Blankenship, J. and Bresnick, E.,** Effects of phenobarbital on the phosphorylation of acidic nuclear proteins of rat liver, *Biochim. Biophys. Acta*, 340, 218, 1974.
263. **Schwartz, S. A.,** Rat embryo nonhistone chromosomal proteins: interaction in vitro with normal and bromodeoxyuridine-substituted DNA, *Biochemistry*, 16, 4101, 1977.
264. **Seeber, S., Meshkov, T., Brucksch, K. P., Käding, J., Schmidt, C. G., and Busch, H.,** Comparative studies on phenol-soluble nonhistone chromatin proteins in normal and leukaemic human leukocytes, *Klin. Wochenschr.*, 57, 257, 1979.
265. **Fraser, I. H., Gordon, A. S., and Liew, C. C.,** Studies of nuclear proteins in the skeletal muscle of the cardiomyopathic Syrian hamster, *J. Neurol. Sci.*, 42, 303, 1979.
266. **Liew, C. C., Suria, D., and Gornall, A. G.,** Effects of aldosterone on acetylation and phosphorylation of chromosomal proteins, *Endocrinology*, 93, 1025, 1973.
267. **Kilianska, Z. and Klyszejko-Stefanowicz, L.,** Electrophoretic studies of non-histone proteins from pig lymph nodes, *Biochimie*, 58, 676, 1976.
268. **Krajewska, W., Wierzbicki, R., and Klyszejko-Stefanowicz, L.,** Non-histone chromatin proteins during maturation of avian erythroid cells, *Acta Biochim. Polon.*, 23, 177, 1976.
269. **Graczynski, M., Kilianska, Z., Gaczyk, G., and Klyszejko-Stefanowicz, L.,** Specificity of non-histone proteins from hamster Kirkman-Robbins hepatoma, *Neoplasma*, 27, 653, 1980.
270. **LeStourgeon, W. M. and Rusch, H. P.,** Nuclear acidic protein changes during differentiation in *Physarum polycephalum*, *Science*, 174, 1233, 1971.
271. **LeStourgeon, W. M. and Rusch, H. P.,** Localization of nuclear and chromatin residual acidic protein. Changes during differentiation in *Physarum polycephalum*, *Arch. Biochem. Biophys.*, 155, 144, 1973.
272. **Rubin, R. W., Hill, M. C., Hepworth, P., and Boehmer, J.,** Isolation and electrophoretic analysis of nucleoli, phenol-soluble nuclear proteins, and outer cyst walls from *Acanthamoeba castellanii* during encystation initiation, *J. Cell. Biol.*, 68, 740, 1976.
273. **Schmidt, G., Billwitz, H., and Lindigkeit, R.,** Characterization of the phenol soluble and insoluble nonhistone proteins of in vivo ^{32}P-labelled rat liver nuclei by high-resolution gel electrophoresis, *Acta Biol. Med. Germ.*, 39, 21, 1980.
274. **Simon, J. H. and Becker, W. M.,** A polyethylene glycol/dextran procedure for the isolation of chromatin proteins (histone and nonhistone) from wheat germ, *Biochim. Biophys. Acta*, 454, 154, 1976.
275. **Lapeyre, J. N. and Bekhor, I.,** Chromosomal protein interactions in chromatin and with DNA, *J. Mol. Biol.*, 104, 25, 1976.
276. **Legraverend, M. and Glazer, R. J.,** Characterization of a non-histone chromosomal protein which stimulates RNA polymerase II, *Biochim. Biophys. Acta*, 607, 92, 1980.
277. **Bhorjee, J. S. and Pederson, T.,** Chromatin: its isolation from cultured mammalian cells with particular reference to contamination by nuclear ribonucleoprotein particles, *Biochemistry*, 12, 2766, 1973.
278. **Jeppesen, P. G. N., Bankier, A. T., and Sanders, L.,** Non-histone proteins and the structure of metaphase chromosomes, *Cell Res.*, 115, 293, 1978.
279. **Krauze, M. O., Kleinsmith, L. J., and Stein, G. S.,** Properties of the genome in normal and SV40 transformed WI38 human diploid fibroblasts, I. Composition and metabolism of nonhistone chromosomal proteins, *Exp. Cell Res.*, 92, 164, 1975.
280. **Sonnenbichler, J. and Nobis, P.,** The so-called non-histones from acid-treated calf thymus chromatin, *Eur. J. Biochem.*, 16, 60, 1970.
281. **Carusi, E. A. and Sinsheimer, R. L.,** The physical properties of a protein isolated from bacteriophage ΦX 174, *J. Mol. Biol.*, 7, 388, 1963.
282. **Ohlenbusch, H. H., Olivera, B. M., Tuan, D., and Davidson, N.,** Selective dissociation of histone from calf thymus nucleoprotein, *J. Mol. Biol.*, 25, 299, 1967.
283. **Fambrough, D. and Bonner, J.,** Selective dissociation of pea bud nucleohistone, *Biochim. Biophys. Acta*, 154, 601, 1968.
284. **Bornkamm, G. W., Nobis, P., and Sonnenbichler, J.,** Concerning the specificity of histone and nonhistone dissociation from calf thymus chromatin by salt, *Biochim. Biophys. Acta*, 278, 258, 1972.

285. **Bidney, D. L. and Reeck, G. R.,** Purification from cultured cells of two nonhistone chromatin proteins with preferential affinity for single-stranded DNA: apparent analogy with calf thymus HMG proteins, *Biochem. Biophys. Res. Commun.,* 85, 1211, 1978.
286. **Martin, R. G. and Ames, B. N.,** A method for determining the sedimentation behaviour of enzymes: application to protein mixtures, *J. Biol. Chem.,* 236, 1327, 1961.
287. **Lee, D., Castro, C. E., Jagodzinski, L. L., and Sevall, J. S.,** Functional significance of rat liver nonhistone protein-DNA interactions: RNA hybridizaton of protein-bound DNA, *Biochemistry,* 18, 3160, 1979.
288. **Steele, W. J. and Busch, H.,** Studies on acidic nuclear proteins of the Walker tumor and liver, *Cancer Res.,* 23, 1153, 1963.
289. **Dingman, C. W. and Sporn, M. B.,** Studies on chromatin. I. Isolation and characterization of nuclear complexes of deoxyribonucleic acid, ribonucleic acid, and protein from embryonic and adult tissues of the chicken, *J. Biol. Chem.,* 239, 3483, 1964.
290. **Frenster, J. H., Allfrey, V. G., and Mirsky, A. E.,** Repressed and active chromatin isolated from interphase lymphocytes, *Proc. Natl. Acad. Sci. U.S.A.,* 50, 1026, 1963.
291. **Frenster, J. H.,** Nuclear polyanions as de-repressors of synthesis of ribonucleic acid, *Nature (London),* 206, 680, 1965.
292. **Chanda, S. K. and Cherian, M. G.,** Isolation and partial characterization of a mercury-binding nonhistone protein component from rat kidney nuclei, *Biochem. Biophys. Res. Commun.,* 50, 1013, 1973.
293. **Montagna, R. A., Rodriquez, L. V., and Becker, F. F.,** A comparative study of the nonhistone proteins of rat liver euchromatin and heterochromatin, *Arch. Biochem. Biophys.,* 179, 617, 1977.
294. **Rickwood, D. and MacGillivray, A. J.,** Quantitative aspects of the binding of nuclear non-histone proteins to DNA as determined by centrifugation in metrizamide gradients, *Exp. Cell Res.,* 104, 287, 1977.
295. **MacGillivray, A. J., Carrol, D., and Paul, J.,** The heterogeneity of the non-histone chromatin proteins from mouse tissues, *FEBS Lett.,* 13, 204, 1971.
296. **Albertsson, P. A.,** Partition methods for fractionation of cell particles and macromolecules, in *Methods of Biochemical Analysis,* Vol. 10, Glick, D., Ed., John Wiley & Sons, New York, 1962, 229.
297. **Turner, G. and Hancock R.,** Fractionation of chromosomal deoxyribonucleoproteins by partition in two-phase aqueous polymer systems, *Biochem. Biophys. Res. Commun.,* 58, 437, 1974.
298. **Alberts, B. M.,** Fractionation of nucleic acids by dextranpolyethylene glycol two-phase systems, in *Methods in Enzymology,* Vol. 12A, Grossman, L. and Moldave, K., Eds., Academic Press, New York, 1967, 566.
299. **Spelsberg, T. C., Mitchell, W. M., and Chytil, F.,** Structural alterations of acidic proteins by acid treatment of chromatin, *Mol. Cell. Biochem.,* 1, 243, 1973.
300. **Catino, J. J., Yeoman, L. C., Mandel, M., and Busch, H.,** Characterization of DNA binding protein from rat liver chromatin which decreases during growth, *Biochemistry,* 17, 983, 1978.
301. **Wilson, E. and Spelsberg, T. C.,** Rapid isolation of total acidic proteins from chromatin of various chick tissues, *Biochim. Biophys. Acta,* 322, 145, 1973.
302. **Briggs, R. C., Campbell, A., Chiu, J. F., Hnilica, L. S., Lincoln, G., Stein, J., and Stein, G.,** Specificity of DNA-associated nuclear antigens in HeLa cells and distribution during the cell cycle, *Cancer Res.,* 39, 3683, 1979.
303. **Chytil, F. and Spelsberg, T. C.,** Tissue differences in antigenic properties of non-histone protein-DNA complexes, *Nature New Biol.,* 233, 215, 1971.
304. **Krajewska W. M., Briggs, R. C., and Hnilica, L. S.,** Cell-specific antigens in chicken erythroid nuclei: species specificity, *Biochemistry,* 18, 5720, 1979.
305. **Adolph, K. W., Cheng, S. M., Paulson, J. R., and Laemmli, U. K.,** Isolation of a protein scaffold from mitotic HeLa cell chromosomes, *Proc. Natl. Acad. Sci. U.S.A.,* 74, 4937, 1977.
306. **Smart, J. E. and Bonner, J.,** Selective dissociation of histone from chromatin by sodium deoxycholate, *J. Mol. Biol.,* 58, 651, 1971.
307. **Ornstein, L.,** Disc electrophoresis. 1. Background and theory, *Ann. N.Y. Acad. Sci.,* 121, 321, 1964.
308. **Gronow, M., Lewis, F. A., and Thackrach, T. M.,** Studies on the degradation of HeLa non-histone proteins, *Biochim. Biophys. Acta,* 606, 157, 1980.
309. **LeStourgeon, W. M. and Beyer, A. L.,** The rapid isolation, high-resolution electrophoretic characterization, and purification of nuclear proteins, in *Methods in Cell Biology,* Vol. 16, Stein, G. S., Stein, J., and Kleinsmith, L. J., Eds., Academic Press, New York, 1977, 387.
310. **Hnilica, L. S., Grimes, S. R., and Chiu, J. F.,** Electrophoretic fractionation of histones utilizing starch gel and sodium dodecyl sulfate-urea gels, in *Methods in Cell Biology,* Vol. 17, Stein, G. S., Stein, J., and Kleinsmith, L. J., Eds., Academic Press, New York, 1978, 211.
311. **Margolis, J. and Kenrick, K. G.,** Polyacrylamide gel electrophoresis in a continuous molecular sieve gradient, *Anal. Biochem.,* 25, 347, 1968.
312. **Laemmli, U. K.,** Cleavage of structural proteins during the assembly of the head of bacteriophage T4, *Nature (London),* 227, 680, 1970.

313. **Slater, G. G.,** Pore-limit electrophoresis on a gradient of polyacrylamide gel, *Anal. Biochem.,* 24, 215, 1968.
314. **Tombs, M. P.,** The interpretation of gel electrophoresis, *Anal. Biochem.,* 13, 121, 1965.
315. **Margolis, J. and Kenrick, K. G.,** Polyacrylamide gel-electrophoresis across a molecular sieve gradient, *Nature (London),* 214, 1334, 1967.
316. **Chambrach, A.,** Electrophoresis and electrofocusing on polyacrylamide gel in the study of native macromolecules, *Mol. Cell. Biochem.,* 29, 23, 1980.
317. **Reisefeld, R. A., Lewis, V. J., and Williams, D. E.,** Disk electrophoresis of basic proteins and peptides on polyacrylamide gels, *Nature (London),* 195, 281, 1962.
318. **Waehneldt, T. V. and Mandel, P.,** Proteins of rat brain myelin. Extraction with sodium dodecyl sulphate and electrophoresis on analytical and preparative scale, *FEBS Lett.,* 9, 209, 1970.
319. **Summers, D. F., Maizel, J. V., Jr., and Darnell, J. E., Jr.,** Evidence for virus-specific non-capsid proteins in polio-virus-infected HeLa cells, *Proc. Natl. Acad. Sci. U.S.A.,* 54, 505, 1965.
320. **Maizel, J. V., Jr.,** Acrylamide-gel electrophorograms by mechanical fractionation: radioactive adenovirus proteins, *Science,* 151, 988, 1966.
321. **Shapiro, A. L., Scharff, M. D., Maizel, J. V., Jr., and Uhr, J. W.,** Polyribosomal synthesis and assembly of the H and L chains of gamma globulin, *Proc. Natl. Acad. Sci. U.S.A.,* 56, 216, 1966.
322. **Vinuela, E., Algranati, I. D., and Ochoa, S.,** Synthesis of virus-specific proteins in *Escherichia coli* infected with the RNA bacteriophage MS2, *Eur. J. Biochem.,* 1, 3, 1967.
323. **Yeoman, L. C., Taylor, C. W., and Busch, H.,** Two-dimensional polyacrylamide gel electrophoresis of acid extractable nuclear proteins of normal rat liver and Novikoff hepatoma ascites cells, *Biochem. Biophys. Res. Commun.,* 51, 956, 1973.
324. **Platz, R. D. and Kleinsmith, L. J.,** Heterogeneity and specificity of nonhistone phosphoproteins, *Comp. Biochem. Physiol.,* 55B, 9, 1976.
325. **Fleischer-Lambropoulos, H. and Pollow, K.,** Comparison of nonhistone chromosomal proteins from neuronal and glial chromatin by isoelectric focusing and microdisc electrophoresis, *Biochem. Biophys. Res. Commun.,* 80, 773, 1978.
326. **Glass, W. F., II, Briggs, R. C., and Hnilica, L. S.,** Identification of tissue-specific nuclear antigens transferred to nitrocellulose from polyacrylamide gels, *Science,* 211, 70, 1981.
327. **Schmidt, W. N., Stryjecka-Zimmer, M., Glass, W. F., II, Briggs, R. C., and Hnilica, L. S.,** Tissue specificity and cellular distribution of Novikoff hepatoma antigenic proteins p39, p49 and p56, *J. Biol. Chem.,* 256, 8117, 1981.
328. **Svensson, H.,** Preparative electrophoresis and ionophoresis, in *Advances in Protein Chemistry,* Vol. 4, Anson, M. L. and Edsall, J. T., Eds., Academic Press, New York, 1948, 251.
329. **Svensson, H.,** Isoelectric fractionation, analysis, and characterization of ampholytes in natural pH gradients. I. The differential equation of solute concentrations at a steady state and its solution for simple cases, *Acta Chem. Scand.,* 15, 325, 1961.
330. **Svensson, H.,** Isoelectric fractionation, analysis, and characterization of ampholytes in natural pH gradients. III. Description of apparatus for electrolysis in columns stabilized by density gradients and direct determination of isoelectric points, *Arch. Biochem. Biophys. Suppl.,* 1, 132, 1962.
331. **Vesterberg, O. and Svensson, H.,** Isoelectric fractionation, analysis and characterization of ampholytes in neutral pH gradients. IV. Further studies on the resolving power in connection with separation of myoglobins, *Acta Chem. Scand.,* 20, 820, 1966.
332. **Wrigley, C. W.,** Analytical fractionation of plant and animal proteins by gel electrofocusing, *J. Chromatogr.,* 36, 362, 1968.
333. **Hunter, L.,** Equilibrium isoelectric focussing in acrylamide gel slabs — reduction of cathodic drift, *Anal. Biochem.,* 89, 279, 1978.
334. **Caspers, M. L., Posey, Y., and Brown, R. K.,** Separator isoelectric focusing: an improved method of protein analysis and purification, *Anal. Biochem.,* 79, 166, 1977.
335. **Gronow, M. and Griffiths, G.,** Rapid isolation and separation of the nonhistone proteins of rat liver nuclei, *FEBS Lett.,* 15, 340, 1971.
336. **Wrigley, C.,** Gel electrofocusing as technique for analysing multiple protein samples by isoelectrofocusing, *Sci. Tools,* 15, 17, 1968.
337. **Finlayson, G. R. and Chrambach, A.,** Isoelectric focusing in polyacrylamide gel and its preparative application, *Anal. Biochem.,* 40, 292, 1971.
338. **MacGillivray, A. J. and Rickwood, D.,** The heterogeneity of mouse-chromatin nonhistone proteins as evidenced by two-dimensional polyacrylamide gel electrophoresis and ion-exchange chromatography, *Eur. J. Biochem.,* 41, 181, 1974.
339. **Margolis, J. and Kenrick, K. G.,** Two-dimensional resolution of plasma proteins by combination of polyacrylamide disc and gradient gel electrophoresis, *Nature (London),* 221, 1056, 1969.

340. **Martini, O. H. W. and Gould, H. J.,** Enumeration of rabbit reticulocyte ribosomal proteins, *J. Mol. Biol.*, 62, 403, 1971.
341. **Orrick, L. R., Olson, M. O. J., and Busch, H.,** Comparison of nucleolar proteins of normal rat liver and Novikoff hepatoma ascites cells by two-dimensional polyacrylamide gel electrophoresis, *Proc. Natl. Acad. Sci. U.S.A.*, 70, 1316, 1973.
342. **Yeoman, L. C., Taylor, C. W., Jordan, J. J., and Busch, H.,** Two-dimensional polyacrylamide gel electrophoresis of chromatin proteins of normal rat liver and Novikoff hepatoma ascites cells, *Biochem. Biophys. Res. Commun.*, 53, 1067, 1973.
343. **Busch, G. I., Yeoman, L. C., Taylor, C. W., and Busch, H.,** Modified two-dimensional polyacrylamide gel electrophoresis systems for higher molecular weight nonhistone chromatin proteins from normal rat liver and Novikoff hepatoma ascites cells, *Physiol. Chem. Phys.*, 6, 1, 1974.
344. **Hirsch, F. W., Nall, K. N., Busch, F. N., Morris, H. P., and Busch, H.,** Comparison of abundant cytosol proteins in rat liver, Novikoff hepatoma, and Morris hepatoma by two-dimensional gel electrophoresis, *Cancer Res.*, 38, 1514, 1978.
345. **Gronow, M.,** Solubilization and partial fractionation of the sulphur-containing nuclear proteins of Hepatoma 223 ascites cells, *Eur. J. Cancer,* 5, 497, 1969.
346. **Loeb, J. E. and Creuzet, C.,** Electrophoretic comparison of acidic proteins of chromatin from different animal tissues, *FEBS Lett.*, 5, 37, 1969.
347. **O'Farrel, P. H.,** High resolution two-dimensional electrophoresis of proteins, *J. Biol. Chem.*, 250, 4007, 1975.
348. **Bray, D.,** The protein explosion, *Nature (London),* 267, 481, 1977.
349. **Anderson, N. G. and Anderson, N. L.,** Molecular anatomy, *Behring Inst. Mitt.*, 63, 169, 1979.
350. **Edwards, J. J., Anderson, N. G., Nance, S. L., and Anderson, N. L.,** Red cell proteins. I. Two-dimensional mapping of human erythrocyte lysate proteins, *Blood,* 53, 1121, 1979.
351. **Peterson, J. L. and McConkey, E. H.,** Non-histone chromosomal proteins from HeLa cells. A survey by high resolution, two-dimensional electrophoresis, *J. Biol. Chem.*, 251, 548, 1976.
352. **Fairbanks, G., Steck, T. L., and Wallach, D. F. M.,** Electrophoretic analysis of the major polypeptides of the human erythrocyte membrane, *Biochemistry,* 10, 2606, 1971.
353. **Bonner, W. M. and Laskey, R. A.,** A film detection method for tritium-labelled proteins and nucleic acids in polyacrylamide gels, *Eur. J. Biochem.*, 46, 83, 1974.
354. **Jackle, H.,** Visualization of proteins after isoelectric focusing during two-dimensional gel electrophoresis, *Anal. Biochem.*, 98, 81, 1979.
355. **Gaczynski, M.,** personal communication.
356. **Elgin, S. C. R., Boyd, J. B., Hood, L. E., Wray, W., and Wu, F. C.,** A prolog to the study of the nonhistone chromosomal proteins, *Cold Spring Harbor Symp. Quant. Biol.*, 38, 821, 1973.
357. **Stumph, W. E., Elgin, S. C. R., and Hood, L.,** Antibodies to proteins dissolved in sodium dodecyl sulfate, *J. Immunol.*, 113, 1752, 1974.
358. **Silver, L. M. and Elgin, S. C. R.,** A method for determination of the in situ distribution of chromosomal proteins, *Proc. Natl. Acad. Sci. U.S.A.*, 73, 423, 1976.
359. **Silver, L. M. and Elgin, S. C. R.,** Production and characterization of antisera against three individual NHC proteins; a case of a generally distributed NHC proteins, *Chromosoma Berl.*, 68, 101, 1978.
360. **Panyim, S. and Chalkley, R.** The heterogeneity of histones. I. A quantitative analysis of calf histones in very long polyacrylamide gels, *Biochemistry,* 8, 3972, 1969.
361. **Hardison, R. and Chalkley, R.,** Polyacrylamide gel electrophoretic fractionation of histones, in *Methods in Cell Biology,* Vol. 17, Stein, G. S., Stein, J., and Kleinsmith, L. J., Eds., Academic Press, New York, 1978, 235.
362. **Yeoman, L. C., Jordan, J. J., Busch, R. K., Taylor, C. W., Savage, H. E., and Busch, H.,** A fetal protein in chromatin of Novikoff hepatoma and Walker 256 carcinosarcoma tumors that is absent from normal and regenerating rat liver, *Proc. Natl. Acad. Sci. U.S.A.*, 73, 3258, 1976.
363. **Knecht, M. E. and Busch, H.,** Isolation of an electrophoretically homogeneous nonhistone nucleolar protein, *Life Sci.*, 10(Part II), 1297, 1971.
364. **Olson, M. O. J. and Busch, H.,** Nucleolar proteins, in *Methods in Cell Biology,* Vol. 17, Stein, G. S., Stein, J., and Kleinsmith, L. J., Eds., Academic Press, New York, 1978, 163.
365. **Krüger, C. and Pollow, K.,** A preparative and analytical microscale procedure for nonhistone chromosomal proteins, *Anal. Biochem.*, 88, 388, 1978.
366. **Gronow, M. and Thackrah, T. M.,** Large scale isoelectric focusing analysis of the non-histone proteins of the nucleus: isolation of components with alkaline isoelectric points, *Experientia,* 34, 430, 1978.
367. **MacGillivray, A. J.,** Fractionation of nonhistone chromosomal proteins utilizing isoelectric focusing techniques, in *Methods in Cell Biology,* Vol. 16, Stein, G. S., Stein, J., and Kleinsmith, L. J., Eds., Academic Press, New York, 1977, 381.

368. **Radola, B. J.,** Isoelectric focusing in layers of granulated gels. I. Thin-layer isoelectric focusing of proteins, *Biochim. Biophys. Acta,* 295, 412, 1973.
369. **Boffa, L. C., Vidali, G., and Allfrey, V. G.,** Selective synthesis and accumulation of nuclear nonhistone proteins during carcinogenesis of the colon induced by 1,2-Dimethylhydrazine, *Cancer,* 36, 2356, 1975.
370. **Bekhor, I., Anne, L., Kim, J., Lapeyre, J. N., and Stambaugh, R.,** Organ discrimination through organspecific nonhistone chromosomal proteins, *Arch. Biochem. Biophys.,* 161, 11, 1974.
371. **Samal, B. and Bekhor, I.,** Nonhistone chromosomal protein interaction with DNA-histone complexes. Circular dichroism, *Arch. Biochem. Biophys.,* 179, 527, 1977.
372. **Bekhor, I. and Samal, B.,** Nonhistone chromosomal protein interaction with DNA-histone complexes. Transcription, *Arch. Biochem. Biophys.,* 179, 537, 1977.
373. **Gates, M. D. and Bekhor, I.,** DNA sequence selection by tightly-bound nonhistone chromosomal proteins, *Nucl. Acids Res.,* 6, 1617, 1979.
374. **Gates, M. D. and Bekhor, I.,** DNA-binding activity of tightly-bound nonhistone chromosomal proteins in chicken liver chromatin, *Nucl. Acids Res.,* 6, 3411, 1979.
375. **Bekhor, I. and Mirell, C. J.,** Simple isolation of DNA hydrophobically complexed with presumed gene regulatory proteins M_3, *Biochemistry,* 18, 609, 1979.
376. **Yeoman, L. C., Taylor, C. W., Jordan, J. J., and Busch, H.,** Differences in chromatin proteins of growing and non-growing tissues, *Exp. Cell Res.,* 91, 207, 1975.
377. **Takami, H., Busch, F. N., Morris, H. P., and Busch, H.,** Comparison of salt-extractable nuclear proteins of regenerating liver, fetal liver, and Morris hepatomas 9618A and 3924A, *Cancer Res.,* 39, 2096, 1979.
378. **Gronow, M.,** The solubilization and partial fractionation of the thiol components of isolated nuclei, *Biochem. J.,* 109, 25P, 1968.
379. **Offenbacher, S. and Kline, E. S.,** Differential phosphorylation of rat liver nuclear non-histone proteins *in vitro, Biochem. Biophys. Res. Commun.,* 66, 375, 1975.
380. **Chiu, J. F., Fujitani, H., and Hnilica, L. S.,** Methods for selective extraction of chromosomal nonhistone proteins, in *Methods in Cell Biology,* Vol. 16, Stein, G. S., Stein, J., and Kleinsmith, L. J., Eds., Academic Press, New York. 1977, 283.
381. **Yaneva M., Beltchev, B., and Tsanev, R.,** Tissue specificity of tightly bound non-histone chromosomal proteins, *Cell Diff.,* 9, 351, 1980.
382. **Wakabayashi, K. and Hnilica, L. S.,** The immunospecificity of nonhistone protein complexes with DNA, *Nature New Biol.,* 242, 153, 1973.
383. **Spelsberg, T. C., Steggles, A. W., Chytil, F., and O'Malley, B. W.,** Progesterone-binding components of chick oviduct. V. Exchange of progesterone-binding capacity from target to non target tissue chromatins, *J. Biol. Chem.,* 247, 1368, 1972.
384. **Spelsberg, T. C., Mitchell, W. M., Chytil, F., Wilson, E. M., and O'Malley, B. W.,** Chromatin of the developing chick oviduct: changes in the acidic proteins, *Biochim. Biophys. Acta,* 312, 765, 1973.
385. **Gevorkian, E. S.,** Binding of hydrocortisone-receptor complex with various groups of non-histone proteins, *Vopr. Med. Chimii,* 26, 656, 1980.
386. **Spelsberg, T. C., Stake, E., and Witzke, D.,** Fractionation of chromosomal nonhistone proteins using chromatin-cellulose resins: purification of steroid hormone acceptor proteins, in *Methods in Cell Biology,* Vol. 17, Stein, G. S., Stein, J., and Kleinsmith, L. J., Eds., Academic Press, New York, 1978, 303.
387. **Stein, G. S., Stein, J. L., Kleinsmith, L. J., Thomson, J. A., Park, W. D., and Jansing, R. L.,** Role of nonhistone chromosomal proteins in the regulation of histone gene expression, *Cancer Res.,* 36, 4307, 1976.
388. **Kish, V. M. and Kleinsmith, L. J.,** Nuclear protein kinases, *J. Biol. Chem.,* 249, 750, 1974.
389. **Hiremath, S. T., Maciewicz, R. A., and Wang, T. Y.,** The loosely bound nonhistone chromosomal proteins of rat prostate in androgen action, *Biochim. Biophys. Acta,* 653, 130, 1981.
390. **Wang, T. Y.,** The role of non-histone chromosomal proteins in the interaction of prostate chromatin with androgen·receptor complex, *Biochim. Biophys. Acta,* 518, 81, 1978.
391. **Rodriguez, L. V. and Becker, F. F.,** Rat liver chromatin. Fractionation into eu- and heterochromatin with localization of ribosomal genes, *Arch. Biochem. Biophys.,* 173, 428, 1976.
392. **Rakowicz-Szulczynska, E. M. and Horst, A,,** Non-histone proteins in fractionated chromatin before and after DNAase II treatment, *Biochim. Biophys. Acta,* 653, 69, 1981.
393. **Blüthmann, H. and Illmensee, K.,** Nuclear nonhistone proteins in mouse Teratocarcinomas. I. Cell lineage specificity, *Wilhelm Roux' Arch.,* 190, 374, 1981.
394. **Grogan, D. E., Desjardins, R., and Busch, H.,** Nucleolar proteins of rat liver and Walker tumor, *Cancer Res.,* 5, 775, 1966.
395. **Hasuma, T., Yukioka, M., Nakajima, S., Morisawa, S., and Inoue, A.,** Preferential phosphorylation of basic non-histone proteins by nuclear protein kinase NII from rat liver, *Eur. J. Biochem.,* 109, 349, 1980.

396. **Medvedev, Zh. A. and Medvedeva, M. N.,** A group of H1 histone satellite acid-soluble non-histone chromatin proteins, *FEBS Lett.,* 112, 35, 1980.
397. **Platz, R. D., Kish, V. M., and Kleinsmith, L. J.,** Tissue specificity of non-histone chromatin phosphoproteins, *FEBS Lett.,* 12, 38, 1970.
398. **Kleinsmith, L. J.,** Acidic nuclear phosphoproteins, in *Acidic Proteins of the Nucleus,* Cameron, I. L. and Jeter, J. R., Jr., Eds., Academic Press, New York, 1974, 103.
399. **Murphy, R. F. and Bonner, J.,** Alkaline extraction of non-histone proteins from rat liver chromatin, *Biochim. Biophys. Acta,* 405, 62, 1975.
400. **Gronow, M., Thackrah, T. M., and Lewis, F. A.,** Instability of rat liver chromatin and other nuclear non-histone proteins in alkaline solution, *Biochem. J.,* 157, 507, 1976.
401. **Ganpath, N., Prestayko, A. W., and Busch, H.,** Comparison of nuclear nonhistone phosphoproteins of rat liver and Novikoff hepatoma, *Cancer Res.,* 37, 1290, 1977.
402. **Dayal, V. K., Saha, A., and Lea, M. A.,** Partial characterization of nonhistone nuclear proteins which are decreased in hepatomas of the rat, *Int. J. Biochem.,* 10, 759, 1979.
403. **Medvedev, Zh. A., Medvedeva, M. N., and Robson, L.,** Age-related changes of the pattern of non-histone chromatin proteins from rat and mouse liver chromatin, *Gerontology,* 25, 219, 1979.
404. **Hnilica, L. S.,** Proteins of the cell nucleus, *Progr. Nucleic Acid Res. Mol. Biol.,* 7, 25, 1967.
405. **Kraut, W. and Werner, D.,** Analysis of the most tightly bound proteins in eukaryotic DNA, *Biochim. Biophys. Acta,* 564, 390, 1979.
406. **Capesius, I., Kraut, W., and Werner, D.,** Proteinase K-resistant and alkali-stable bound proteins in higher plant DNA, *FEBS Lett.,* 110, 184, 1980.
407. **Hershey, H. V. and Verner, D.,** Evidence for non-deoxynucleotide links in Ehrlich ascites tumour cell DNA, *Nature (London),* 262, 148, 1976.
408. **Riley, D. E., Keller, J. M., and Byers, B.,** The isolation and characterization of nuclear ghosts from cultured HeLa cells, *Biochemistry,* 14, 3005, 1975.
409. **Riley, D. E. and Keller, J. M.,** The polypeptide composition and ultrastructure of nuclear ghosts isolated from mammalian cells, *Biochim. Biophys. Acta,* 444, 899, 1976.
410. **Adolph, K. W., Cheng, S. M., and Laemmli, U. K.,** Role of nonhistone proteins in metaphase chromosome structure, *Cell,* 12, 805, 1977.
411. **Tsutsui, Y. and Hoshino, M.,** Redistribution of DNA-bound nonhistone chromosomal proteins during mitosis, *Cell Struct. Function,* 4, 321, 1979.
412. **Plagens, U.,** Effect of salt-treatment on manually isolated polytene chromosomes from *Chironomus tentans, Chromosoma,* 68, 1, 1978.
413. **Kamiyama, M. and Dastugue, B.,** Rat liver non-histone proteins: correlation between protein kinase activity and activation of RNA synthesis, *Biochem. Biophys. Res. Commun.,* 44, 29, 1971.
414. **Kamiyama, M., Dastugue, B., and Kruh, J.,** Action of phosphoproteins and protein kinase from rat liver chromatin on RNA synthesis, *Biochem. Biophys. Res. Commun.,* 44, 1345, 1971.
415. **MacGillivray, A. J., Rickwood, D., Cameron, A., Carrol, D., Ingles, C. J., Krause, R. J., and Paul, J.,** Methods for isolation and characterization of chromosomal nonhistone proteins, in *Methods in Enzymology,* Vol. 40, O'Malley, B. W. and Hardman, J. G., Eds., Academic Press, New York, 1975, 160.
416. **MacGillivray, A. J., Johnston, C., MacFarlane, R., and Rickwood, D.,** The isolation and partial characterization of low-molecular-weight phosphorylated component of the non-histone proteins of mouse nuclei, *Biochem. J.,* 175, 35, 1978.
417. **Stein, G. S., Park, W. D., and Stein, J. L.,** Fractionation of nonhistone chromosomal proteins, in *Methods in Cell Biology,* Vol. 17, Stein, G. S., Stein, J., and Kleinsmith, L. J., Eds., Academic Press, New York, 1978, 293.
418. **Paul, J. and Gilmour, R. S.,** Organ-specific restriction of transcription in mammalian chromatin, *J. Mol. Biol.,* 34, 305, 1968.
419. **Righetti, P. and Drysdale, J. W.,** Isoelectric focusing in polyacrylamide gels, *Biochim. Biophys. Acta,* 236, 17, 1971.
420. **Tiselius, A., Hjerten, S., and Levin, O.,** Protein chromatography on calcium phosphate columns, *Arch. Biochem. Biophys.,* 65, 132, 1956.
421. **Bernardi, G. and Cook, H.,** Separation and characterization of the two high density lipoproteins of egg yolk, α- and β-lipovitellin, *Biochim. Biophys. Acta,* 44, 96, 1960.
422. **Bernardi, G.,** Chromatography of nucleic acids on hydroxylapatite. I. Chromatography of native DNA, *Biochim. Biophys. Acta,* 174, 423, 1969.
423. **Bernardi, G.,** Chromatography on nucleic acids on hydroxylapatite columns, in *Methods in Enzymology,* Vol. 21, Part D, Grossman, L. and Moldave, K., Eds., Academic Press, New York, 1971, 95.
424. **Bernardi, G. and Kawasaki, T.,** Chromatography of polypeptides and proteins on hydroxylapatite columns, *Biochim. Biophys. Acta,* 160, 301, 1968.

425. **Faulhaber, I. and Bernardi, G.,** Chromatography of calf-thymus nucleoprotein on hydroxylapatite columns, *Biochim. Biophys. Acta,* 140, 561, 1967.
426. **MacGillivray, A. J.,** Fractionation of nonhistone chromosomal proteins utilizing hydroxylapatite chromatography, in *Methods in Cell Biology,* Vol. 16, Stein, G. S., Stein, J., and Kleinsmith, L. J., Eds., Academic Press, New York, 1977, 329.
427. **MacGillivray, A. J. and Rickwood, D.,** Physical methods for fractionation of chromatin, in *The Cell Nucleus,* Vol. 6, Busch, H., Ed., Academic Press, New York, 1978, 263.
428. **Appels, R., Bolund, L., and Ringertz, N. R.,** Biochemical analysis of reactivated chick erythrocyte nuclei isolated from chick-HeLa heterokaryons, *J. Mol. Biol.,* 87, 339, 1974.
429. **Blüthmann, H.,** Changes in nonhistone chromosomal proteins in phytohemaglutinin stimulated lymphocytes, *Mol. Biol. Rep.,* 4, 97, 1978.
430. **Blüthmann, H.,** Specific binding of a nonhistone chromosomal protein from lymphocyte to DNA, *Eur. J. Biochem.,* 70, 233, 1976.
431. **Blüthmann, H.,** DNA-binding properties of a nonhistone chromosomal protein from lymphocytes, *Mol. Cell. Biochem.,* 19, 147, 1978.
432. **Lapeyre, J. N. and Bekhor, I.,** Effects of 5-bromo-2-deoxyuridine and dimethyl sulfoxide on properties and structure chromatin, *J. Mol. Biol.,* 89, 137, 1974.
433. **Bloom, K. S. and Anderson, J. N.,** Fractionation and characterization of chromosomal proteins by the hydroxylapatite dissociation method, *J. Biol. Chem.,* 253, 4446, 1978.
434. **Yon, R. J.,** Chromatography of lipophilic proteins on adsorbents containing mixed hydrophobic and ionic groups, *Biochem. J.,* 126, 765, 1972.
435. **Hjerten, S.,** Some general aspects of hydrophobic interaction chromatography, *J. Chromatogr.,* 87, 325, 1973.
436. **Shaltiel, S.,** Hydrophobic chromatography, in *Methods in Enzymology,* Vol. 34, Jacoby, W. B. and Wilchek, M., Eds., Academic Press, New York, 1974, 126.
437. **Ochoa, J. L.,** Hydrophobic interaction chromatography, *Biochimie,* 60, 1, 1978.
438. **Hjerten, S., Rosengren, J., and Pahlman, S.,** Hydrophobic interaction chromatography. The synthesis and the use of some alkyl and aryl derivates of agarose, *J. Chromatogr.,* 101, 281, 1974.
439. **Cuatrecasas, P.,** Protein purification by affinity chromatography. Derivatizations of agarose and polyacrylamide beads, *J. Biol. Chem.,* 245, 3059, 1970.
440. **Hofstee, B. H. J.,** Hydrophobic affinity chromatography of proteins, *Anal. Biochem.,* 52, 430, 1973.
441. **Livanova, N. B., Eronina, T. B., Silonova, G. V., and Ramensky, E. V.,** ω-Aminohexyl-Sepharose in purification of liver glycogen phosphorylase b, *FEBS Lett.,* 69, 95, 1976.
442. **Helene, C., Dimicoli, J. L., and Brun, F.,** Binding of tryptamine and 5-hydroxytryptamine serotonine to nucleic acids. Fluorescence and proton magnetic resonance studies, *Biochemistry,* 10, 3802, 1971.
443. **Lancelot, G., Mayer, R., and Helene, C.** Models of interaction between nucleic acids and proteins. Hydrogen bonding of arginine with nucleic acid bases, phosphate groups and carboxylic acids, *Biochim. Biophys. Acta,* 564, 181, 1979.
444. **Calendi, E., Di Marco, A., Reggiani, M., Scarpinato, B., and Valentini, L.,** On physico-chemical interactions between daunomycin and nucleic acids, *Biochim. Biophys. Acta,* 103, 25, 1965.
445. **Kikuchi, H. and Watanabe, M.,** Significance of use of amino acids and histamine for the elution of nonhistone proteins in copper-chelate chromatography, *Anal. Biochem.,* 115, 109, 1981.
446. **Cuatrecasas, P. and Anfinsen, C. B.,** Affinity chromatography, *Annu. Rev. Biochem.,* 40, 259, 1971.
447. **Cuatrecasas, P., Wilchek, M., and Anfinsen, C. B.,** Selective enzyme purification by affinity chromatography, *Proc. Natl. Acad. Sci. U.S.A.,* 61, 636, 1968.
448. **Litman, R. M.,** A deoxyribonucleic acid polymerase from *Micrococcus luteus (Micrococcus lysodeikticus)* isolated on deoxyribonucleic acid-cellulose, *J. Biol. Chem.,* 243, 6222, 1968.
449. **Kleinsmith, L. J., Heidema, J., and Carrol, A.,** Specific binding of rat liver nuclear proteins to DNA, *Nature (London),* 226, 1025, 1970.
450. **Allfrey, V. G.,** DNA-binding proteins and transcriptional control in prokaryotic and eukaryotic systems, in *Acidic Proteins of the Nucleus,* Cameron, I. L. and Jeter, J. R., Jr., Eds., Academic Press, New York, 1974, 1.
451. **Allfrey, V. G., Inoue, A., Karn, J., Johnson, E. M., Good, R. A., and Hadden, J. W.,** Sequence-specific binding of DNA by non-histone proteins and their migration from cytoplasm to nucleus during gene activation, in *The Structure and Function of Chromatin,* Ciba Found. Symp. 28 (new series), Associated Scientific Publishers, New York, 1975, 199.
452. **Allfrey, V. G. and Inoue, A.,** Affinity chromatography of DNA-binding proteins on DNA covalently attached to solid supports, in *Methods in Cell Biology,* Vol. 17, Stein, G. S., Stein, J., and Kleinsmith, L. J., Eds., Academic Press, New York, 1978, 253.
453. **Kaplowitz, P. B., Platz, R. D., and Kleinsmith, L. J.,** Nuclear phosphoproteins. III. Increase in phosphorylation during histone-phosphoprotein interaction, *Biochim. Biophys. Acta,* 229, 739, 1971.

454. **Conner, B. J. and Comings, D. E.,** Isolation of non-histone chromosomal high mobility group protein from mouse liver nuclei by hydrophobic chromatography, *J. Biol. Chem.*, 256, 3283, 1981.
455. **Thompson, S. T., Cass, K. H., and Stellwagen, E.,** Blue Dextran-Sepharose: an affinity column for the dinucleotide fold in proteins, *Proc. Natl. Acad. Sci. U.S.A.*, 72, 669, 1975.
456. **Wilson, J. E.,** Applications of Blue Dextran and Cibacron Blue F3GA in purification and structural studies of nucleotide-requiring enzymes, *Biochem. Biophys. Res. Commun.*, 72, 816, 1976.
457. **Klyszejko-Stefanowicz, L.,** Heterogeneity and specificity of non-histone proteins, *Post. Biochem.*, 25, 287, 1979.
458. **Chan, P. K. and Liew, C. C.,** Purification of a phosphoprotein from chromatin of rat liver, *Biochem. J.*, 183, 147, 1979.
459. **Ohtsuki, K., Shiraishi, H., Yamada, E., Nakamura, M., and Ishida, N.,** A non-histone chromatin protein that is a specific phosphate acceptor of nuclear cAMP-independent protein kinase from mouse spleen cells, *J. Biol. Chem.*, 255, 2391, 1980.
460. **Ohtsuki, K., Yamada, E., Nakamura, M., and Ishida, N.,** Mouse spleen cell nuclear kinases and the stimulating effect of dsDNA on NHP phosphorylation by cyclic AMP-independent protein kinase in vitro, *J. Biochem.*, 87, 35, 1980.

Chapter 2

THE HIGH MOBILITY GROUP (HMG) PROTEINS

G. H. Goodwin and E. W. Johns

TABLE OF CONTENTS

I. INTRODUCTION

The discovery[1] and subsequent initial characterization[2,3] of the group of chromosomal proteins called the HMG proteins has created considerable interest because of the suggestion that they may be associated with active genes. This suggestion, made initially[4] solely on the basis of the quantity of HMG proteins relative to the histones, has gained acceptability (at least in the case of two of the HMG proteins) because of subsequent experimental work (see Section IV. D). Consequently much effort has been, and is now being expended on all the problems associated with a relatively new group of proteins; i.e., tissue and species specificity, turnover, synthesis, modification, structure, and localization within chromatin. We have incomplete answers to most of the problems, and to some only guesses. However, there is much solid information with regard to the structure of these proteins, and as was the case with the histones, a groundwork of good protein chemistry in terms of fractionation, isolation, characterization, and structure laid the foundation for the subsequent functional work.

The HMG proteins are chromatin-associated proteins which can be extracted from chromatin with 0.35 *M* NaCl and are soluble in 2% trichloroacetic acid (TCA).[2] They are additionally characterized by having high contents of acidic and basic amino acids. Although the four main mammalian and avian HMG proteins, HMG 1, 2, 14, and 17 were originally considered to be related members of a single group of proteins on the basis of the above properties, the more recent structural and functional studies have, as we shall see, clearly demonstrated that we are dealing with two types of structurally (and probably functionally) distinct proteins. HMG1 and 2 are clearly related to one another but are quite different from the smaller HMG proteins, HMG14 and 17 (and the other low molecular weight proteins in birds and fish) which are structurally and functionally very similar to one another. Thus, in this chapter we have chosen to review the HMG proteins in this light and we have in effect written two reviews; one on the HMG1 and 2 group proteins (the high molecular weight HMG proteins) and one on the HMG14 and 17-type proteins (the low molecular weight HMG proteins). The various proteins that we will be dealing with are tabulated in Table 1; we have deliberately focused our attention on the proteins from mammals, birds, and fish since so little is known about the various HMG-like proteins which have been isolated from invertebrates, plants, and lower eukaryotes that it is difficult to relate them to the vertebrate HMG proteins. This review is not an exhaustive literature survey on what is becoming a rapidly expanding field — we have, rather, concentrated on dealing with the more recently published data. The reader is referred to several previous reviews for more detailed accounts of the earlier literature,[5,6] and a book on HMG proteins comprehensively deals with all aspects of HMG protein biochemistry,[7] including an account of HMG-like proteins in invertebrates, plants, and lower eukaryotes.[8]

Although a good deal is known about the structures of the main HMG proteins it has to be admitted at the outset that we do not know the nuclear function (or functions) of the high molecular weight HMG proteins and we are only just beginning to understand the role of the low molecular weight HMG proteins in the structure of transcriptionally active chromatin. Earlier studies[5] have established that all four HMG proteins are associated with nucleosomes, but it is important to keep in mind when we come to discuss in detail the functions of the various HMG proteins and their association with nucleosomes that these proteins are present in the cell nucleus in much smaller quantities than the histones. If we count HMG1 and 2 together as one protein we calculate from the yields of these two proteins from calf thymus that in the cell nucleus there is about one molecule of HMG1 (or HMG2) per 3000 base pairs of DNA. Similarly, there is about one molecule of HMG14 (or HMG17) per 2000 base pairs of DNA. If one assumes that most of the DNA in the cell nucleus is packaged into nucleosomes then a maximum of 1 in every 15 nucleosomes will have HMG1 or 2

Table 1
THE TWO CATEGORIES OF HMG PROTEINS IN VERTEBRATES[a]

	Mammals (calf, pig, rabbit rat, mouse)	Birds (chicken, duck)	Fish (trout, winter flounder)
High mol wt proteins	HMG1	HMG1	HMGT[b]
	HMG2 (five subfractions)	HMG2b } five subfractions	HMGT2
		HMG2a }	HMGT1
Low mol wt proteins	HMG14	HMG14	D
	HMG17	HMG17	C
	(HMG19A, B?)	HMGY	H6

[a] The columns show the proteins in order of increasing mobility on SDS polyacrylamide electrophtresis gels.
[b] The electrophoretic mobility of HMGT relative to HMGT1 and T2 is not known.

bound and 1 in 10 will have HMG14 or 17 bound. Scans of polyacrylamide gels of thymus acid extracts have given a somewhat higher content of HMG1 and 2, suggesting that there might be as much as one molecule of HMG1 or 2 for every five or six nucleosomes.[9] These calculations assume one HMG binding site per nucleosome; if there are two or more (see below) then the proportion of HMG-containing nucleosomes in the cell nucleus will be correspondingly less.

II. ISOLATION

There are basically two methods for the extraction of the four HMG proteins from the nucleus or chromatin. The first is the 0.35 *M* NaCl extraction method followed by 2% TCA precipitation.[3,10] The other is based on the finding that 5% perchloric acid (PCA) selectively extracts the HMG proteins together with histone H1. The H1 can then be subsequently removed by fractional acetone precipitation.[1,5,11,12]

The four HMG proteins can be separated from one another and obtained in a pure form by the procedures outlined in Figure 1 which involves (in the final stages), ion-exchange chromatography on CM-Sephadex® or CM-cellulose columns at pH 8.8 to 9.0.[3,5,10,12]

The HMG proteins prepared by these methods are recovered in a denatured state due to the exposure to acids. This is not a problem with HMG14 and 17 (being random coil proteins anyway), but HMG1 and 2 proteins may be irreversibly affected since highly they are proteins which have highly ordered structures (see Section III. C). Thus a method has been developed recently for isolating nondenatured proteins by avoiding the use of acids.[13] In this method the HMG proteins are extracted from chromatin with 0.35 *M* NaCl and the extract is fractionated by phosphocellulose chromatography at pH 7.5.

III. THE HIGH MOLECULAR WEIGHT HMG PROTEINS

A. Occurrence and Composition

The nuclei of a wide variety of mammalian, avian, and fish tissues contain a series of proteins of about 29,000 mol wt which have high contents of acidic and basic amino acids. Originally two such proteins, HMG1 and 2, were isolated by salt or PCA extractions of calf thymus chromatin,[1,3,10] but later it was shown that HMG2 was composed of a family of at least four subfractions which differed in their isoelectric points.[14] Figure 2 is an example of a more recent analysis and shows that HMG2 is composed of five subfractions, A to E.[15] (HMG1 does not focus as a discrete band in the isoelectric focusing dimension of Figure 2,

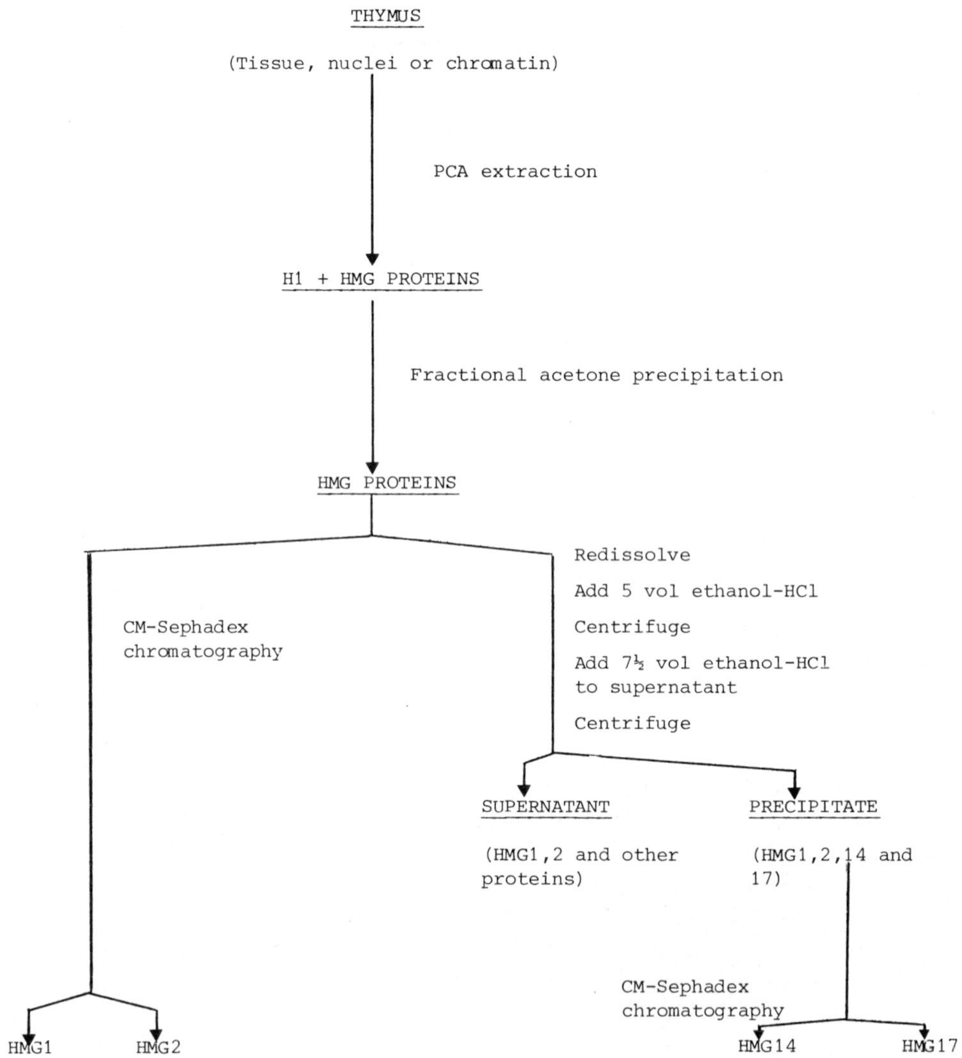

FIGURE 1. Schematic diagram of the isolation of the four calf HMG proteins.

probably because of aggregation.) These six proteins (HMG1, 2A, 2B, 2C, 2D, 2E) are apparently present in all calf tissues[15,16] and very similar proteins have been found in other mammalian tissues.[5,8] The amino acid compositions of calf thymus HMG1 and four of the five HMG2 subfractions are given in Table 2. These analyses demonstrate the highly charged nature of the proteins and the close similarities of the members of the group of proteins within a species.

Avian cells have proteins very like calf HMG1 and 2[17-21] and again the avian HMG2 has five subfractions.[20] The individual HMG2 subfractions have not been isolated but they do partially separate on CM-Sephadex® columns into two peaks, HMG2a and2b.[20] (Note: these two peaks have also been termed HMGE and HMG2, respectively, by Sterner et al.[18] and HMG2A and HMG2B by Gordon et al.)[21] These two fractions (HMG2a and 2b) differ from one another somewhat more than do the individual calf HMG2 subfractions. Thus HMG2a has more aspartic acid but less glutamic acid than HMG2b. Also, while the calf subfractions all have the same N-terminal glycine, the avian HMG2 proteins have glycine or alanine N-

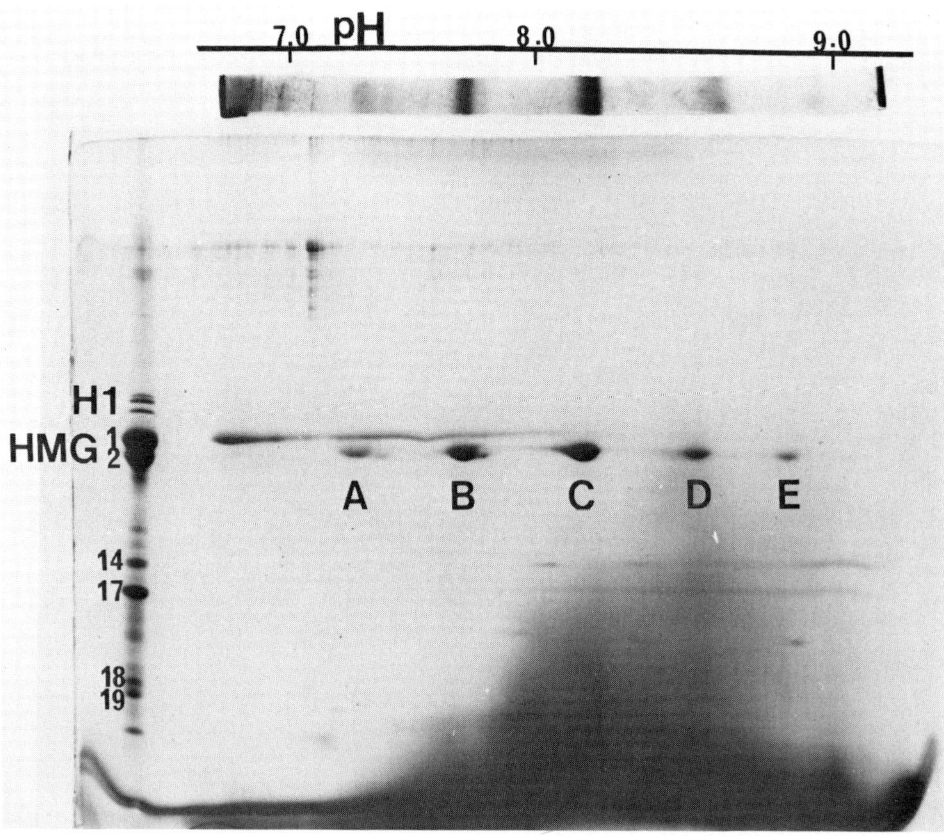

FIGURE 2. Two-dimensional analysis of PCA-extracted proteins from calf thymus showing HMG1, the five HMG2 subfractions, HMG14, and HMG17[15] (first dimension — isoelectric focusing; second dimension — SDS electrophoresis, top to bottom).

terminals.[18,20] Like histone H1, quantitative variations in the subfractions are seen in different avian tissues.[18,20,21] Although HMGE (or HMG2a) was originally thought to be erythrocyte specific (hence the E),[18] it was later shown that this protein fraction is present in other avian tissues.[20,21]

In fish there are again several proteins with similar properties to the avian and mammalian proteins.[22-25] Three proteins, HMGT, HMGT1, and HMGT2, with different isoelectric points, have so far been described in trout testis and liver.[22-24] It is not clear exactly how these proteins relate to the six calf proteins; nevertheless they clearly fall into the same category of proteins since they have similar amino acid compositions (Table 2), N-terminal amino acid sequences (see below), and molecular weights. The trout proteins differ from the mammalian and avian proteins in that they have less acidic amino acids, apparently lacking the highly acidic peptide near the C-terminal end of the HMG1 and 2 molecules (see below), and they do not fold into ordered secondary and tertiary structures like HMG1 and 2 (see below).

B. Primary Structures

Most of the amino acid sequence of calf HMG1 has been published together with that of the unfractionated HMG2 (Figure 3)[26] and has shown that these two proteins are very closely related in their primary structures. Since no obvious evidence of sequence heterogeneity or postsynthetic modifications appeared during the sequencing of HMG2, the sequence given

Table 2
AMINO ACID COMPOSITIONS (MOL %) OF CALF HMG1 AND HMG2 SUBFRACTIONS, AND TROUT PROTEIN T, T1, AND T2

Amino acid	Calf thymus					Trout testis		
	HMG1	HMG2A	HMG2B	HMG2C	HMG2D	HMGT	HMGT1	HMGT2
Asp	10.7	9.3	10.0	9.7	9.0	13.7	15.2	12.4
Thr	2.5	2.1	2.0	2.0	2.5	3.1	3.0	4.0
Ser	5.0	7.2	7.6	7.5	7.8	5.5	8.0	6.6
Glu	18.1	17.3	18.4	17.6	17.3	10.6	12.9	9.7
Pro	7.0	10.6	8.3	9.4	8.3	8.2	7.4	9.9
Gly	5.3	5.9	6.6	6.2	6.8	7.8	8.0	9.3
Ala	9.0	7.0	7.7	7.7	7.9	10.1	8.8	11.5
Val	1.9	1.5	1.4	1.8	2.2	3.9	3.8	3.6
Cys	tr	—	0.2	—	—	0.8	n.d.	n.d.
Met	1.5	0.3	0.6	0.6	0.7	2.0	0.9	1.5
Ile	1.8	1.7	1.1	1.1	1.5	1.6	2.0	1.1
Leu	2.2	2.3	1.6	1.7	2.3	2.3	3.0	1.6
Tyr	2.9	3.2	2.4	2.3	2.5	2.3	2.2	1.7
Phe	3.6	3.1	3.1	3.1	3.5	3.5	3.2	3.5
Lys	21.3	19.1	19.8	20.1	18.6	14.2	14.2	16.8
His	1.7	2.1	2.1	2.1	2.3	1.6	1.5	1.7
Arg	3.9	4.6	4.3	4.6	4.7	5.5	5.7	5.0
Trp	n.d.	n.d.	n.d.	n.d.	n.d.	0.4	n.d.	n.d.

in Figure 3 is either the sequence of the major subfraction (subfraction B, see Figure 2), or all the subfractions have the same sequences in the regions that have been sequenced. The most striking feature of the sequences of both HMG1 and 2 is the continuous tract of 35 or 41 aspartic and glutamic acids in the C-terminal region. Examination of the sequences of the two proteins reveal that the basic amino acids and the rest of the acidic amino acids are distributed fairly evenly throughout the rest of the molecule, making the N-terminal two thirds of the proteins quite basic. Hydrophobic amino acids are also distributed fairly evenly throughout the sequences outside the highly acidic regions.

The N-terminal sequences of the trout proteins T[22] and T2[24] are the only other published sequences in this group of proteins (Figure 4). It is clear from these sequences that all three are related to the calf HMG1 and 2 proteins. Since the avian HMG proteins have very similar compositions and isoelectric focusing properties to the calf proteins it is expected that they, too, will have the highly acidic sequence in the C-terminal. The HMGT proteins probably do not have it since they have less acidic amino acids and no such highly acidic peptide has been found in HMGT.[27]

C. Secondary, Tertiary, and Quaternary Structures

Proteins HMG1 and 2, as originally prepared by PCA extraction or salt extraction and TCA fractionation, are obtained in a denatured state. Nevertheless, both proteins can refold in solution at pHs around neutrality or by the addition of salt to give highly structured polypeptides with 40 to 50% α-helix.[28,29] Tertiary folding of the protein is also evident when analyzed by NMR[28] and fluorescence anisotrophy.[29] However, comparison of such "renatured" proteins with HMG1 and 2 proteins made under nondenaturing conditions[13] shows that the acid-treated proteins are not folded to quite the same degree as the native proteins,[30] thus for functional studies it may be important to isolate HMG1 and 2 under nondenaturing conditions. This may be especially relevant to the isolation of the HMGT proteins since the

10 20

HMG 1: Gly-Lys-Gly-Asp-Pro-Lys-Lys-Pro-Arg-Gly-Lys-Met-Ser-Ser-Tyr-Ala-Phe-Phe-Val-Gln-Thr-Ser-Arg-Glu-Glu-His-
HMG 2: Gly-Lys-Gly-Asp-Pro-Asn-Lys-Pro-Arg-Gly-Lys-Met-Ser-Ser-Tyr-Ala-Phe-Phe-Val-Gln-Thr-Ser-Arg-Glu-Glu-His-

30 39

HMG 1: Lys-Lys-Lys-His-Pro-Asp-Ala-Ser-Val-Asn-Phe-Ser-Glu-(Phe-Ser-Lys-Lys-Cys-Ser-Glu, Ser, Gly, Ala, Tyr,
HMG 2: Lys-Lys-Lys-His-Pro-Asp-Ala-Ser-Val-Asx-Phe-Ser-Glu-(Phe,Ser,Lys,Lys,Cys, Glu, Val, Gly, Ala, Tyr,

60 70

HMG 1: Lys_2,Glu_2)-Arg-Trp-Lys-Thr-Met-Ser-Ala-Lys-Glu-Lys-Gly-Lys-Phe-Glu-Asp-Met-Ala-Lys-Ala-Asp-Lys-Ala-Arg-
HMG 2: Lys ,)-Arg-Trp-Lys-Thr-Met-Ser-Ala-Lys-Glu-Lys-Ser-Lys-Phe-Glu-Asp-Met-Ala-Lys-Ser-Asp-Lys-Ala-Arg-

80 90 100

HMG 1: Tyr-Glu-Arg-Glu-Met-Lys-Thr-Tyr-Ile-Pro-Pro-Lys-Gly-Glu-Thr-Lys-Lys-Lys-Phe-Lys-Asp-Pro-Asn-Ala-Pro-Lys-
HMG 2: Tyr-Asp-Arg-Glu-Met-Lys-Asn-Tyr-Val-Pro-Pro-Lys-Gly-Asp-Lys-Lys-Gly-Lys-Lys-Lys-Asp-Pro-Asn-Ala-Pro-Lys-

145 160

110 120

HMG 1: Arg-Pro-Pro-Ser-Ala-Phe-Phe-Leu-Phe-Ala-Ser-Glu-Tyr-Arg-Pro-Lys-Ile-Lys-Gly-Glu-His-Pro-Gly-Leu-Ser-Ile-
HMG 2: Arg-Pro-Pro-Ser-Ala-Phe-Phe-Leu-Phe-Ser-Ala-Glu-His-Arg-Pro-Lys-Ile-Lys-Ala-Glu-His-Pro-Gly-Leu-Ser-Ile-

130 140 150

HMG 1: Gly-Asp-Val-Ala-Lys-Lys-Leu-Gly-Glu-Met-Trp-Asn-Asn-Thr-Ala-Ala-Asp-Asp-Lys-Gln-Pro-Tyr-Glu-Lys-Lys-Ala-
HMG 2: Gly-Asp-Thr-Ala-Lys-Lys-Leu-Gly-Glu-Met-Trp-Ser-Gln-Gln-Ser-Ala-Lys-Asp-Lys-Gln-Pro-Tyr-Glu-Gln-Lys-Ala-

160 170 180

HMG 1: Ala-Lys-Leu-Lys-Glu-Lys-Tyr-Glu-Lys-(?)-Ala-Ala-Tyr-Arg-Ala-Lys-Gly-Lys-Pro-Asp-Ala-Ala-Lys-Lys-Gly-Val-
HMG 2: Ser-Lys-Leu-Lys-Glu-Lys-Tyr-Glu-Lys-(?)-Ala-Ala-Tyr-Arg-Ala-Lys-Gly-Lys-Ser-Glu-Ala-Gly-Lys-Lys-Gly-Pro-

190 200

HMG 1: Val-Lys-Ala-Glu-Lys-Ser-Lys-Lys-Lys-Lys-Glu-Glu-Glu-Glu-Asp-Glu-Glu-Asp-Asp-Glu-(Glu_{22},Asp_9, Lys)-
HMG 2: Gly-Arg-Pro-Thr-Gly-Ser-Lys-Lys-Lys-Asn-Glu-Pro-Glu-Asp-Glu-Glu-Glu-Glu-Glu-Glu-(Glu_{21},Asp_5,Pro,Lys)-

259

HMG 1: (Asp-Ile-Ala-Ala-Tyr,Thr,Pro,Ala,Leu,Phe,Arg,Ser_2,Glu_3,Gly_2,Lys_5)-Phe-Ala-Lys.
HMG 2: (Asp,Ile,Ala,Ala,Tyr,Thr,Pro, Leu, Arg,Ser_2,Glu_4,Gly_3,Lys_5)-Phe-Ala-Lys.

FIGURE 3. The primary structures of calf thymus HMG1 and HMG2.[26]

```
                                                  10                                                 20
HMG1    Gly - Lys - Gly - Asp - Pro - Lys - Lys - Pro - Arg - Gly - Lys - Met - Ser - Ser - Tyr - Ala - Phe - Phe - Val - Gln -
HMG2    Gly - Lys - Gly - Asp - Pro - Asn - Lys - Pro - Arg - Gly - Lys - Met - Ser - Ser - Lys - Ala - Phe - Phe - Val - Gln -
HMGT    Pro - Gly - Lys - Asp - Pro - Asn - Lys - Pro - Lys - Gly - Lys - Thr - Ser - Ser - Ser - Ala - Phe - Phe - Val - Ala -
HMGT2   Pro - Gly - Lys - Asp - Pro - Asn - Lys - Pro - Lys - Gly - Lys - Thr - Ser - Ser - Tyr - Ala - Phe - Phe - Val - Ala -

                                                  30
HMG1    Thr - Ser - Arg - Glu - Glu - His - Lys - Lys - Lys - His
HMG2    Thr - Ser - Arg - Glu - Glu - His - Lys - Lys - Lys - His
HMGT    Gln - Arg - Arg - Glx - Glx - His
```

FIGURE 4. N-terminal sequences of calf HMG1 and 2, and trout testis HMGT[22] and T2.[24]

acid-treated proteins are found not to refold in solution[84] (whether or not the HMGT proteins are naturally in a folded state remains to be established). Gel filtration[30] and sedimentation studies[31,32] show that both renatured and "native" HMG1 and 2 are monomeric proteins in solution although in chromatin they may be associated as oligomers (see below).

D. Synthesis and Postsynthetic Modifications

The synthesis of HMG1 and HMG2 does not appear to be exclusively confined to any one phase of the cell cycle. In the study by Kuehl,[33] using regenerating liver, the synthesis of these two proteins was found to drop to a minimum 4 hr after hepatectomy, then rise again before the onset of DNA synthesis, and to peak a few hours after the peak of DNA synthesis. Similar results were obtained by Craddock and Henderson,[34] but the minimum at 4 hr was not observed. Unlike the radioactive labeling of nuclear histones, the synthesis of HMG1 and 2 proteins is thus not closely coordinated with DNA synthesis. Also inhibitors of DNA synthesis have little effect on HMG1 and 2 synthesis,[33] but dimethylnitrosamine, a potent carcinogen, causes a considerable reduction in the synthesis of these two proteins.[34]

The lack of coordination between HMG1 and 2 synthesis and DNA synthesis is also seen in synchronized HeLa cells,[35] though in this case the peak of HMG1 and 2 synthesis is in early G1, well before the peak of DNA synthesis.

Like the histones, proteins HMG1 and 2 are modified by postsynthetic modification and there are data to suggest that they can be modified by acetylation,[36,37] methylation,[38] phosphorylation[39,40] (but see Reference 41), and possibly modified by poly(ADP-ribose) groups.[42] Only in the former two cases have the modifications been chemically characterized in the purified proteins. Two sites of acetylation have been identified as the lysines at residues numbered 2 and 11 of the sequences[37] (see Figure 3), and dimethylarginine has been detected in the amino acid analyses of purified calf HMG1 and 2 (3 to 4% of the arginines in HMG1 and 8 to 9% of the arginines in HMG2 are so modified). These modifications could account for the microheterogeneity seen in calf HMG2 though alkaline phosphatase and neutral hydroxylamine treatment (to remove phosphate and ADP-ribose groups, respectively) had no effect on the isoelectric focusing pattern.[43] In the case of avian HMG2, HMG2a and 2b differ in their amino acid composition (see above) and so must differ in their primary structures.

E. Association with DNA, Nucleosomes, and Chromatin

Protein HMG1 and 2 bind to single- and double-stranded DNA in vitro primarily through ionic interactions between basic amino acids (mostly in the N-terminal two thirds of the proteins) and the phosphate of the DNA; the highly acidic C-terminal regions do not bind and are presumably free to interact with other molecules.[28,31,32] Since HMG1 and 2 bind more strongly to single-stranded DNA[44,45] these two proteins can unwind the helix of double-stranded DNA[46] and in this way can induce supercoils in circular DNA.[47] There is some evidence that HMG1 and 2 might actually be bound to single-stranded DNA in vivo.[48] In experiments in which salt dissociation of HMG1 and 2 from single-stranded DNA, double-stranded DNA, and native chromatin are compared, it was found that HMG1 and 2 elute from single-stranded DNA and chromatin at comparable ionic strengths while they elute from double-stranded DNA at significantly lower ionic strengths. (However, the tighter binding of HMG1 and 2 in chromatin as compared to binding to naked double-stranded DNA could be due to the fact that the DNA in the cell nucleus is supercoiled or be due to additional interactions between HMG and other chromosomal constituents.) Support for the idea that HMG1 and 2 are bound to single-stranded DNA regions in chromatin is provided by the isolation of nucleosomes from the cell nucleus which have a high content of single-stranded DNA and which have stoichiometric quantities of HMG1 and 2.[49] These were prepared by digesting nuclei with micrococcal nuclease at near physiological ionic strength — conditions which favor digestion of single-stranded DNA. Nucleosomes solubilized under such conditions were found to have enriched quantities of HMG1 and 2 and the DNA 35 to 40% single-stranded. These nucleosomes lacked histone H1, intimating that it may have been replaced on these nucleosomes by the HMG proteins. (Note: the HMG1- and 2-containing nucleosomes did not appear to have HMG14 and 17 bound to them as well.)

The possible association of HMG proteins with single-stranded DNA might suggest that HMG1 and 2 are acting as DNA helix-unwinding proteins during DNA replication (or even possibly transcription). Unfortunately there is no other evidence to confirm or deny this idea. Examination of the quantities of HMG1 and 2 in tissues with different nucleic acid synthetic rates and cells at different stages of the cell cycle have not furnished support for such a role in nucleic acid synthesis — for example, proliferating tissues such as thymus and appendix[50] do not appear to have dramatically more HMG1 and 2 than nondividing cells such as erythrocytes. The relative quantities of HMG2 vs. HMG1 have been examined in a number of rat tissues[51] and it was found that proliferating tissues have higher levels of HMG2, but this does not appear to be so in chicken[17,18,20,21] and rabbit tissues[50] so the implication of HMG2 in the DNA replication process also remains equivocal.

Digestion of nuclei with all three commonly used nucleases (micrococcal, DNAse I, and DNAse II) results in the rapid release of variable amounts of the HMG1 and HMG2 (and HMGT) from the nucleus.[19,52-58] Much of this released protein is not bound to nucleosomes and so could be due to an association of HMG1 and 2 with some very susceptible DNA (e.g., linker DNA of nucleosomes on transcriptionally active genes).[59] However, in mammalian cells, half of the HMG1 and 2 appears to be loosely bound in the cell nucleus anyway, being extractable with 0.15 *M* NaCl,[56,60] and could therefore have been released by nuclease action simply as a result of fragmenting the nuclear matrix. In fact HMG1 and 2 have been found in the cytoplasm (as well as the nucleus) of some cells[61,62] and it is possible that HMG1 and 2 are in equilibrium between chromatin, nucleoplasm, and cytoplasm and the levels in these compartments may fluctuate according to the metabolic state of the cell or the cell cycle.[63] (A recent report has suggested that the finding of substantial amounts of HMG1 and 2 in the cytoplasm could be an experimental artifact caused by leakage of protein from the nucleus.[64] Also Rechsteiner and Kuehl[65] introduced labeled HMG1 into cells by microinjection and found that most of the protein migrates into the nucleus.)

Chemical cross-linking studies on chromatin have shown that HMG1 may be associated as oligomers or at least sequestered in specific regions of the genome.[66] (It is of interest to note in this connection that both HMG1 and 2 when bound to DNA in vitro formed beaded nucleosome-like structures[67] suggesting an oligomeric association of the proteins, though the isolated proteins themselves do not form oligomers on their own.)[30] As yet little is known about what regions the HMG1 and 2 proteins might be bound to in chromatin. There are some rather indirect hints from nuclease digestion experiments[19,52-58] that HMG1 and 2 might be associated with transcribed DNA sequences, but as yet a clear association of HMG1 and or 2 with expressed sequences has not been demonstrated (for example, the DNA sequences of the HMG1- and 2-containing nucleosomes described above have not as yet been analyzed, though they are depleted in 5-methylcytosine).[49] No (or little) HMG1 and 2 has been found in purified rat liver satellite chromatin.[68] Thus the function of the HMG1 and 2 group of proteins remains obscure, but the most likely role is promoting strand separation for DNA replication. Finally, one possibility that has not been ruled out is that HMG1 and 2 might not be bound to the DNA at all, but to RNA. Superficially, this might seem unlikely in view of the fact that erythrocytes, which have little nuclear RNA, have HMG1 and 2 levels much the same as any other cell — nevertheless, this possibility cannot be discarded and should be borne in mind.

Table 3
AMINO ACID COMPOSITIONS (MOL%) AND MOLECULAR WEIGHTS OF THE LOW MOLECULAR WEIGHT PROTEINS FROM CALF, CHICKEN, AND TROUT

Amino acids	Calf thymus		Chicken erythrocyte and oviduct			Trout liver		
	HMG14	HMG17	HMG14	HMG17	HMGY	C	D	H6
Asp	8.3	12.0	9.3	9.1	6.2	6.0	8.0	6.7
Thr	4.1	1.2	4.6	3.0	5.0	2.6	4.1	1.6
Ser	8.0	2.3	5.2	4.3	4.8	4.4	4.3	5.6
Glu	17.5	10.5	15.6	11.7	13.7	23.8	21.9	6.1
Pro	8.1	12.9	10.5	12.1	10.9	10.9	8.4	12.3
Gly	6.4	11.2	5.6	10.0	8.2	2.8	3.3	7.4
Ala	14.8	18.4	18.0	17.2	14.8	16.0	16.6	25.4
Cys	—	—	—	—	—	—	—	—
Val	4.0	2.0	tr	2.2	1.8	4.2	2.5	3.4
Met	—	—	—	—	—	tr	tr	—
Ile	—	—	—	—	0.3	0.6	—	—
Leu	2.0	1.0	1.1	1.2	1.8	0.8	1.0	1.2
Try	—	—	tr	tr	—	tr	—	—
Phe	—	—	tr	tr	0.4	tr	—	—
His	—	—	1.1	tr	0.8	tr	0.9	—
Lys	21.1	24.3	24.0	23.6	26.2	19.6	23.7	23.1
Arg	5.4	4.2	4.1	4.6	4.2	4.6	4.2	7.2
Mol wt	10,700	9,200	12,800	9,200	—	—	—	7,200

IV. THE LOW MOLECULAR WEIGHT HMG PROTEINS

A. Occurrence and Composition

This group of proteins is coextracted with the high molecular weight HMG proteins when chromatin is treated with 0.35 *M* NaCl or PCA. Although the two groups of proteins have a number of properties in common (e.g., they are present in the chromatin in similar quantities — a few percent by weight of the DNA — and the proteins of both groups have high contents of acidic and basic amino acids), the lower molecular weight proteins are quite different structurally and they probably have a different function.

Mammalian tissues have been found to have two major proteins in this group, HMG14 and HMG17.[5,6] Originally isolated from calf and pig thymus,[10,12,69] these two small basic proteins have subsequently been shown to be present in many other calf tissues[16] and to be present in rabbits, rats, and mice.[33,34,68,70-72] The amino acid compositions and molecular weights of calf thymus HMG14 and 17 are given in Table 3. In addition to the high contents of lysine, aspartic acid, and glutamic acid they have high contents of glycine, alanine, and proline. They have few hydrophobic amino acids, which explains why these proteins do not give tertiary-folded structures in solution and in fact are completely random-coil proteins. Two similar proteins have been found in avian cells[17,75] (and have been termed HMG14 and 17), and just recently a third, smaller protein, HMGY, has been described which has the same N-terminal amino acid (proline) and similar composition to the avian HMG14 and 17[76] (In fact HMGY runs as a doublet on polyacrylamide electrophoretic gels.)

In fish (trout) the major low molecular weight HMG protein is the protein termed H6, first found in trout testis,[77] but later shown to be present in liver[23] and in other organs of

```
                                                                    10
Pro - Lys - Arg - Lys - Val - Ser - Ser - Ala - Glu - Gly -
                                                                    20
Ala - Ala - Lys - Glu - Glu - Pro - Lys - Arg - Arg - Ser -
                                                                    30
Ala - Arg - Leu - Ser - Ala - Lys - Pro - Ala - Pro - Ala -
                                                                    40
Lys - Val - Glu - Thr - Lys - Pro - Lys - Lys - Ala - Ala -
                                                                    50
Gly - Lys - Asp - Lys - Ser - Ser - Asp - Lys - Lys - Val-
                                                                    60
Gln - Thr - Lys - Gly - Lys - Arg - Gly - Ala - Lys - Gly -
                                                                    70
Lys - Gln - Ala - Glu - Val - Ala - Asn - Gln - Glu - Thr -
                                                                    80
Lys - Glu - Asp - Leu - Pro - Ala - Glu - Asn - Gly - Glu -
                                                                    90
Thr - Lys - Asn - Glu - Glu - Ser - Pro - Ala - Ser - Asp -
                                                                   100
Glu - Ala - Glu - Glu - Lys - Glu - Ala - Lys - Ser - Asp.
```

FIGURE 5. The primary structure of calf thymus HMG14.[79]

trout,[70] and in other fish.[25] H6 is smaller than HMG14 and 17 (see Table 2) and is about the same size as HMGY. There are two other HMG14/17-like proteins in trout tissues, proteins C and D,[23] but these are present in smaller amounts than H6. These two proteins are larger than H6 and have electrophoretic mobilities quite close to calf HMG14 and 17 (see Table 1).

In addition to the above-described proteins, a number of other low molecular weight PCA-soluble proteins have been detected in small quantities in mammalian cell nuclei.[78] Two of these proteins, HMG19A and 19B, resemble HMG14 and 17 in their amino acid compositions and a third, HMG18, resembles histones H1 and H5, but further characterization of these proteins is required before we can classify them with confidence in the low molecular weight HMG group.

B. Primary Structures

The complete amino acid sequences of calf HMG14[79] and HMG17,[80] chicken HMG17,[81] and trout H6[82] proteins have been determined (Figures 5, 6, and 7). Partial sequences have been published for some of the other low molecular weight avian (HMG14) and trout proteins (proteins C and D)[83] (Figure 8). These sequences show that all seven proteins are quite closely related and confirms the grouping of these proteins shown in Table 1.

The chicken and calf HMG17 sequences are very similar (Figure 6), differing only at five positions and even then the differences are conservative (e.g., aspartic acid is replaced by glutamic acid at position 64). The same is not true for the HMG14 proteins in the two species (Figure 8), for we find that comparing the N-terminal sequences there are quite a

```
                                                                10
Pro - Lys - Arg - Lys - Ala - Glu - Gly - Asp - Ala - Lys -
                                               (Thr)

                                                                20
Gly - Asp - Lys - Ala - Lys - Val - Lys - Asp - Glu - Pro -

                                                                30
Gln - Arg - Arg - Ser - Ala - Arg - Leu - Ser - Ala - Lys -

                                                                40
Pro - Ala - Pro - Pro - Lys - Pro - Glu - Pro - Lys - Pro -

                                                                50
Lys - Lys - Ala - Pro - Ala - Lys - Lys - Gly - Glu - Lys -
                 (Ala) (Pro)             (Ser)

                                                                60
Val - Pro - Lys - Gly - Lys - Lys - Gly - Lys - Ala - Asp -

                                                                70
Ala - Gly - Lys - Asp - Gly - Asn - Asn - Pro - Ala - Glu -
                 (Glu)

                                                                80
Asn - Gly - Asp - Ala - Lys - Thr - Asn - Gln - Ala - Glu -

                                                          89
Lys - Ala - Glu - Gly - Ala - Gly - Asp - Ala - Lys.
```

FIGURE 6. The primary structure of calf thymus HMG17.[80] The chicken erythrocyte HMG17 sequence is the same except for five substitutions which are shown in the brackets underneath.[81]

```
                                                                10
Pro - Lys - Arg - Lys - Ser - Ala - Thr - Lys - Gly - Asp -

                                                                20
Glu - Pro - Ala - Arg - Arg - Ser - Ala - Arg - Leu - Ser -

                                                                30
Ala - Arg - Pro - Val - Pro - Lys - Pro - Ala - Ala - Lys -

                                                                40
Pro - Lys - Lys - Ala - Ala - Ala - Pro - Lys - Lys - Ala -

                                                                50
Val - Lys - Gly - Lys - Lys - Ala - Ala - Glu - Asn - Gly -

                                                                60
Asp - Ala - Lys - Ala - Glu - Ala - Lys - Val - Gln - Ala -

                                                          69
Ala - Gly - Asp - Gly - Ala - Gly - Asn - Ala - Lys.
```

FIGURE 7. The primary structure of trout testis H6.[82]

```
                                                           10                                                         18
Trout D   Pro-Lys-Arg- o -Lys- o - o -Gln-Gly-Ala-Ala-Gly-Asp- o - o - o - o - o -Val- o - o -Lys-Glu-Glu-Pro-Gln-Arg-Arg-

                                                                                  10                                          20
Trout C   Pro-Lys-Arg-Ser-Lys-Ala- o -Asn- o -Ala- o - o -Asp- o - o - o - o - o -Val-Glu-Ala-Ala- o -Glu-Pro-Lys-Arg-Arg-Ser-Glu-

                                                      10                                                                  20
CTHMG14   Pro-Lys-Arg- o -Lys-Val-Ser-Ser- o -Ala-Glu-Gly- o - o - o - o - o - o -Ala- o -Ala-Lys-Glu-Glu-Pro-Lys-Arg-Arg-Ser-Ala-

                                                           10                                                             20
CEHMG14   Pro-Lys-Arg- o -Lys-Ala-Pro- o - o -Ala-Glu-Gly-Glu- o - o - o - o - o - o - o -Ala-Lys-Glu-Glu-Pro-Lys-Arg-Arg-Ser-Ala-

                                                                   10                                    20
CTHMG17   Pro-Lys-Arg- o -Lys- o - o - o - o -Ala-Glu-Gly-Asp-Ala-Lys-Gly-Asp-Lys-Ala-Lys-Val-Lys-Asp-Glu-Pro-Gln-Arg-Arg-Ser-Ala

                                                                                                  10
Trout H6  Pro-Lys-Arg- o -Lys- o -Ser- o - o -Ala- o - o - o - o - o - o - o - o -Thr-Lys-Gly- o -Asp-Glu-Pro-Ala-Arg-Arg-Ser-Ala-


                                      28
Trout C   ( )-Leu-Ser-Asn-Lys-Pro-Ala-Ile-

                                          30                                      40
CTHMG14   Arg-Leu-Ser-Ala-Lys-Pro-Ala-Pro-Ala-Lys-Val-Glu-Thr-Lys-Pro-Lys-Lys-Ala-Ala-Gly-Lys-Asp-Lys-Ser-Ser- o - o -Asp-Lys-Lys-Val-

                                              30                                      40
CEHMG14   Arg-Leu-Ser-Ala-Lys-Pro-Ala-Pro-Pro-Lys-Pro-Glu-Pro-Lys-Pro-Lys-Lys-Ala-Ala-Pro-Pro-Lys-Glu-Lys-Ala-Ala-Asp-Asp-Lys-Lys-Glu-

                          30                                      40                                          50
CTHMG17   Arg-Leu-Ser-Ala-Lys-Pro-Ala-Pro-Pro-Lys-Pro-Glu-Pro-Lys-Pro-Lys-Lys- o -Ala-Pro-Ala-Lys-Lys-Gly-Glu-Lys-Val-Pro-Lys-Gly-Lys-

                  20                                          30                                      40
Trout H6  Arg-Leu-Ser-Ala-Arg-Pro-Val-Pro- o -Lys-Pro-Ala-Ala-Lys-Pro-Lys-Lys-Ala-Ala-Ala-Pro-Lys-Lys-Ala-Val-Lys-Gly-Lys-Lys-Ala-Ala-
```

FIGURE 8. Comparison of the N-terminal sequences of trout proteins C, D, and H6 and calf thymus and chicken erythrocyte HMG14 and 17.[83] The amino terminal sequence of chicken erythrocyte HMG17 shows only two differences from calf thymus HMG17, having Thr at residue 9 and Ser at residue 48.

```
                                                                                    10
H6       Pro-Lys-Arg-Lys-(-)-(-)-Ser-Ala-Thr-Lys-Gly-(-)-(-)-(-)-(-)-(-)-(-)-Asp-Glu-Pro-Ala-
HMG 17   Pro-Lys-Arg-Lys-Ala-Glu-Gly-Asp-Ala-Lys-Gly-Asp-Gly-Ala-Lys-Val-Lys-Asp-Glu-Pro-Gln-
                                             10                                      20

                                  20                                            30
H6       Arg-Arg-Ser-Ala-Arg-Leu-Ser-Ala-Arg-Pro-Val-Pro-(-)-Lys-Pro-Ala-Ala-Lys-Pro-Lys-Lys-
HMG 17   Arg-Arg-Ser-Ala-Arg-Leu-Ser-Ala-Lys-Pro-Ala-Pro-Pro-Lys-Pro-Glu-Pro-Lys-Pro-Lys-Lys-
                                         30                                      40

                                  40
H6       Ala-Ala-Ala-Pro-Lys-Lys-Ala-(-)-(-)-Val-(-)-Lys-Gly-Lys-(-)-(-)-Lys-Ala-(-)-(-)-(-)-
HMG 17   Ala-Pro-Ala-(-)-Lys-Lys-Gly-Glu-Lys-Val-Pro-Lys-Gly-Lys-Lys-Gly-Lys-Ala-Asp-Ala-Gly-
                                         50                                      60

                                  (ADPR)
                                             50
H6       (-)-(-)-(-)-(-)-(-)-(-)-Ala-Glu-Asn-Gly-Asp-Ala-Lys-Ala-(-)-Glu-Ala-(-)-Lys-Val-Gln-
HMG 17   Lys-Asp-Gly-Asn-Asn-Pro-Ala-Glu-Asn-Gly-Asp-Ala-Lys-Thr-Asn-Gln-Ala-Glu-Lys-Ala-Glu-
                                     70                                      80

             60                                      69
H6       Ala-Ala-Gly-Asp-Gly-Ala-Gly-Asn-Ala-Lys.
HMG 17   (-)-(-)-(-)-(-)-Gly-Ala-Gly-Asp-Ala-Lys.
                                             89
```

FIGURE 9. Comparison of the amino acid sequences of trout testis H6 and calf thymus HMG17.

few differences between the two proteins, and in fact chicken HMG14 is in one respect more like calf (and avian) HMG17 than calf HMG14. This is seen in the chicken erythrocyte HMG14 sequence from residues 26 to 40, a sequence rich in prolines which is a characteristic feature of HMG17 (residues 31 to 44) (see Figure 8). Thus the sequence of chicken HMG14 is more akin to the calf HMG17 than to the calf HMG14 sequence in this region.

The trout protein H6 has considerable sequence homologies with calf HMG17 (Figure 9), including a proline-rich region, but being smaller H6 has a number of deletions. Trout proteins C and D are also similar to the calf HMG14 and 17 proteins in their N-terminal sequences (Figure 8), but until the complete sequences of these proteins have been determined it is not possible to deduce the exact evolutionary relationship between the trout proteins (C, D, and H6) and the avian/mammalian proteins (HMG14, 17, and Y).

The sequences of HMG14, 17, and H6 reveal a number of interesting points. The lysine and arginine residues are sequestered mainly in the N-terminal two thirds of the proteins, making these regions highly basic in nature, while the C-terminal thirds are acidic in nature, weakly so in the HMG17 and H6 cases, but markedly so in HMG14. From interaction studies with DNA (see below) we know that the basic amino acids (particularly the arginines) within the N-terminal two thirds bind to DNA, primarily through ionic interactions, while the acidic C-terminal remains free. Another point of interest is that there are sequence homologies between HMG17 (and less so HMG14) and histones H1 and H5. These sequences are in the basic N-terminal two thirds of HMG17 and are homologous with blocks of sequences in the random coil regions of histones H1 and H5 (Figure 10).

C. Synthesis and Postsynthetic Modifications

Two studies using regenerating rat liver have shown that the synthesis of HMG17 after partial hepatectomy follows very much the same course as HMG1 and 2 and is not linked with DNA synthesis;[33,34] synthesis of HMG17 commences before the onset of DNA synthesis and reaches a peak several hours after the peak of DNA synthesis. Also DNA synthesis

	1 — 17	29
HMG17	Pro-Lys-Arg-Lys-Ala-Glu-Gly	Ala-Lys-Pro-Ala-Pro-Pro-Lys-Pro-
	32 — 38	169
H1	Ala-Lys-Arg-Lys-Ala-Ala-Gly	Ala-Lys-Pro-Ala-Ala- 0 -Lys-Pro-
	39 — 47	52
HMG17	Lys-Pro-Lys-Lys-Ala-Pro- 0 -Ala-Lys-Lys	Pro-Lys-Gly-Lys-Lys
	139 — 148	119
H1	Lys-Pro-Lys-Lys-Ala-Ala-Gly-Ala-Lys-Lys	Pro-Lys-Pro-Lys-Lys-
	22 — 35	
HMG17	Arg-Arg-Ser-Ala-Arg-Leu-Ser-Ala-Lys-Pro-Ala-Pro-Pro-Lys	
	20 — 23	6 — 12
H5	Arg-Arg-Ser-Ala	Leu-Ser- 0 - 0 -Pro-Ala-Pro-Ala-Lys

FIGURE 10. Sequenced homologues between calf HMG17 and rabbit histone H1 and chicken erythrocyte H5.

inhibitors do not affect HMG17 synthesis but dimethylnitrosamine does.[34] No data are available for HMG14 but presumably its synthesis is like that of HMG17.

There is now good evidence that HMG14 and 17 are modified by phosphorylation,[40,41] but as yet the sites of phosphorylation have not been identified. Trout testis H6 has been found to be modified by poly ADP-ribosylation and this has been shown to be attached to the glutamic acid at residue 48.[82] Since calf HMG14 and HMG17 both have a glutamic acid in the C-terminal surrounded by the same sequence as that around the modified residue in H6 we may anticipate that mammalian and avian HMG14 and HMG17 will also be found to have a poly (ADP-ribose) at this position.

Like HMG1 and 2, HMG14 and 17 have been shown to be modified by acetylation and so far one *N*-acetyl lysine residue has been formed at residue 2 of the sequences of the two proteins.[85] These modifications could account for the spread of HMG14 and HMG17 protein in the isoelectric focusing analysis of Figure 2.

D. Association with DNA, Nucleosomes, and Chromatin

HMG14 and HMG17 bind to DNA by ionic interactions between basic amino acids and the DNA phosphate groups.[73,74] Nuclear magnetic resonance analysis of this interaction shows the lysine, and more so the arginine residues, binding to the DNA. The sequence Arg-Arg-Ser-Ala-Arg is found in HMG14, 17, and H6 (and a similar sequence is found in histoneH5) and so this sequence may be crucial for HMG-DNA interactions.

When monomer nucleosomes and core particles are isolated from nuclease-digested nuclei the low molecular weight HMG protproteins are found associated with such particles,[4,56,75,86,87] but because the quantities of HMG14 and 17 in the cell nucleus are much smaller than the histones (only about 2% by weight) then obviously only a proportion of the total nucleosomes in the cell nucleus have HMG protein attached to them — only about 10%, assuming one HMG binding site per nucleosome. Several lines of evidence equate this population of HMG-containing nucleosomes with those nucleosomes specifically associated with transcribed sequences. First, using micrococcal nuclease to fragment chromatin, monomer nucleosomes have been isolated which are enriched in transcribed sequences and which have enriched or stoichiometric quantities of HMG14, 17, HMGY, or H6.[75,76,87,100] Secondly, subnucleosome fragments produced by nuclease degradation, composed of 20 to 30 base pairs of DNA associated with just HMG14 or 17, have been shown to have enriched quantities of transcribed sequences.[71] Thirdly, transcribed sequences in chromatin (usually about 10% of the total) are more sensitive to degradation by DNAse I digestion than nontranscribed sequences,[88] and this has been shown to be due to the presence of HMG14 and 17 (or H6) on those nucleosomes bound to transcribed sequences.[52,89,90-93] Thus the binding of HMG14 and 17 to "active" nucleosomes results in a conformational change in the nucleosomes rendering them more susceptible to DNAse I degradation. Although these results have led to the belief that HMG14 and 17 are exclusively located on transcribed sequences, the situation has been complicated by a number of other findings. First, although HMG14 and 17 bind preferentially to nucleosomes associated with transcribed sequences in reconstitution experiments,[90,94] it has been found that all (or most) of the nucleosomes in the cell nucleus have HMG14 and 17 binding sites.[94-96] (Two binding sites have been mapped on the core particle to two regions near the two ends of the DNA.)[94,96] Secondly, HMG proteins have been found associated (albeit in reduced quantities) with satellite chromatin, which is generally considered to be mostly transcriptionally inert.[68,97] Thus an exclusive location of HMG proteins on transcribed sequences may be an oversimplification and a more realistic view might be that there are low- and high-affinity HMG binding sites; the former being on nucleosomes bound to nontranscribed sequences and the latter on nucleosomes bound to transcribed sequences. The HMG proteins would then be distributed on the two in such a way that most (over half, say) are associated with "active" nucleosomes and the balance distributed throughout the rest of the genome. This would explain the presence of reduced quantities of HMG proteins in satellite chromatin.[68] Only when bound to "active" nucleosomes do the HMG proteins induce a DNAse I-sensitive configuration.

The nature of the interactions between the HMG proteins and the nucleosomal components in the two types of binding interaction are not known, but presumably in both there is binding of the basic N-terminal sequence to the DNA while the more acidic C-terminal sequence interacts with the core histones; in the high-affinity interaction the C-terminal sequence interaction may well be important in inducing the DNAse I-sensitive configuration of the nucleosomes, though this has not yet been tested experimentally. In what way do transcriptionally active nucleosomes differ from inactive nucleosomes to account for this preferential binding of HMG proteins? In order to answer this question, methods have been developed for purifying HMG-containing nucleosomes by salt fractionation[98] and/or gel electrophoresis techniques,[75,86,95,97,99,100] or, by using the HMG proteins themselves to isolate those nucleosomes with high-affinity sites for HMG proteins.[101] Nucleosomes soluble in 0.1 *M* NaCl or solutions containing divalent metal ions have been found to be enriched in expressed sequences,[87,98,102] and HMG proteins.[75,87,98] These salt-soluble nucleosomes can be further fractionated by preparative gel electrophoresis to obtain nucleosomes with one or two low molecular weight HMG proteins per nucleosome.[75,87,100] These nucleosomes have a full complement of the four core histones but in many cases lack histone H1 (they also have little HMG1 and 2). It is difficult to decide whether this lack of H1 reflects the in vivo situation since H1 binding to monomer nucleosomes appears to be so labile,[103,104] but it is certain this putative absence of H1 is not the reason for tighter HMG binding to active nucleosomes since in reconstitution experiments using total unfractionated H1- and HMG-depleted nucleosomes, the preferential DNase I-sensitivity of active genes can still be regained by adding back HMG14 and 17 alone.[90] (Note also that chemical cross-linking experiments by Ring and Cole[105] suggest that HMG14 and 17 might be juxtaposed to H1.) Although the salt-soluble H6-containing nucleosomes from trout testis have large amounts of highly acetylated forms of the histones[99,106] this does not seem to be true for HMG-containing nucleosomes from other tissues.[75,76,86,95,97,100] Nucleosomes prepared by HMG-affinity chromatography have been shown to be enriched in transcribed sequences and depleted in nontranscribed sequences.[101] These also do not have high enough levels of the highly acetylated histones to account for the preferential binding of one or more HMG proteins per nucleosome. However, such nucleosomes have a lower content of 5-methylcytosine in their DNA and this agrees with a number of studies that have shown that, generally speaking, expressed genes are undermethylated as compared with inactive genes.[107-109] Thus methylation of cytosine at certain positions on the nucleosome could inhibit HMG binding to inactive genes and hence would account for the depleted content of HMG14 and 17 seen in satellite chromatin, the DNA of which is more highly methylated than main band DNA.[110] However, methylation or rather nonmethylation of DNA cannot be the sole distinguishing feature of active nucleosomes to account for high-affinity binding sites and HMG-induced DNAse I-sensitivity of active nucleosomes. Thus nucleosomes reconstituted from nonmethylated DNA, poly(dA-dT), and histones are not rendered DNAse I-sensitive when HMG 14 and 17 are added.[96] A further problem is that undermethylation of active genes is not confined to the transcribed sequences themselves but extends several thousands of base pairs into neighboring flanking sequences. Although flanking sequences are DNAse I-sensitive this apparently is not caused by HMG14 and 17.[111] (An exception to this is found in the α^D-α^A globin gene region in chick erythrocytes.[101] In this case, the coding regions, the spacer DNA joining these two regions, and about 1 Kb of DNA to the 3′ side of the α^A gene are all DNAse I-sensitive due to HMG binding.) It should also be pointed out that not all sites in a gene that are found to be methylated in various nonexpressing tissues are unmethylated in the tissue which is expressing the gene.[108,109] Thus the reason for HMG binding more tightly to active gene sequences is not clear and how these proteins induce DNAse I-sensitivity is still a matter of conjecture.

Although HMG14 and 17 have been shown to induce a DNAse I-sensitive configuration on active nucleosomes it is possible that they have other functions in addition. Thus, because the bulk of the HMG14 and 17-containing monomer nucleosomes when isolated are devoid of H1, and because HMG17 (and less so HMG14) have sequence homologies with H1 (Figure 10), it is possible that these two HMG proteins displace H1 or at least weaken the binding of H1 on active nucleosomes and hence alter the stability of higher order chromatin structures. HMG14 and 17 have been shown to inhibit the histone deacetylase in vitro[72] and since histone acetylation may be important in the process of RNA synthesis (for example, in unfolding higher order chromatin structures to allow progress of the RNA polymerase) it is conceivable that HMG14 and 17 promote RNA transcription by helping to maintain elevated levels of the highly acetylated histones.

Both HMG14 and 17 can substitute equally well in the reconstitution of DNAse I-sensitive nucleosomes. Why the cell utilizes two such proteins (and maybe more, e.g., HMGY, proteins C and D) when apparently one would do is food for thought. It could be, for example, that genes transcribed at different rates or at different stages of the cell cycle have different HMG proteins bound to them. The possibility that postsynthetic modifications of the HMG14 and 17 proteins play a role in modulating gene transcription will also be an important line of investigation in future studies.

A number of puzzling findings in relation to the association of HMG proteins and transcribed sequences have yet to be resolved.

1. Although the ovalbumin gene sequences are DNAse I-sensitive when oviduct nuclei are digested with DNAse I, they do not retain this sensitivity when isolated monomer nucleosomes are digested with DNAse I.[112]
2. In trout testis not all transcribed sequences appear to be associated with HMG(H6)-containing nucleosomes, since Levy and Dixon[113] have found that salt-insoluble H1-containing monomer nucleosomes lacking H6 are also enriched in transcribed sequences.

3. Salt-soluble nucleosomes from chicken erythrocytes, containing a 10- to 15-fold enrichment of HMG14 and HMG17, were found to be at most 2-fold enriched in globin gene sequences.[114]

ACKNOWLEDGMENTS

The authors would like to thank all those who have worked in their laboratory over the years and have contributed to the work described in this chapter. The authors also thank Audrey Inglefield and Carol Wright for typing and helping to prepare the manuscript. Grants from the Medical Research Council, and the Cancer Research Campaign are gratefully acknowledged.

REFERENCES

1. **Johns, E. W.,** Studies on histones, VII. Preparative methods for histone fractions from calf thymus, *Biochem. J.*, 92, 55, 1964.
2. **Goodwin, G. H., Sanders, C., and Johns, E. W.,** A new group of chromatin-associated proteins with a high content of acidic and basic amino acids, *Eur. J. Biochem.*, 38, 14, 1973.
3. **Goodwin, G. H. and Johns, E. W.,** Isolation and characterisation of two calf-thymus chromatin non-histone proteins with high contents of acidic and basic amino acids, *Eur. J. Biochem.*, 40, 215, 1973.
4. **Goodwin, G. H., Woodhead, L., and Johns, E. W.,** The presence of high mobility group non-histone chromatin proteins in isolated nucleosomes, *FEBS Lett.*, 73, 85, 1977.
5. **Goodwin, G. H., Walker, J. M., and Johns, E. W.,** The high mobility group (HMG) non-histone chromosomal proteins, in *The Cell Nucleus,* Vol. 6, Busch, H., Ed., Academic Press, New York, 1978, chap. 6.
6. **Walker, J. M., Goodwin, G. H., Smith, B. J., and Johns, E. W.,** The chromosomal proteins, in *Comprehensive Biochemistry,* in press, 1981.
7. **Johns, E. W., Ed.,** The HMG Chromosomal Proteins, Academic Press, London, 1982.
8. **Mayes, E. L. V.,** Species and tissue specificity, in *The HMG Chromosomal Proteins,* Johns, E. W., Ed., Academic Press, London, 1982, chap. 2.
9. **Smith, B. J., Toogood, C. I. A., and Johns, E. W.,** Quantitative staining of submicrogram amounts of histone and high-mobility group proteins on sodium dodecylsulphate-polyacrylamide gels, *J. Chromatogr.*, 200, 200, 1980.
10. **Goodwin, G. H., Nicolas, R. H., and Johns, E. W.,** An improved large scale fractionation of high mobility group non-histone chromatin proteins, *Biochim. Biophys. Acta* 405, 280, 1975.
11. **Sanders, C. and Johns, E. W.,** A method for the large-scale preparation of two chromatin proteins, *Biochem. Soc. Trans.*, 2, 547, 1974.
12. **Sanders, C.,** A method for the fractionation of the high mobility group non-histone chromosomal proteins, *Biochem. Biophys. Res. Commun.*, 78, 1034, 1977.
13. **Isackson, P. J., Debold, W. A., and Reeck, G. R.,** Isolation and separation of chicken erythrocyte HMG non-histone chromatin protein by chromatography on phosphocellulose, *FEBS Lett.*, 119, 337, 1980.
14. **Goodwin, G. H., Nicolas, R. H., and Johns, E. W.,** Microheterogeneity in a non-histone chromosomal protein, *FEBS Lett.*, 64, 412, 1976.
15. **Nicolas, R. H. and Goodwin, G. H.,** Isolation and analysis, in *The HMG Chromosomal Proteins,* Johns, E. W., Ed., Academic Press, London, 1982, chap. 3.
16. **Rabbani, A., Goodwin, G. H., and Johns, E. W.,** Studies on the tissue specificity of high mobility group non-histone chromosomal proteins from calf, *Biochem. J.*, 173, 497, 1978.
17. **Rabbani, A., Goodwin, G. H., and Johns, E. W.,** HMG non-histone chromosomal proteins from chicken erythrocytes, *Biochem. Biophys. Res. Commun.*, 81, 351, 1978.
18. **Sterner, R., Boffer, L. C., and Vidali, G.,** Comparative structural analysis of high mobility group proteins from a variety of sources. Evidence for a high mobility group protein unique to avian erythrocyte nuclei, *J. Biol. Chem.*, 253, 3830, 1978.

19. **Teng, C. S., Andrews, G. K., and Teng, C. T.,** Studies on the high-mobility group non-histone protein from hen oviduct, *Biochem. J.*, 181, 585, 1979.
20. **Mathew, C. G.P., Goodwin, G. H., Gooderham, K., Walker, J. M., and Johns, E. W.,** A comparision of the high mobility group non-histone chromatin protein HMG2 in chicken thymus and erythrocytes, *Biochem. Biophys. Res. Commun.*, 87, 1243, 1979.
21. **Gordon, J. S., Rosefeld, D. I., Kaufman, R., and Williams, D. L.,** Evidence for a quantitative tissue-specific distribution of the high mobility group chromosomal proteins, *Biochemistry*, 19, 4395, 1980.
22. **Watson, D. C., Peters, E. H., and Dixon, G. H.,** The purification, characterisation and partial sequence determination of a trout testis non-histone protein HMG-T, *Eur. J. Biochem.*, 74, 53, 1977.
23. **Rabbani, A., Goodwin, G. H., Walker, J. M., Brown, E., and Johns, E. W.,** Trout liver high mobility group non-histone chromosomal proteins, *FEBS Lett.*, 109, 294, 1980.
24. **Brown, E., Goodwin, G. H., Mayes, E. L. V., Hastings, J. R. B., and Johns, E. W.,** Heterogeneity of protein resembling high mobility group protein HMGT in trout testis nuclei, *Biochem. J.*, 191, 661, 1980.
25. **Kennedy, B. P. and Davies, P. L.,** Acid-soluble nuclear proteins of the testis during spermatogenesis in the winter flounder. Loss of the high mobility group proteins, *J. Biol. Chem.*, 255, 2533, 1980.
26. **Walker, J. M., Gooderham, K., Hastings, J. R. B., Mayes, E., and Johns, E. W.,** The primary structure of non-histone chromosomal proteins HMG1 and 2, *FEBS Lett.*, 122, 264, 1981.
27. **Dixon, G. H.,** personal communication, 1980.
28. **Cary, P. D., Crane-Robinson, C., Bradbury, E. M., Javaherian, K., Goodwin, G. H., and Johns, E. W.,** Conformational studies of two non-histone chromosomal proteins and their interactions with DNA, *Eur. J. Biochem.*, 62, 583, 1976.
29. **Smerdon, M. J. and Isenberg, I.,** Conformational changes in subfractions of calf thymus histone H1, *Biochemistry*, 15, 4233, 1976.
30. **Cockerill, P., Goodwin, G. H., Turner, C.,and Cary, P.,** *Biochim. Biophys. Acta*, in press.
31. **Shooter, K. V., Goodwin, G. H., and Johns, E. W.,** Interactions of a purified non-histone chromosomal protein with DNA and histones, *Eur. J. Biochem.*, 47, 263, 1974.
32. **Goodwin, G. H., Shooter, K. V., and Johns, E. W.,** Interaction of a non-histone chromatin protein (high mobility group protein 2) with DNA, *Eur. J. Biochem.*, 54, 427, 1975.
33. **Kuehl, L.,** Synthesis of high mobility group proteins in regenerating rat liver, *J. Biol. Chem.*, 254, 7276, 1979.
34. **Craddock, V. M. and Henderson, A.R.,** The effect of dimethylnitrosamine on the synthesis of high mobility group non-histone proteins in regenerating rat liver, *Carcinogenesis*, 1, 445, 1980.
35. **Goldknopf, I., Sudhakov, S., Rosenbaum, F., and Busch, H.,** Timing of ubiquitin synthesis and conjugation into protein A24 during the HeLa cell cycle, *Biochem. Biophys. Res. Commun.*, 95, 1253, 1980.
36. **Sterner, R., Vidali, G., Heinrikson, R. L., and Allfrey, V. G.,** Postsynthetic modification of high mobility group proteins. Evidence that high mobility group proteins are acetylated, *J. Biol. Chem.*, 253, 7601, 1978.
37. **Sterner, R., Vidali, G., and Allfrey, V. G.,** Studies of acetylation and deacetylation in high mobility group proteins. Identification of the sites of acetylation in HMG-1, *J. Biol. Chem.*, 254, 11577, 1979.
38, **Boffa, L. C., Sterner, R., Vidali, G., and Allfrey, V. G.,** Post synthetic modifications of nuclear proteins are methylated, *Biochem. Biophys. Res. Commun.*, 89, 1322, 1979.
39. **Fonagy, A., Ord, M. G., and Stocken, L. A.,** Phosphorylation of rat thymus histones, its control and the effects thereon of gamma-irradiation, *Biochem. J.*, 162, 171, 1977.
40. **Schroter, H., Haase, E., and Arfmann, H. A.,** *Proc. 2nd Int. Cong. Cell Biology*, Abstr. 6241, 1980, 81.
41. **Saffer, J. D. and Glazer, R. I.,** The phosphorylation of high mobility group proteins 14 and 17 from Ehrlich ascites and L1210 *in vitro*, *Biochem. Biophys. Res. Commun.*, 93, 1280, 1980.
42. **Caplan, A., Ord, M. G., and Stocken, L. A.,** Chromatin structure through the cell cycle. Studies with regenerating rat liver, *Biochem. J.*, 174, 475, 1978.
43. **Goodwin, G. H. and Johns, E. W.,** Further characterisation of HMG non-histone proteins in the organisation and expression of the eukaryotic genome, Bradbury, E. M. and Javaherian, K., Eds., Academic Press, London, 1977, 44.
44. **Bidney, D. L. and Reeck, G. R.,** Purification from cultured hepatoma cells of two non-histone chromatin proteins with preferential affinity for single-stranded DNA: apparent analogy with calf thymus HMG proteins, *Biochem. Biophys. Res. Commun.*, 85, 1211, 1978.
45. **Isackson, P. J., Fishback, J. L., Bidney, D. L., and Reeck, G. R.,** Preferential affinity of high molecular weight high mobility group non-histone chromatin proteins for single-stranded DNA, *J. Biol. Chem.*, 254, 5569, 1979.

46. **Javaherian, K., Sadeghi, M., and Liu, L. F.,** Non histone proteins HMG1 and HMG2 unwind DNA double helix, *Nuc. Acids Res.*, 6, 3569, 1979.
47. **Javaherian, K., Liu, E. F., and Wang, J. C.,** Non histone proteins HMG1 and HMG2 change the DNA helical structure, *Science,* 199, 1345, 1978.
48. **Isackson, P. J., Clow, L. G., and Reeck, G. R.,** Comparison of the salt dissociations of high molecular weight HMG chromatin proteins from double stranded DNA and from chromatin, *FEBS Lett.* 125, 30, 1981.
49. **Jackson, J. B., Pollock, J. M., and Rill, R. L.,** Chromatin fractionation procedure that yields nucleosomes containing near-stoichiometric amounts of high mobility group non histone chromosomal proteins, *Biochemistry,* 18, 3739, 1979.
50. **Kay, E. M. and Goodwin, G. H,.** Unpublished observations, 1981.
51. **Seyedin, S. M. and Kistler, W. S.,** Levels of chromosomal protein high mobility group 2 parallel the proliferative activity of testis, skeletal muscle and other organs, *J. Biol. Chem.*, 254, 11264, 1979.
52. **Levy, W. B., Wong, N., and Dixon, G. H.,** Selective association of the trout-specific H6 protein with chromatin regions susceptible to DNAse I and DNAse II: possible location of HMG-T in the spacer region between core nucleosomes, *Proc. Natl. Acad. Sci. U.S.A.*, 74, 2810, 1977.
53. **Vidali, G., Boffa, L. C., and Allfrey, V. G.,** Selective release of chromosomal proteins during limited DNAse I digestion of avian erythrocyte chromatin, *Cell,* 12, 409, 1977.
54. **Goodwin, G. H. and Johns, E. W.,** Are the high mobility group non-histone chromosomal proteins associated with 'active' chromatin?, *Biochim. Biophys. Acta,* 519, 279, 1978.
55. **Levy, W.B. and Dixon, G. H.,** A study of the localisation of high mobility group proteins in chromatin, *Can. J. Biochem.*, 56, 480, 1978.
56. **Mathew, C. G. P., Goodwin, G. H., and Johns, E. W.,** Studies on the association of the high mobility group non-histone chromatin proteins with isolated nucleosomes, *Nucl. Acids Res.*, 6, 167, 1979.
57. **Kuehl, L., Lyness, T., Dixon, G. H., and Levy, W. B.,** Distribution of high mobility group proteins among domains of trout testis chromatin differing in their susceptibility to micrococcal nuclease, *J. Biol. Chem.*, 255, 1090, 1980.
58. **Davie, J. R. and Candido, E. P. M.** DNAse I sensitive chromatin is enriched in the acetylated species of histone H4, *FEBS Lett.*, 110, 164, 1980.
59. **Peter, E. H., Levy, W. B., and Dixon, G. H.,** Evidence for the location of high mobility group protein T in the internucleosomal linker regions of trout testis chromatin, *J. Biol. Chem.*, 254, 3358, 1979.
60. **Goodwin, G. H.,** unpublished results.
61. **Bustin, M. and Neihart, H. K.,** Antibodies against chromosomal HMG proteins stain the cytoplasm of mammalian cells, *Cell,* 16, 181, 1979.
62. **Isackson, P. J., Bidney, D. L., Reeck, G. R., Neihart, H. K., and Bustin, M.,** High mobility group chromosomal proteins isolated from nuclei and cytosol of cultured hepatoma cells are similar, *Biochemistry,* 19, 4466, 1980.
63. **Smith, B. J., Robertson, D., Birbeck, M. S. C., Goodwin, G. H., and Johns, E. W.** Immunochemical studies of high mobility group nonhistone chromatin proteins HMG1 and HMG2, *Exp. Cell Res.*, 115, 420, 1978.
64. **Gordon, J. S., Bruno, J., and Lucas, J. J.,** Heterogeneous binding of high mobility group chromosomal proteins to nuclei, *J. Cell Biol.*, 88, 373, 1981.
65. **Rechsteiner, M. and Kuehl, L.,** Microinjection of non-histone chromosomal protein HMG1 into bovine fibroblasts and HeLa cells, *Cell,* 16, 901, 1979.
66. **Itkes, A. V., Glotov, B. O., Nikolaev, L. G., and Severin, E. S.,** Clusters of non-histone chromosomal protein HMG1 molecules in intact chromatin, *FEBS Lett.*, 118, 63, 1980.
67. **Mathis, D. J., Kindelis, A., and Spadafora, C.,** HMG proteins (1 & 2) form beaded structures when complexed with closed circular DNA, *Nucl. Acids Res.*, 8, 2577, 1980.
68. **Mathew, C. G. P., Goodwin, G. H., Igo-Kemenes, T., and Johns, E. W.,** The protein composition of rat satellite chromatin, *FEBS Lett.*, 125, 25, 1981.
69. **Goodwin, G.H., Rabbani, A., Nicolas, R. H., and Johns, E. W.,** The isolation of the high mobility group non-histone chromosomal protein HMG14, *FEBS Lett.*, 80, 413, 1977.
70. **Rabbani, A.,** Studies on the High Mobility Group Non-Histone Chromosomal Proteins, Ph.D. thesis, London University, 1979.
71. **Bakayev, V. V., Schmatchenko, V. V., and Georgiev, G. P.,** Subnucleosome particles containing high mobility group proteins HMG-E and HMG-G originate from transcriptionally active chromatin, *Nucl. Acids Res.*, 7, 1525, 1979.
72. **Reeves, R. and Candido, E. P. M.,** Partial inhibition of histone deacetylase in active chromatin by HMG14 and HMG17, *Nucl. Acids Res.*, 8, 1947, 1980.

73. **Abercrombie, B. D., Kneale, G. G., Crane-Robinson, C., Bradbury, E. M., Goodwin, G. H., Walker, J. M., and Johns, E. W.,** Studies on the conformational properties of the high-mobility group chromosomal protein HMG17 and its interaction with DNA, *Eur. J. Biochem.,* 84, 173, 1978.
74. **Cary, P. D., King, D. S., Crane-Robinson, C., Bradbury, E. M., Rabbani, A., Goodwin, G. H., and Johns, E. W.,** Structural studies on two high mobility group proteins from calf thymus, HMG14 and ubiquitin, and their interaction with DNA, *Eur. J. Biochem.,* 112, 577, 1980.
75. **Goodwin, G. H., Mathew, C. G. P., Wright, C. A., Venkov, C., and Johns, E. W.,** Analysis of HMG proteins associated with salt-soluble nucleosomes *Nucl. Acids Res.,* 7, 1815, 1979.
76. **Goodwin, G. H., Wright, C. A., and Johns, E. W.,** The characterisation of ISF monomer nucleosomes from hen oviduct and the partial characterisation of a third HMG14/17-like protein in such nucleosomes, *Nucl. Acids Res.,* in press, 1981.
77. **Huntley, G. H. and Dixon, G. H.,** The primary structure of the NH_2O-terminal region of histone T, *J. Biol. Chem.,* 247, 4916, 1972.
78. **Goodwin, G. H., Brown, E., Walker, J. M., and Johns, E. W.,** The isolation of three new high mobility group nuclear proteins, *Biochim. Biophys. Acta,* 623, 329, 1980.
79. **Walker, J. M., Goodwin, G. H., and Johns, E. W.,** The primary structure of the nucleosome-associated chromosomal protein HMG14, *FEBS Lett.,* 100, 394, 1979.
80. **Walker, J. M., Hastings, J. R. B.,and Johns, E. W.** The primary structure of a non-histone chromosomal protein, *Eur. J. Biochem.,* 76, 461, 1977.
81. **Walker, J. M., Stearn, C., and Johns, E. W.,** The primary structure of non-histone chromosomal protein HMG17 from chicken erythrocyte nuclei, *FEBS Lett.,* 112, 207, 1980.
82. **Watson, D. C., Wong, N. C. W., and Dixon, G. H.,** The complete amino-acid sequence of a trout-testis non-histone protein, H6, localised in a subset of nucleosomes and its similarity to calf thymus nonhistone proteins, HMG14 and HMG17, *Eur. J. Biochem.,* 95, 193, 1979.
83. **Walker, J. M., Brown, E., Goodwin, G. H., Stearn, C., and Johns, E. W.,** Studies on the structures of some HMG-like non-histone chromosomal proteins from trout and chicken tissues. Comparison with calf thymus HMG14 and 17, *FEBS Lett.,* 113, 253, 1980.
84. **Brown, E., Goodwin, G. H., and Crane-Robinson, C.,** unpublished results, 1981.
85. **Allfrey, V. G.,** Post-synthetic modifications, in *The HMG Chromosomal Proteins,* Johns, E. W., Ed., Academic Press, London, 1982, chap. 7.
86. **Bakayev, V. V., Bakayeva, T. G., Schmatchenko, V. V., and Georgiev, P.,** Non-histone proteins in mononucleosomes and subnucleosomes, *Eur. J. Biochem.,* 91, 291, 1978.
87. **Levy, W. B., Conner, W., and Dixon, G. H.,** A subset of trout testis nucleosomes enriched in transcribed DNA sequences contains high mobility group proteins as major structural components, *J. Biol. Chem.,* 254, 609, 1979.
88. **Weintraub, H. and Groudine, M.,** Chromosomal subunits in active genes have an altered conformation, *Science,* 193, 848, 1976.
89. **Weisbrod, S. and Weintraub, H.,** Isolation of a subclass of nuclear proteins responsible for conferring a DNAse 1-sensitive structure on globin chromatin, *Proc. Natl. Acad. Sci. U.S.A.,* 76, 630, 1979.
90. **Weisbrod, S., Groudine, M., and Weintraub, H.,** Interaction of HMG14 and 17 with actively transcribed genes, *Cell,* 19, 289, 1980.
91. **Levy, W. B., Kuehl, L., and Dixon, G. H.,** The release of high mobility group protein H6 and protamine gene sequences upon selective DNAse 1 degradation of trout testis chromatin, *Nucl. Acids Res.,* 8, 2859, 1980.
92. **Gazit, B., Panet, A., and Cedar, H.,** Reconstitution of a deoxyribonuclease I-sensitive structure on active genes, *Proc. Natl. Acad. Sci. U.S.A.,* 77, 1787, 1980.
93. **Gazit, B. and Cedar, H.,** Nuclease sensitivity of active chromatin, *Nucl. Acids Res.,* 8, 5143, 1980.
94. **Sandeen, G., Wood, W. I., and Felsenfeld, G.,** The interaction of high mobility proteins HMG14 and 17 with nucleosomes, *Nucl. Acids Res.,* 8, 3757, 1980.
95. **Albright, S. C., Wiseman, J. M., Lange, R. A., and Garrard, W. T.,** Subunit structures of different electrophoretic forms of nucleosomes, *J. Biol. Chem.,* 255, 3673, 1980.
96. **Mardian, J. K. W., Paton, A. E., Bamick, G. J., and Olins, D. E.,** Nucleosome cores have two specific binding sites for non-histone chromosomal proteins HMG14 and HMG17, *Science,* 209, 1534, 1980.
97. **Levinger, L., Barsoum, J., and Varshavsky, A.,** Two dimensional hybridization mapping of nucleosomes, *J. Mol. Biol.,* 146, 287, 1981.
98. **Levy, W. B. and Dixon, G. H.,** Partial purification of transcriptionally active nucleosomes from trout testis cells, *Nucl. Acids Res.,* 5, 4155, 1978.
99. **Hutcheon, T., Dixon, G. H., and Levy, W. B.,** Transcriptionally active mononucleosomes from trout testis are heterogeneous in composition, *J. Biol. Chem.,* 255, 681, 1980.
100. **Albanese I. and Weintraub, H.,** Electrophoretic separation of a class of nucleosomes enriched in HMG14 and 17 and actively transcribed globin genes, *Nucl. Acids Res.,* 8, 2787, 1980.

101. **Weisbrod, S. and Weintraub, H.,** Isolation of actively transcribed nucleosomes using immobilized HMG14 and HMG17 and an analysis of α-globin chromatin, *Cell,* 23, 391, 1981.
102. **Bloom, K. S. and Anderson, J. M.,** Fractionation of hen oviduct chromatin into transcriptionally active and inactive regions after selective micrococcal nuclease digestion, *Cell,* 15, 141, 1978.
103. **Gaubatz, J. W. and Chalkley, R.,** Distribution of H1 histone in chromatin digested by micrococcal nuclease, *Nucl. Acids Res.,* 4, 3281, 1977.
104. **Renz, M., Nehl, P., and Hozier, J.,** Involvement of histone H1 in the organisation of the chromosome fiber, *Proc. Natl. Acad. Sci. U.S.A.,* 74, 1879, 1977.
105. **Ring, D. and Cole, R. D.,** Chemical cross-linking of H1 histone to the nuceosomal histones, *J. Biol. Chem.,* 254, 11688, 1979.
106. **Levy, W. B., Watson, D. C., and Dixon, G. H.,** Multiacetylated forms of H4 are found in a putative transcriptionally competent chromatin fraction from trout testis, *Nucl. Acids Res.,* 6, 259, 1979.
107. **McGhee, J. D. and Ginder, G. D.,** Specific DNA methylation sites in the vicinity of the chicken β-globin genes, *Nature (London),* 280, 419, 1979.
108. **Van der Ploeg, L. H. T. and Flavell, R. A.,** DNA methylation in the Human γδβ-globin locus in crythroid and non erythroid tissues, *Cell,* 19, 947, 1980.
109. **Mandel, L. and Chambon, P.,** DNA methylation: correlation with DNAse 1 sensitivity of chicken ovalbumin and conalbumin chromatin, *Nucl. Acids Res.* 7, 2081, 1979.
110. **Salomon, R., Kay, A. M., and Herzberg, M.,** Mouse nuclear satellite DNA: 5-methylcytosine content, pyrimidine isoplith distribution and electron microscopic appearance, *J. Mol. Biol.,* 43, 581, 1969.
111. **Stalder, J., Larsen, A., Engel, J. D., Dolan, M., Groudine, M., and Weintraub, H.** Tissue specific DNA cleavages in the globin domain introduced by DNAse 1, *Cell,* 20, 451, 1980.
112. **Garel, A. and Axel, R.,** Selective digestion of transcriptionally active ovalbumin genes from oviduct nuclei, *Proc. Natl. Acad. Sci. U.S.A.,* 73, 3960, 1976.
113. **Levy, W. B. and Dixon, G. H.,** Limited action of micrococcal nuclease on trout testis nuclei genetics: two mononucleosome subsets enriched in transcribed DNA sequences, *Proc. Natl. Acad. Sci. U.S.A.,* 76, 1682, 1979.
114. **Mathew, C. G. P., Goodwin, G. H., Wright, C. A., and Johns, E. W.,** The high mobility group proteins and transcribed nucleosomes, *Cell Biol. Int. Rep.,* 5, 37, 1981.

Chapter 3

NONHISTONE NUCLEAR PHOSPHOPROTEINS

Mark O. J. Olson

TABLE OF CONTENTS

INTRODUCTION

Of the numerous posttranslational covalent modifications of proteins, phosphorylation is the one most frequently found in eukaryotic cells. Nature's extensive use of the phosphoryl group in proteins may be due to several unique features:

1. The introduction of a phosphoryl group into a site in a protein adds between one and two negative charges to that site.[1,2] This alteration, if strategically placed, may change interactions within the molecule as well as interactions of the protein with other molecules.
2. The enzyme system (Figure 1) for the phosphorylation reaction is relatively simple and may be regulated by metal ions, cyclic nucleotides, nucleoside triphosphate concentrations, or phosphoprotein kinase specificity.
3. The phosphoryl group may be removed by a hydrolytic process which does not require energy input.

Clearly, the convenience and simplicity of the system makes phosphorylation a particularly attractive posttranslational modification for regulatory processes.

Phosphorylated proteins are distributed throughout eukaryotic cells.[4-6] However, the proteins with the highest phosphate content are found in the cell nucleus.[7,8] The major nuclear nonhistone phosphoprotein fraction contains approximately 1.3% phosphorous by weight and accounts for as much as 9% of the total dry weight of the cell nucleus.[9] The level of phosphorylation of proteins in this fraction is equivalent to an average of 4 to 5 phosphorylated amino acids per 100 residues.

The presence of the highly phosphorylated fraction of nuclear proteins in chromatin suggested to the early workers in the field that these proteins are involved with the regulation of transcription.[7-9] Subsequently, numerous studies revealed that in addition to being in chromatin, protein-bound phosphate is present in virtually all subcomponents of the nucleus. For example phosphorylated proteins have been found in hnRNP (heterogeneous nuclear ribonucleoprotein) particles and nuclear envelope lamina.[10-12] Therefore, additional roles such as transport of gene products, RNP assembly, and regulation of nuclear substructure throughout the cell cycle should be included in the list of nuclear phosphoprotein function.

This chapter will cover only the nonhistone nuclear phosphoproteins. For simplicity, the nonhistone class of nuclear proteins will include all proteins which cannot be clearly classified as histones. The objectives of this review are (1) to provide a general overview of the subject, (2) to emphasize material that has not been covered in recent reviews (see reviews by Kleinsmith,[13] Jungman and Kranias,[14] and Ahmed and Wilson[15], and (3) to bring to the attention of the reader recent information on the structure of nonhistone phosphoproteins.

II. GENERAL CHEMICAL CHARACTERISTICS OF NONHISTONE PHOSPHOPROTEINS

A. Identification of Phosphoproteins

Those nonhistone proteins which are classified as phosphoproteins are identified as such by one or several methods. The most common initial approach is to label cells or nuclei with ^{32}P orthophosphate and then search for the comigration of ^{32}P radioactivity with stained protein bands displayed on electrophoretic separation systems. With this method it is important to insure that the radioactivity is not due to tightly bound nucleic acid fragments or phospholipids. To eliminate this possibility the proteins may be treated with hot trichloroacetic acid to remove nucleic acids, and chloroform/methanol to liberate bound lipid ma-

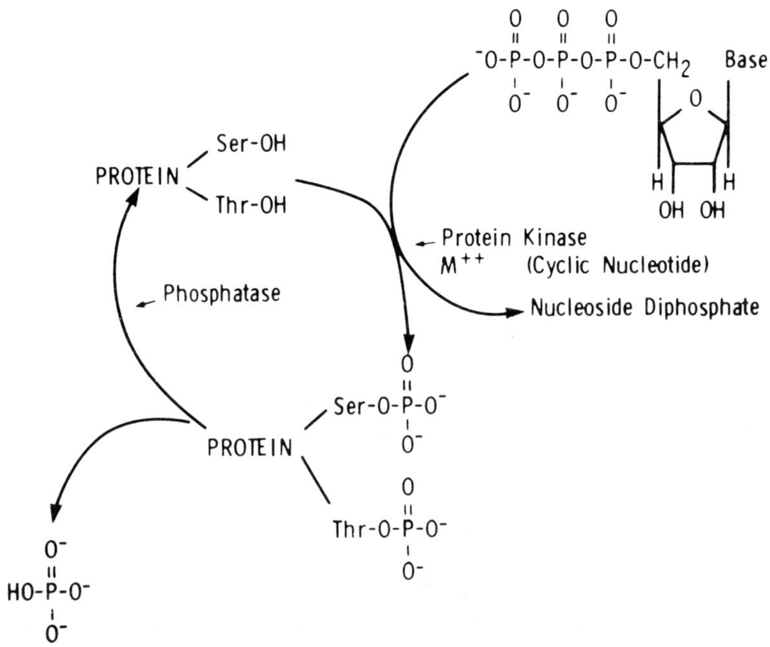

FIGURE 1. Summary diagram of the enzyme systems for phosphorylation and dephosphorylation of proteins. Multiple phosphoprotein kinases, either cyclic nucleotide dependent or independent, may be present. All protein kinases require nucleoside triphosphates and divalent cations. Phosphoprotein phosphatases may or may not require divalent cations. (From Olson, M. O. J., *Advances in Inorganic Biochemistry,* Vol. 3, Eichhorn, G. and Marzilli; L., Eds., Elsevier, New York, 1981. With permission.)

terial.[8] Alternatively, the gels on which proteins have already been separated may be treated with hot trichloroacetic acid to eliminate contaminating nonprotein ^{32}P radioactivity.[16]

B. Analyses of Phosphorylated Amino Acids

Once a protein or protein fraction is identified as phosphorylated, an attempt may be made to quantitate the degree of phosphorylation. The overall level of phosphorylation may be determined as alkali-labile phosphate. The release of phosphate from proteins by base depends not on hydrolysis, but on β-elimination.[17] This procedure is valid if phosphoryl groups are linked to the protein only by phosphoester bonds. Although Schiltz and Sekeris[18] have presented evidence for in vitro phosphorylation of lysine and arginine in nuclear proteins, and 3-phosphohistidine and ε-phospholysine have been found in histones,[18-20] the presence of the P-N bond has not been confirmed in nonhistone nuclear proteins. Therefore, as a method for determining total phosphate content in nonhistone proteins, liberation of phosphate by alkali appears to be valid. The released phosphate is generally measured as ^{32}P-labeled orthophosphate selectively extracted as the phosphomolybdate complex.[21] Phosphate may be quantitated colorimetrically after reduction of the molybdate complex with $SnCl_2$.[22]

It is also desirable to know quantitatively and qualitatively which phosphoamino acids are contained in the nonhistone proteins under consideration. Phosphoserine and phosphothreonine are readily released by acid hydrolysis. Unfortunately, they are unstable under the conditions required for complete hydrolysis and their yields are difficult to predict. However, if partial hydrolysis conditions are used (2 *M* HCl, 8 to 10 hr at 110°) sufficient quantities of phosphoamino acids survive so that qualitative analyses may be performed. This approach has been generally successful in determining whether proteins contain radioactively labeled phosphoamino acids. In fact, ^{32}P-labeled bands may be simply cored from polyacrylamide gels in order to identify the phosphoamino acids present in individual proteins.[23] The ^{32}P labeled phosphoamino acids are readily analyzed by ion exchange chromatography,[22,24] or by paper electrophoresis.[25] A word of caution is appropriate regarding paper electrophoretic analyses. Recently, phosphotyrosine was found in various sarcoma virus transforming gene products.[26] Phosphotyrosine comigrates with phosphothreonine during electrophoresis at pH 1.9. Therefore, phosphothreonine could easily be confused with phosphotyrosine, unless a second dimension of chromatography or electrophoresis is run. To this author's knowledge, phosphotyrosine has not been identified in nuclear proteins.

Phosphoserine and phosphothreonine may also be determined using the amino acid analyzer after appropriate modifications of the buffer system. Theoretically, it should be possible to correct for losses of phosphoserine and phosphothreonine during hydrolysis. However, the rates of destruction of phosphoamino acids are dependent on the neighboring amino acid residues in the peptide or protein.[27] Thus, quantitation of phosphoserine and phosphothreonine by application of correction factors for destruction during hydrolysis will provide only an approximation.

In the author's hands, the most reliable method of quantitation of phosphoamino acids is that devised by Richardson et al.[28] In this procedure the protein sample is subjected to β-elimination with NaOH followed by reduction with $NaBH_4$. This treatment completely destroys phosphoester-containing amino acids. The β-eliminated sample and an untreated sample are subjected to complete acid hydrolysis followed by amino acid analyses. The reductions in amounts of serine and threonine in the β-eliminated sample are used to calculate the phosphoserine and phosphothreonine contents of the protein by utilization of formulas provided by Richardson et al.[28]

Because of the aforementioned difficulties in phosphoamino acid analysis, the precise level of phosphorylation of most nonhistone phosphoproteins is not known. As more phosphorylated nuclear proteins are purified and chemically characterized, this information should become available.

III. BIOLOGY AND BIOCHEMISTRY OF NONHISTONE NUCLEAR PHOSPHOPROTEINS

A. Nonhistone Phosphoproteins of Unspecified Nuclear Locations

1. General Considerations

In the late 1960s and early 1970s numerous workers examined the compositions of nuclear protein fractions in various cell types under varying physiological conditions. They found a highly heterogeneous population of proteins, many of them containing covalently bound phosphate. Although the definition of chromatin spans a wide range, most of the nonhistone proteins were presumed to be chromatin associated. In general, the presumption of chromatin association of most of the nonhistone phosphoproteins is still valid, although many of the nuclear nonhistone proteins have been shown to be present in regions of the nucleus which cannot strictly be considered as chromatin. In any event, the heterogeneity, species specificity, tissue specificity, and chromatin association of the nonhistone phosphoproteins led to the belief that these proteins were involved in the regulation of gene expression. The evidence for this remains largely circumstantial, although in certain areas recent data have strengthened the case for regulation of specific genes by nonhistone proteins and their phosphorylation.

Numerous tissues under various physiological conditions have been analyzed primarily by one- or two-dimensional polyacrylamide gel electrophoretic systems. Nearly all of the studies revealed changes in nuclear phosphoproteins which correlated with changes in development or changes induced by drugs, hormones, or transforming agents. The list of these

Table 1
OBSERVATIONS ON NONHISTONE PHOSPHOPROTEINS

Observation	Tissue(s) or system(s)	Ref.
Tissue specificity of nonhistone phosphoproteins	Various rat tissues	29
	Various bovine tissues	30
	Various mouse tissues	31, 32
Tissue and species specificity	Rat, chicken, mouse turtle, frog, monkey, and opossum tissues	33
Changes in phosphoproteins during differentiation	*Physarum polycephelum*	34
	Sea urchin embryos	35
	Perinatal rat hepatocytes	36
Specific DNA binding of nonhistone phosphoproteins	Rat liver	37, 38
	Calf thymus	39
	Ehrlich ascites cells	40
	Bovine lymphocytes	41
Nonhistone protein phosphorylation is cell cycle dependent	HeLa cells	42, 43
Phosphorylation of nonhistone proteins is altered by changes in physiological conditions	Regenerating rat liver	44—46
	Aging rat liver	47
Normal and tumor tissues differ in nonhistone phosphoproteins	Morris hepatomas, Novikoff hepatoma, and Walker tumor	45, 48—51
Phosphorylation changes after administration of hormones		
Testosterone	Rat prostate	52, 53
Corticosterone	Rat liver	54, 55
Aldosterone	Rat kidney	56
Estradiol	Rat uterus	57
Estradiol	Hen oviduct	58
Cyclic AMP	Rat liver	59
Acetylcholine, cyclic GMP	Horse lymphocytes	60
Phosphorylation altered by chemical agents		
Azo dyes	Rat liver	61, 62
Concanavalin A	Horse lymphocytes	63
Isoproterenol	Rat submandibular gland	64, 65
α 1,2,3,4,5,6-Hexa-chlorocyclohexane	Rat liver	66
Phenobarbital	Rat liver	67
Bleomycin and cytarabine	Rat salivary gland	68

studies is extensive and this area has been well covered in a recent review by Kleinsmith.[13] Therefore, the major studies are summarized in Table 1 and only brief mentions of them will be made in the text.

2. Tissue and Species Specificity

A recurring observation in the studies on nonhistone proteins has been that the electrophoretic profiles of these proteins were more or less tissue-specific (Table 1). Although many polypeptides are common to different tissues within an organism, several others are unique to certain tissues.[29-33] This is in direct contrast to observations on the histones, where only four different major polypeptides are found in the nucleosome core and these are found in all tissues of a given organism. Furthermore, the finding that many of the polypeptides in the fraction were phosphorylated enhanced the tissue specificity of the profile. Additional evidence for tissue specificity has been provided by the examination of developing tissues. These systems offer the advantage of observing changes in the nonhistone phosphoproteins as the tissue changes during development. For example, changes in the nonhistone phosphoproteins have been reported during differentiation of *Physarum polycephalum,*[34] sea urchin embryos,[35] and in perinatal rat hepatocytes.[36] These observations have led to the tentative conclusion that the expression of the gene complement of a given tissue is specified by the nonhistone protein content of the chromatin of that tissue. The proteins may serve as repressors or activators of specific genes and phosphorylation may be a further level of fine control in the system. Of course, tissue specificity of the nonhistone phosphoproteins is only consistent with their proposed involvement in gene regulation and it does not prove the hypothesis.

3. DNA Binding

If nonhistone phosphoproteins regulate transcription of specific genes they would be expected to recognize and bind to specific sites on DNA. Several studies have presented evidence for this. Teng et al.[29] found that the greatest degree of binding of nonhistone phosphoproteins was with DNA of the same species. For example, rat phosphoproteins complexed best with rat DNA, to a lesser extent with mouse DNA, and not at all with calf, human, or dog DNA. Kleinsmith showed that the DNA binding fraction accounted for approximately 1% of the total nonhistone protein in chromatin.[37] Saturation of binding was achieved at a ratio of 1 μg of nonhistone phosphoprotein to 100 μg of homologous DNA. Even with the 100-fold enrichment the fraction of bound protein was still a highly heterogeneous mixture.

Kostraba et al.[40] reported similar findings with loosely bound chromatin proteins. The DNA binding components of this protein fraction were also enriched in phosphate. In addition, the DNA binding proteins specifically stimulated transcription of DNA with homologous RNA polymerase but not with bacterial RNA polymerase. Prestayko et al.[38] also found that many rat liver nuclear proteins which were soluble at low ionic strength were capable of binding DNA. Several of these, including proteins designated C18, CN′, and B33 were phosphorylated.

Bluthmann identified a predominant DNA binding phosphoprotein from bovine lymphocytes.[41] This protein (NH 30,000), so designated because of its molecular weight, bound DNA from lymphocytes much more effectively than DNA from *Escherichia coli.* Protein NH 30,000 was capable of binding as much as 30% of the lymphocyte DNA, or one binding site per 1700 base pairs. Therefore, this protein does not recognize DNA with a high degree of selectivity. The amino acid composition (Table 2) indicates that the protein has a high content of acidic amino acids and resembles HMG proteins. However, it has an amino terminal arginine residue which distinguishes it from any of the HMG proteins. It also contains 1% phosphate, or an average of two phosphate residues per molecule.

The above studies indicate that some of the nonhistone phosphoproteins meet another criterion for their proposed gene regulatory role: specific DNA binding. However, precise recognition of specific genes has been demonstrated only in isolated cases, such as the specific binding of a 70,000 mol wt phosphoprotein to the palindromic ribosomal DNA of *Physarum* (see Section III.C.1.c.)[69] Greater progress in the study of proteins that regulate specific genes should now be possible with the cloned genes and nitrocellulose blotting techniques for detection of DNA binding proteins.[70,71]

4. Nonhistone Phosphoproteins and the Cell Cycle

Synchronously dividing cell populations are also useful for studying the possible role of protein phosphorylation in gene regulation. RNA synthesis is generally suppressed during mitosis,[72,73] and histone mRNA synthesis occurs primarily in late G_1 and early S phases of the cell cycle.[74-76] Thus, changes in rates of transcription may be monitored during the cell cycle and these may be correlated with changes in nuclear phosphoproteins. Platz et al.[43] reported that phosphorylation of nuclear acidic proteins was suppressed during mitosis in HeLa S-3 cells. This observation was expanded by Karn et al.,[42] who found that uptake of phosphate into most major phosphoprotein species increased during early G_1 and early S

Table 2
AMINO ACID COMPOSITIONS OF ISOLATED NUCLEAR PHOSPHOPROTEINS

Amino acid	NH 30,000	HMG1	HMG14	Protein B_2	Protein D-55	INHP	13K NHCP	Histone deacetylase	Nucleoplasmin
Lys	10.2	16.8	19.0	7.8	7.0	6.7	6.3	6.6	11.3
His	2.2	1.7	0.3	2.7	2.6	1.2	2.3	2.3	2.2
Arg	5.7	5.5	5.6	13.2	5.1	7.6	2.6	3.0	2.2
Asx	9.6	11.0	8.1	8.9	9.2	9.1	8.2	12.3	9.2
Thr	4.5	2.9	4.2	3.3	5.0	4.9	4.3	5.5	5.4
Ser	6.0	5.3	7.8	6.7	7.3	7.5	17.9	5.9	6.7
Glx	15.5	19.0	17.1	13.0	16.0	12.9	18.0	17.5	20.0
Pro	6.9	6.0	8.5	7.0	2.7	7.3	0	3.6	7.3
Gly	6.3	5.9	6.5	11.5	8.4	11.6	17.2	10.2	8.2
Ala	7.4	8.6	14.5	9.3	7.2	10.8	10.4	7.8	7.2
Cys	0	1.9	6.7	ND[a]	2.1	ND	0	0	ND
Val	6.0	2.3	4.2	1.9	5.2	7.7	3.7	7.2	6.2
Met	0	1.1	0	ND	2.2	ND	0	1.0	1.0
Ileu	4.2	2.0	0.5	2.5	4.4	2.8	2.5	4.3	3.2
Leu	9.4	2.4	2.0	6.9	8.6	5.1	3.5	7.4	6.2
Tyr	1.7	3.1	0.4	2.5	2.6	1.3	0.6	2.3	2.1
Phe	3.3	5.0	0.6	2.9	2.9	3.5	2.3	3.0	2.3
Trp	0.9	tr	ND	ND	1.5	ND	ND	1% (w/w)	ND
PO_4 (w/w%)	1.0	ND	ND	0.3	0.17	2.7	ND	1.3	ND
% Charged A.A.	43.2	52	50.1	45.6	39.9	37.5	35.1	41.7	44.9
Asx + Glx/ Lys + His + Arg	1.4	1.2	1.0	0.93	1.7	1.4	2.9	2.5	1.9
Ref.	41	91	91	105	109	115	117	186	192

[a] ND indicates not determined.

phases and was minimal during the late G_2 to M period. Furthermore, the turnover of phosphate in different proteins varied according to the phase of the cell cycle, i.e., some proteins lost phosphate much more rapidly than others. A general conclusion from these studies is that protein phosphorylation is highest when the rate of RNA synthesis is maximal.

5. Hormones and Chemical Agents

Various hormones and chemical agents which stimulate transcription have been used to correlate changes in nonhistone phosphoproteins with changes in gene activities (Table 1). In general, there is a correlation between quantitative changes in nonhistone proteins or their phosphorylation and increases in transcription induced by the agent. In most of the investigations the agent employed produced a general stimulation of phosphorylation affecting a broad range of proteins. This may reflect a linkage of genes being activated or coupling of processes in the nucleus after induction by the agent. In a few studies effects on specific proteins have been observed. For example, when lymphocytes are stimulated with concanavalin A, phosphorylation of a polypeptide of 52,000 mol wt is differentially enhanced.[63] The effect is seen as early as 15 min after addition of concanavalin A; 2 hr after concanavalin addition, the rate of phosphorylation of acidic proteins increases fourfold and reaches a maximum after 8 hr. Since concanavalin A induces RNA synthesis in lymphocytes, followed by enlargement of the cells, DNA synthesis, and mitosis, phosphorylation may be important in the control of cell proliferation.

Phosphorylation of nonhistone proteins may be involved at pretranscriptional levels of hormone action. Klysejko-Stefanowicz et al.[77] found that binding of the 5 α-dihydroxytestosterone-receptor complex to rat prostate chromatin was enhanced by phosphorylation of the chromatin. In another system, Nielsen et al.[78] showed that the glucocorticoid receptor was inactivated by exposure to alkaline phosphatase. These studies suggest that protein phosphorylation affects hormone action by modulating the interaction of the hormone with its receptor or the receptor with its chromatin acceptor, prior to transcription.

6. Reconstituted Chromatin Systems

Because of the limitations of studies which simply correlate rates of RNA synthesis with protein phosphorylation, attempts have been made to test the effects of the phosphoproteins on transcription more directly. Ultimately, one would like to be able to reconstruct chromatin from DNA, histones, and nonhistone proteins in phosphorylated or dephosphorylated forms.

A number of initial studies showed simple addition of nonhistone phosphoproteins to either DNA or chromatin increases the rate of RNA synthesis.[7,29,40,79-84] At least two of the studies correlated the phosphate content of the proteins with their ability to stimulate transcription.[80,81] However, this approach was less than satisfactory because most of the studies did not deal with transcription of specific genes and because the modified chromatin probably bore little physiological resemblance to chromatin in living cells.

The problem was approached more directly by Kleinsmith et al.[85] where chromatin from HeLA S-phase cells was reconstituted from DNA, histones, and either control nonhistone proteins or dephosphorylated nonhistone proteins. The dephosphorylation was achieved by incubating nonhistone proteins with phosphoprotein phosphatase derived from a nuclear fraction. This was covalently linked to agarose, to facilitate the removal of the enzyme after dephosphorylation. Chromatin reconstitution was carried out by gradient dialysis. The number of initiation sites in the dephosphorylated chromatin was reduced by 50% compared to the control chromatin. Furthermore, when the transcription products were analyzed by hybridization to DNA complementary to histone messenger RNA, the percentage of transcribed histone mRNA sequences fell by nearly 70% in the dephosphorylated chromatin. Thus, it was concluded that the level of phosphorylation of nonhistone proteins determines the availability of the template for transcription and also does this in a manner which is selective for certain genes. This system may be less than physiological due to use of heterologous polymerase and because of the uncertainty regarding the fidelity of chromatin reconstitution. However, this is the first case where phosphorylation of nonhistone proteins has been shown to have a direct effect on the transcription of specific genes.

B. Chromatin Associated Nonhistone Phosphoproteins

1. HMG Proteins

The high mobility group (HMG) of proteins is a subclass of nonhistone proteins defined by their extractability from chromatin by 0.35 *M* NaCl, their solubility in 2% TCA, and their relatively high electrophoretic mobility.[86] Four major proteins (HMG-1, 2, 14, and 17) are present in this group.[87] Isolated nucleosomes contain HMG proteins,[87,88] and HMG 14 and 17 are associated with transcriptionally active or DNase I-sensitive regions of chromatin.[89,90]

The HMG proteins have roughly equal contents of acidic and basic amino acids which together account for approximately 50% of the total residues (Table 2).[91] Moreover, the bulk of the acidic residues are found in the carboxyl terminal portions of the HMG protein molecules.[92-95] This unusual combination of acidic and basic residues makes the HMG proteins interesting "hybrids" of histones and acidic proteins.

Since these proteins were discovered relatively recently, their postsynthetic modification has not received extensive attention. However, the acetylation[96] and methylation[97] of HMG-1 and HMG-2 have been reported. Very recently, the phosphorylation of HMG-14 and HMG-17 in cell suspensions and by isolated protein kinase was observed. Saffer and Glazer[98] incubated Ehrlich ascites and L1210 cells with ^{32}P phosphate and noted that HMG 14 and 17, but not HMG 1 and 2 were labeled. They also found that the kinetics of labeling and the turnover rates were similar to that of histone H1. In another study, employing CHO cell suspensions, all HMG proteins were phosphorylated in interphase cells, with HMG-17 incorporated 80% of the radioactivity.[99] In contrast, in metaphase cells the phosphate content of HMG 1, 2, and 14 increased threefold and the phosphate content of HMG-17 dropped to half of its interphase value. Thus, the HMG proteins are not only phosphorylated, but the phosphorylation appears to be cell-cycle dependent.

Additional studies have examined the phosphorylation of HMG proteins by nuclear protein kinases. Inoue et al.[100] reported that HMG-17 and probably HMG-14, but not HMG 1 and 2, were phosphorylated by rat liver cyclic nucleotide independent kinase NII. Protein kinase NI had little or no activity in phosphorylating the HMG proteins. Similarly, Arfmann and Baydoun found that nuclear cyclic AMP independent kinase preferentially phosphorylated HMG-17.[101] On the other hand, the cytoplasmic kinase phosphorylated HMG 1 and 2 to a higher degree than HMG 14 and 17. HMG-1 also serves as a substrate for cyclic AMP dependent kinase using [γ^{35}S]ATP as a thiophosphate donor.[102]

Since the phosphorylation of HMG proteins was only recently observed, little can be said about the function of this modification. The fact that the phosphorylation is, in part, cell-cycle dependent suggests that phosphorylation may play a role in chromatin assembly and chromosome condensation and dispersal.

2. *Nucleosome-Associated Phosphoproteins*

Although chromatin subunits or nucleosome core particles are primarily composed of histones and DNA, they also contain small quantities of nonhistone proteins. Liew and Chan[104] noted the presence of specific nonhistone proteins associated with mononucleosomes from nuclease digested chromatin. Further analyses indicated that one of these proteins, designated B2, was highly phosphorylated.[104] Two other proteins, A4 (39,000 mol wt, pI 5.3 to 6.0) and A5 (31,000 mol wt, pI 5.1 to 5.8) were also highly phosphorylated, but were released from chromatin by micrococcal nuclease and were not associated with nucleosomes. Protein B2 has a molecular weight of 68,000 and an isoelectric point of 6.5 to 8.2.

Chan and Liew have recently purified protein B2 and the amino acid composition (Table 2) indicates that the protein has many similarities to the HMG proteins.[105] Protein B2, as well as the HMG proteins, has nearly a one-to-one ratio of acidic-to-basic amino acids. Similarly, nearly half of its amino acids contain charged side chains. This is a recurring theme in nearly all of the nuclear nonhistone phosphoproteins that have been isolated to date. Protein B2 may be similar or identical to a 70,000 mol wt phosphoprotein observed in chromatin subunits of mouse ascites cells.[106]

A number of phosphorylated proteins as well as phosphoprotein kinase activity were seen in mononucleosomes of rat liver nuclei by Defer et al.[107] The same group recently expanded this work to show that two predominant polypeptides, a (41,000 mol wt) and b (31,000 mol wt) were present in nucleosomes of hepatoma tissue culture cells.[108] Phosphoprotein a was found in all of the monomer particles released by micrococcal nuclease. Phosphoprotein b was absent in the smallest class of nucleosomes, but was present in larger monomers and dimer particles. It was suggested by the authors that the two types of phosphoproteins are associated with different parts of chromatin, and that phosphoprotein b is bound to internucleosomal linker DNA.

An interesting phosphorylated nucleosome binding protein has been isolated from calf thymus by Saffer and Coleman.[109] This protein (D-55) has a molecular weight of 55,000 and incorporates 1 mol of phosphate per mole of protein if incubated with endogeneous protein kinase and [γ-^{32}P]ATP. The amino acid composition (Table 2) indicates that it is potentially rich in acidic amino acids. The unphosphorylated form binds DNA, histones, and nucleosomes. Phosphorylation of the protein enhances its binding to histones and nucleosomes, but does not affect its DNA binding properties. Protein D-55, when bound to nucleosomes, enhances transcription of nucleosome DNA by *E. coli* RNA polymerase about 100-fold. This is a level of transcription four times that of naked DNA. Even more interesting is the fact that this enhancement of transcription is abolished by phosphorylation of protein D-55. This is an opposite effect from what is usually observed; phosphorylation is commonly thought to enhance transcription. Although the physiological role of protein D-55 is unknown, it seems likely that proteins of this nature are needed to facilitate transcription through nucleosome-bound DNA.

3. Low Molecular Weight Chromatin-Associated Phosphoproteins

Workers in several laboratories have observed phosphorylated polypeptides of relatively low molecular weight (10,000 to 16,000 daltons) in chromatin. MacGillivray et al.[110] found that after labeling mouse liver nuclei with [γ-^{32}P]ATP as much as 10 to 20% of the radioactivity was incorporated into a low molecular weight component of saline-soluble nuclear and chromatin fractions. This polypeptide, which they designated component 10, had a pI near 4.5 and a molecular weight of approximately 10,000. The protein contained both phosphoserine and phosphothreonine and yielded two phosphorylated peptides upon tryptic digestion and phosphopeptide mapping. Although nothing is known about the function of this protein, it was observed in all tissues that were examined by MacGillivray and coworkers,[31,32,111] Similar phosphorylated low molecular weight components were reported in rat liver nuclei[56,112,113] and in fibroblast chromatin.[114]

Kostraba and Wang[115] isolated a nonhistone phosphoprotein from Ehrlich ascites tumor chromatin which had a molecular weight of approximately 11,000 daltons (Table 2).[115] The protein designated I-NHP, contained 2.7% phosphorous, was acidic, and bound to DNA, but unlike many other phosphoproteins, it inhibited transcription of DNA by RNA polymerase. Kostraba et al. further presented evidence that the site of inhibition was at RNA chain initiation. They also isolated an analagous protein from calf thymus chromatin.[116] The calf thymus protein preferentially inhibited transcription of homologous DNA and was effective in the inhibition of transcription of heterologous DNAs. However, the calf thymus protein contained two subunits with estimated molecular weights of 16,500 and 13,000, in contrast to the Ehrlich ascites protein which had a single polypeptide of 11,000 mol wt.

In a recent study, Ohtsuki et al.[117] isolated a phosphate acceptor polypeptide from mouse spleen chromatin. This protein (13,000 mol wt) served as a very effective substrate for purified nuclear cyclic AMP independent protein kinase. The phosphorylation of the protein was stimulated by double stranded DNA and inhibited by histones. The amino acid composition (Table 2) was rather unusual, with high contents of serine, glycine, and glutamic acid and a complete absence of proline. In addition, the protein had nearly a three-to-one ratio of potentially acidic-to-basic amino acids. However, the isoelectric point was about 9.0, suggesting a high proportion of acidic amino acids in the amidated form.

The above studies indicate that chromatin contains several relatively low molecular weight phosphoproteins. Although they each have distinctly different amino acid compositions, they have in common a high content of potentially charged amino acids. Little is known about their functions. Because of their small size it is conceivable that they are breakdown products of larger proteins in a manner similar to the way that ubiquitin is cleaved from the A24 protein during stages of development in chicken erythrocytes.[118]

C. Nucleolar Phosphoproteins

1. General Studies

The nucleolus is the site of synthesis of ribosomal RNA as well as the location of ribosome precursor assembly and processing.[119] Since the genes for preribosomal RNA are located exclusively in the nucleolus, it contains a specific fraction of chromatin enriched in rRNA cistrons. Therefore, this subnuclear organelle offers unique opportunities for studying the biochemical and morphological events accompanying gene activation, ribosome precursor assembly and processing, and the mechanisms involved in their control.

Because of the multiple processes taking place in the nucleolus, it contains several classes of proteins.[11] These include histones and nonhistone proteins from the chromatin component, and ribosomal proteins from the preribosomal particles. The preribosomal RNP particles also contain nonribosomal or "nucleolar stable" proteins which are recycled within the nucleolus.[120] In addition, several enzymes are found in the nucleolus, some of them specifically localized to the organelle.[11] The most common examples of nucleolar enzymes are those that are involved in specific nucleolar functions such as RNA polymerase I, methylases which act upon preribosomal RNA, and nucleases for processing preribosomal RNA.

In the early 1970s the nucleolus began to be analyzed for its content of proteins. Orrick et al.[121] separated more than 100 nucleolar acid-soluble proteins from normal rat liver and Novikoff hepâtoma nucleoli (Figure 2a). In subsequent studies it was found that approximately 40 of the polypeptides separated by two-dimensional electrophoresis were phosphorylated under in vivo labeling conditions (Figure 2b).[23] Most, but not all of the proteins labeled in vivo, were also labeled in vitro when isolated nucleoli were incubated with [γ-^{32}P]ATP in the presence of divalent cations.[122] Two polypeptides of molecular weights of approximately 40,000 and 100,000 were highly labeled. These were designated proteins B23 and C23, respectively, according to their relative electrophoretic mobilities.[121] These early studies provided a basis for localization and characterization of nucleolar proteins and the nucleolar phosphorylating system.

Two-dimensional electrophoretic analyses showed that proteins from both ribosomal subunits were present in nucleolar preribosomal particles.[123,124] Among the proteins present in the preribosomal particles were the two proteins previously shown to be highly phosphorylated, B23 and C23. Olson et al.[125] found that approximately 19 proteins including B23 and C23 were phosphorylated in vivo. Four of these were ribosomal proteins, but they were different from the proteins phosphorylated in cytoplasmic ribosomes. These results suggested that phosphorylation of proteins of the nucleolar preribosomes is independent of phosphorylation of cytoplasmic ribosomal proteins. It was also proposed that protein phosphorylation is involved in the maturation process of preribosomal particles.

Nucleolar chromatin has also been examined for phosphorylated proteins. In the acid-insoluble or nonhistone fraction of chromatin, protein C18 was the major phosphorylated protein.[126] This also appears to be a uniquely nucleolar protein since its concentration is substantially reduced in extranucleolar chromatin. More recently it was found that protein C18 has physical and chemical characteristics that are virtually identical to protein C23.[127] Thus, the protein probably exists in two forms: a tightly bound chromatin-associated form and a nonchromatin form complexed, in part, with RNA.

A nucleolus-specific phosphoprotein from mouse ascites tumor cells was also purified by Kawashima and Izawa.[128] This protein has a molecular weight somewhat higher (120,000) than protein C23, but also contains rapidly turning over phosphate. The protein may be functionally analagous to or the same protein as protein C23.

More recently, Tsutsui et al.[129] purified a high molecular weight (110,000) acid-soluble protein from mouse ascites sarcoma cell nuclei. This protein was shown to be phosphorylated and it appeared to exist in two different forms on the basis of its elution from DEAE Sephadex®. The 110,000 mol wt protein was found in several cell types including Ehrlich ascites, mouse leukemia, XC, and SR-CBH/He cells, and mouse and rat liver. It is interesting that, of the tissues examined, the rapidly growing ones contained the greatest amounts of the 110,000 mol wt protein. On the basis of molecular weight, isoelectric point, and amino acid composition (Table 3) it is probably the same protein as C23.

Atmar et al.[130] found somewhat analagous phosphoproteins in the lower eukaryote *Phy-*

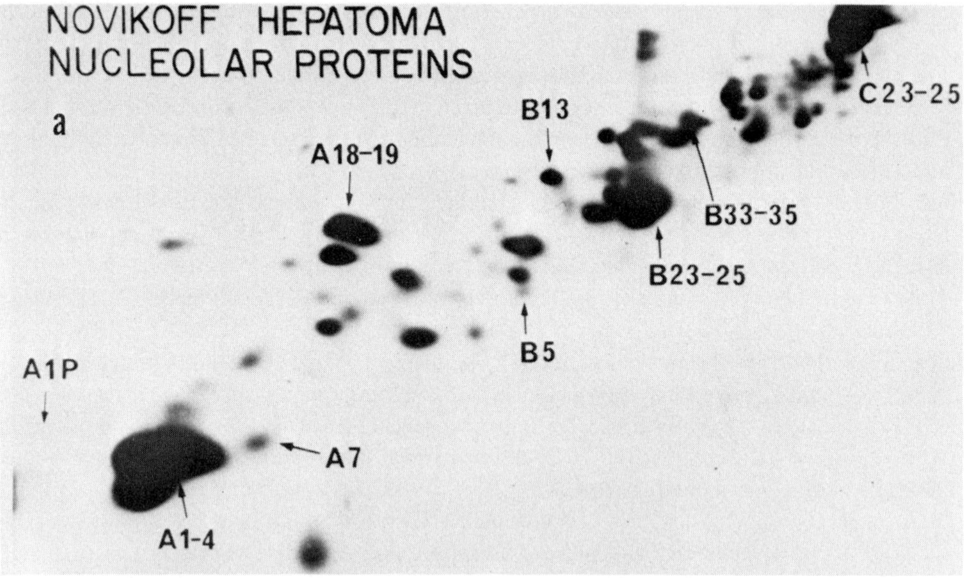

FIGURE 2a. Two-dimensional polyacrylamide gel electrophoresis of 250 μg of Novikoff hepatoma nucleolar proteins. Samples were first loaded on tube gels of 10% acrylamide-6 *M* urea and run in the first dimension for 5 hr at 120 V (constant voltage). For the second dimension, a 12% acrylamide, 0.1% sodium dodecylsulfate slab gel was run for 14 hr at 50 ma constant amperage and then stained with Coomassie® blue. (From Olson, M. O. J., Orrick, L. R., Jones, C. E., and Busch, H., *J. Biol. Chem.*, 249, 2823, 1974. With permission.)

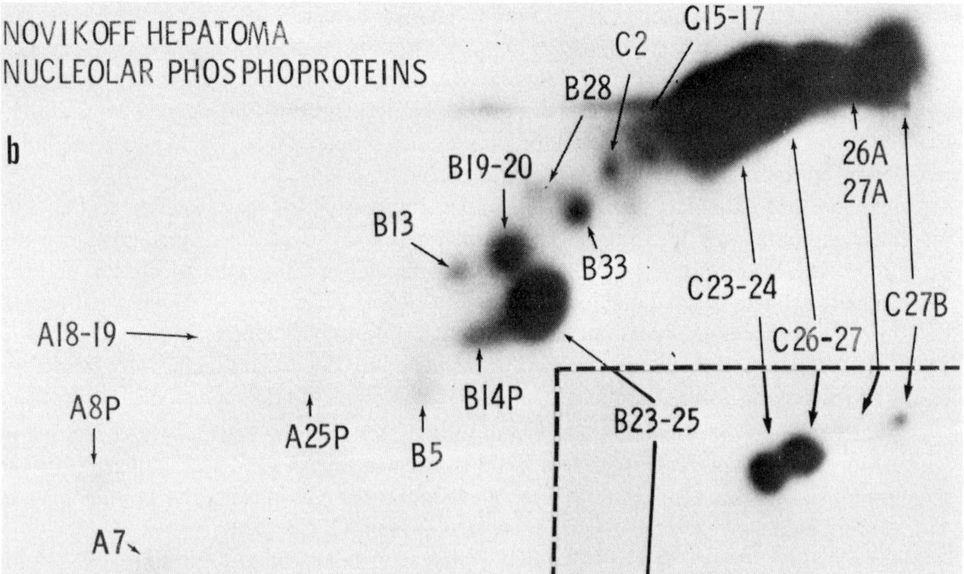

FIGURE 2b. Visualization of ^{32}P-labeled nucleolar proteins by autoradiography on X-ray film. The nucleolar proteins were labeled by injection of 40 mCi per rat intraperitoneally into rats with 6-day-old transplants of Novikoff hepatoma ascites tumors. Proteins were prepared and subjected to electrophoresis as in (a). The numbers indicate radioactive spots coinciding with numbered spots. Numbers designated with P indicate radioactive spots coinciding with numbered spots. *Inset,* 24-hr exposure time to resolve radioactive spots in *Region C.* (From Olson, M. O. J., Orrick, L. R., Jones, C. E.,and Busch, H., *J. Biol. Chem.*, 249, 2823, 1974. With permission.)

Table 3
AMINO ACID COMPOSITIONS OF ISOLATED NUCEOLAR PHOSPHOPROTEINS

Amino acid	C23	110,000 mol wt protein	C18	B23	*Physarum* 70,000 mol wt phosphoprotein
Lys	11.8	11.0	10.5	9.4	8.3
His	0.5	0.8	0.6	1.3	3.2
Arg	2.9	3.1	3.8	3.7	4.2
Asx	12.0	10.0	11.2	12.9	11.0
Thr[a]	5.6(5.5)	5.3	5.1	4.2(3.1)	4.5
Ser[b]	5.9(2.4)	8.5	6.1	7.9(6.1)	7.2
Glx	18.5	17.5	17.9	14.6	16.2
Pro	5.3	4.8	5.6	4.1	4.1
Gly	10.0	8.9	10.3	9.1	12.2
Ala	10.3	10.8	9.3	6.4	7.7
1/2 Cys	0	ND[c]	0	0	0.2
Val	6.0	5.9	5.5	5.3	5.0
Met	1.2	0.9	1.0	2.2	0.8
Ileu	2.5	3.1	2.7	3.2	3.5
Leu	5.6	5.3	5.7	7.5	7.4
Try	0.8	1.1	0.5	1.7	1.7
Phe	3.5	3.0	3.6	2.4	2.9
Trp	0.	ND	ND	0	0
P-Ser[d]	3.5	ND	ND	1.8	ND
P-Thr[d]	0.1	ND	ND	1.1	ND
% Charged AA	45.7	42.5	44.0	41.9	42.9
Asx + Glx/Lys + His + Arg	2.0	1.8	2.0	1.9	1.7
Ref	144	129	146	146	69

[a] In protein C23, C18, and B23 threonine was corrected for 7% destruction during hydrolysis. The number in parenthesis indicates the threonine value after subtracting the phosphothreonine content.
[b] In proteins C23, C18, and B23 serine was corrected for 10% destruction. The number in parenthesis indicates value after phosphoserine was subtracted.
[c] ND indicates not determined.
[d] Phosphoserine + phosphotheonine were determined by the change in serine and threonine after β elimination.[31]

sarum polycephalum. Nucleoli of this slime mold contain two polypeptides of 52,000 and 70,000 mol wt which are the predominant phosphorylated proteins of the organelle. Their in vitro phosphorylation is stimulated by the presence of polyamines. The 70,000 mol wt polypeptide exists as a dimer of 139,000 and is part of the nucleolar ribosomal deoxyribonucleoprotein complex.[69] This protein also binds with high affinity and specificity to the palindromic ribosomal DNA. Furthermore, the protein stimulates rRNA synthesis as much as fivefold when added to the purified rDNP complex which contains RNA polymerase I. Dephosphorylation of the protein by alkaline phosphatase treatment abolished the specific DNA binding ability as well as the stimulatory effects on transcription. Thus, this protein may be a specific regulator of transcription of ribosomal DNA and phosphorylation of the protein may regulate its activity.

In a number of studies, the phosphorylation of nucleolar proteins was found to vary according to the physiological state of the cells from which the proteins are derived. For example, the uptake of ^{32}P into total nucleolar acid-soluble proteins in vivo or in vitro was found to be at least twofold greater in Novikoff hepatoma than in normal rat liver.[131] One of the proteins showing the greatest differences in labeling was protein C23.

Another study which employed in vitro labeling examined the changes in nucleolar protein phosphorylation during liver regeneration.[132] Uptake of ^{32}P from [γ-^{32}P]ATP into total acid-soluble nucleolar proteins increased 22-fold after 24 hr of liver regeneration, with uptake into protein C23 accounting for most of the difference. Several other proteins increased in ^{32}P uptake while others decreased in labeling during the time course of liver regeneration. These increases in uptake of ^{32}P in proliferating systems are consistent with the increases in protein kinase activity also observed in regenerating liver and Novikoff hepatoma nucleoli.[133] The regenerating liver and tumor systems reflect increases in ribosome output by the nucleolus which are concomitant with enhanced levels of RNA polymerase I.[134] Thus, there appears to be a general correlation between the level of phosphorylation of proteins in the nucleolus and its rate of output of ribosomes.

2. The Nucleolar Phosphorylating and Dephosphorylating System

The presence of phosphoprotein kinase activity in the nucleolus has been demonstrated by several groups.[133,135-139] Although multiple kinases have been observed,[133,135] the nucleolus, like the nucleus, appears to contain only two major ionic classes of cAMP-independent protein kinase activity.[133,136,137] Some of the studies suggest that nucleolus associated protein kinases are different from those in the extranucleolar portion of the nucleus.[137,139] Three fractions of protein kinase from Novikoff hepatoma nucleoli are capable of phosphorylating nucleolar protein C23.[133]

In general, nucleolus-associated protein kinases are of the cyclic nucleotide independent variety. However, Kawashima and Izawa reported that the phosphorylation of the 120K nucleolus-specific phosphoprotein of mouse ascites cells was stimulated by cyclic GMP.[135] In the *Physarum* system the phosphorylation of acidic nucleolar proteins is not stimulated by cyclic nucleotides, but is enhanced by the addition of polyamines.[130] Very recently a polyamine-activated protein kinase was purified.[140] These studies provide some indication that nucleolar protein phosphorylation may be at least partially controlled by small molecules.

Phosphoprotein phosphatase is also present in nucleoli. Olson and Guetzow partially purified and characterized a nucleolus-associated phosphoprotein phosphatase from Novikoff hepatoma cells.[141,142] In this regard, the phosphatase may be as important as the protein kinase in determining the overall pattern of phosphorylation. As discussed earlier, Kang et al.[122] found that the profile of in vitro labeling was dependent on the type of divalent cation present. For example, protein B23 was not labeled in the presence of 5 m*M* $MgCl_2$, but was labeled when 5 m*M* $ZnCl_2$ was added to the system. The greatest overall uptake of ^{32}P from [γ-^{32}P]ATP into total proteins was also found in the presence of $ZnCl_2$. It is interesting to note that Zn^{2+} was also a very effective inhibitor of phosphoprotein phosphatase.[142] Differential metal ion effects may not have their basis at the kinase level, but certain metals may selectively inhibit phosphatases, thereby modulating the turnover of phosphate on specific proteins.

The above studies were cited in order to illustrate the complexity of protein phosphorylation in the nucleolus. The net content of phosphate in some proteins appears to be the result of the opposing forces of phosphorylation and dephosphorylation. The enzymes may in turn be regulated by the concentrations of cyclic nucleotides, polyamines, or metal ions.

3. Characteristics of Nucleolar Phosphoproteins

Although very little is known about the physical and chemical properties of isolated nonhistone nuclear phosphorylated proteins, a few nucleolar phosphoproteins have been isolated and partially characterized. These proteins have a number of unique properties.

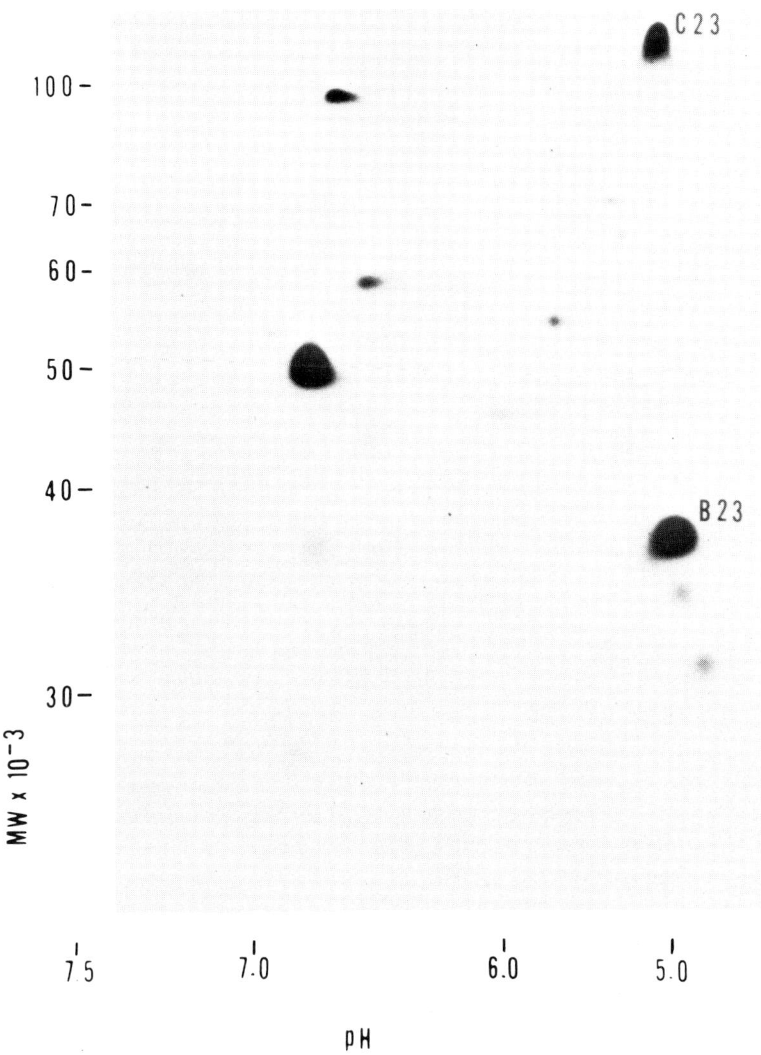

FIGURE 3. Two-dimensional polyacrylamide gel electrophoresis of saline-soluble nucleolar proteins. Nucleoli from Novikoff hepatoma were extracted first with 0.075 *M* NaCl, 0.025 *M* EDTA, pH 7.5 and then with 0.01 *M* Tris, pH 7.5. The combined extracts were run on a two-dimensional gel system as described in Reference 145. Molecular weights are indicated on the ordinate. Isoelectric ponits (pH) are indicated on the abscissa. Proteins C23 and B23 are indicated by numbers on photo.

a. Protein C23/C18

Protein C23 was initially isolated from 0.4 *N* H_2SO_4 extracts of Novikoff hepatoma nucleoli by preparative polyacrylamide gel electrophoresis followed by hydroxylapatite chromatography.[143,144] The purified protein appears as an elongated spot on two-dimensional gel electrophoresis and has a homogeneity of greater than 99%. The molecular weight was determined to be approximately 110,000 and the pI is near 5.5 (see Figure 3).[145,146] The protein contains about 30% acidic amino acids as well as nearly 20% basic amino acids (Table 3). In addition, it contains more than 30 phosphoserine residues. Thus, it is an acidic protein with approximately half of its residues carrying charged side chains.

Protein C23 contains a large number of its acidic residues in clusters of high net negative charge. At least four clusters containing from 10 to 30 carboxyl groups plus 1 to 3 phosphoserine residues have been identified.[143,144] One of these regions, C23-Ca, has been sequenced:[144]

Ala-Ala-Pro-Ala-Ala-Pro-Ala-Ser(P)-Glu-Asp-Glu-Asp-Glu-Glu-
Asp-Asp-Asp-Asp-Glu-Asp-Asp-Asp-Asp-Asp-Ser(P)-Gln-
Glu-Ser(P)-Glu-Glu-Glu-Asp-Glu-Glu-Val-Met-Glu-Ile-
Thr-Pro-Ala-Lys.

Preliminary studies suggest that most of the acidic clusters are located in one end of the molecule.[147] The NH_2-terminal sequence of C23 is relatively basic: Val-Lys-Leu-Ala-Lys-Ala-Gly-Lys-Thr-. Thus, this protein has the potential for ionic interactions with basic regions of either histones or ribosomal proteins. In addition, ionic interactions with acidic groups of DNA, RNA, or acidic proteins should be possible through the amino terminal end of protein C23.

Protein C18 was previously shown to be the major nucleolar chromatin nonhistone phosphoprotein.[126] this protein was purified from nucleoli after preextraction with acid and dilute salt solutions. The protein was released by DNase digestion and then put through the electrophoresis-hydroxylapatite procedure as above. The properties of protein C18 (molecular weight, isoelectric point, and amino terminal sequence) were found to be virtually identical to those of protein C23.[146] The amino acid composition (Table 3) is also similar to that of protein C23. Thus, protein C23 has a tightly bound counterpart, C18. The reasons for this distribution are unclear. However, a certain fraction of the protein may be involved in specific chromatin organizing functions.

b. Protein B23

This protein was also isolated from nucleolar acid extracts by the above procedure.[143,146] Protein B23 has a pI of 4.9 to 5.4 and a molecular weight of 38,000. This protein also contains acidic clusters. However, at least two of the clusters contain phosphothreonine instead of phosphoserine. The N-terminus appears to be blocked, but a single C-terminal leucine was detected by hydrazinolysis. Thus, at least two distinctly different proteins, C23 and B23, contain the unusual, highly negatively charged clusters of amino acids.

Another unusual property of these two proteins is their ability to stain with silver. Lischwe et al.[145] extracted proteins from nucleoli by various methods and separated them by two-dimensional polyacrylamide gel electrophoresis. The gels were subjected to a silver staining technique that selectively stains the nucleolus organizer regions of chromosomes.[148,149] In all of the extracts, the major silver staining proteins were proteins C23 and B23. The basis of the selectivity of silver staining of active nucleolus organizer regions is not known and it remains to be determined whether proteins C23 or B23 are nucleolus organizer proteins. However, the highly negatively charged regions of these proteins would be expected to attract relatively large quantities of Ag^+. Upon reduction the deposited silver may produce a densely stained region. In the same manner, the negatively charged regions could conceivably interact with polycationic regions of histones in chromatin to organize the nucleolus. The role of the acidic clusters and the sites of phosphorylation contained therein remain a subject for speculation and should be an area of fruitful research in the future.

c. The 70,000 Mol Wt Phosphoprotein of Physarum Nucleoli

Kuehn et al.[69] purified by nondenaturing methods a phosphorylated nonhistone protein from *P. polycephalum* nucleoli. The protein has a subunit molecular weight of 70,000 but apparently exists as a dimer of 139,000. The amino acid composition of the protein is presented in Table 3. The 70,000 mol wt protein, like proteins B23 and C23, is an acidic protein with nearly twice as many potentially acidic residues as basic residues. Similarly,

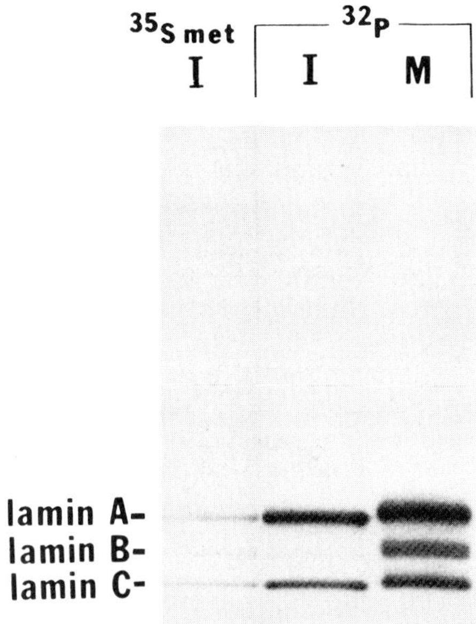

FIGURE 4. SDS gel electrophoresis of the lamins from ^{32}P pulse-labeled interphase or mitotic cells. Samples of ^{35}S-methionine-labeled interphase (I) cells or ^{32}P pulse-labeled exponentially growing interphase or synchronized mitotic (M) cells were immunoprecipitated with antilamin A antibodies and electrophoresed on an SDS gel. Visualization was by autoradiography. A Dupont® Hi-Plus intensifying screen was used to enhance ^{32}P lanes. (From Gerace, L. and Blobel, G., *Cell,* 19, 277, 1980. Copyright M.I.T. Press. With permission.)

it contains no tryptophan and nearly half of its residues have charged side chains. It is conceivable that proteins C23 and B23 play similar roles in higher eukaryotic nucleoli, as does the 70K protein in the slime mold.

D. Nuclear Membrane Phosphoproteins

The nuclear envelope lamina is a supramolecular assembly which is at the interface between the nucleoplasm and the inner nuclear membrane. This structure contains three predominant peripheral membrane polypeptides which have been designated lamins, A, B, and C, having molecular weights of 70, 67, and 60 kdaltons, respectively.[150,151] The lamina is reversibly disassembled during cell division, concurrent with the disappearance of the nuclear envelope at mitosis. During mitosis lamins A and C, dissociate from the membrane and become soluble.[12] However, lamin B appears to remain associated with membrane fragments. The lamins, which are in a polymerized state in the interphase membrane, are monomeric during mitosis. All three lamins are phosphorylated.[12]

Gerace and Blobel suggest that phosphorylation of the lamins is involved in the reversible depolymerization process.[12] All three lamins undergo a lowering of their isoelectric points during mitosis. Furthermore, lamins A and C incorporate three times as much ^{32}P phosphate in mitotic cells than lamins A and C in interphase cells (Figure 4). The augumentation of ^{32}P incorporation in lamin B is even more dramatic, a sixfold increase. In early G1 phase cells the level of phosphorylation of the lamins is intermediate between the mitotic and the exponentially growing cells. Thus, phosphorylation seems to be regulating the state of association and the three-dimensional structures of supramolecular assemblies at the periphery of the nucleus.

Other studies have shown the in vitro phosphorylation of nuclear envelope polypeptides and the presence of protein kinases in nuclear membranes. Lam and Kasper reported the selective phosphorylation of a major membrane polypeptide (68,000 mol wt) when highly purified nuclear envelope was incubated with [γ-^{32}P]ATP.[152] This is likely to be the lamin B polypeptide of Gerace et al.[151] Nuclear envelope-associated protein kinase activity was also observed by Steer et al.[153] and was shown to be activated by Co^{++} ions.[154] Although Gerace and Blobel[12] present evidence that the protein phosphorylation is involved in membrane dissociation during mitosis, Lam and Kasper[152] have suggested that the 68,000-dalton protein and its associated kinase activity may play a role in nucleocytoplasmic transport.

E. Nuclear Matrix Phosphoproteins

The nuclear matrix consists largely of acidic nonhistone proteins with three main polypeptides which range in molecular weight from 60,000 to 70,000 daltons.[155] This structure is believed to provide fixed sites for the attachment of DNA replicating complexes and the newly replicating DNA is suggested to be reeled through these sites.[156] Allen et al.[157] studied the phosphorylation of the nuclear matrix proteins in regenerating liver. They found that the uptake of ^{32}P from [γ-^{32}P]ATP into nuclear matrix proteins was two- to three-fold higher than uptake into total nuclear proteins in either normal rat liver or regenerating liver. Phosphorylation of nuclear matrix proteins was further enhanced during regeneration, reaching a maximum at 12 hr after partial hepatectomy. This preceded the onset of DNA synthesis which occurs 14 to 16 hr after partial hepatectomy. Since the nuclear matrix is involved in the DNA replicative process, it is conceivable that phosphorylation of the matrix proteins plays a role in regulation of DNA synthesis.

F. Nuclear RNP Particle Proteins

Heterogeneous nuclear RNA (hnRNA) which is believed to be the precursor to messenger RNA is packaged in the nucleus as a complex of RNA and protein. These complexes range in size from 30S to 300S depending on the conditions of their isolation.[158] The repeating subunit structure consists of 40S RNP particles which may be released by micrococcal nuclease or other methods of limited cleavage of RNA.[10] The protein mass of the 40S particles is composed largely of six major polypeptides which vary in molecular weights from 32,000 to 44,000.[159] However, additional proteins having molecular weights in the neighborhood of 65,000 are present in the larger intact hnRNP particles.[160]

In the early 1970s there were reports of the presence of phosphorylated proteins in nuclear hnRNP particles from rat brain.[158,161] By in vivo labeling the ^{32}P radioactivity was associated predominantly with two polypeptides in the 30,000 to 40,000 mol wt range. However, these proteins were not characterized further at the time.

A later study found that four proteins associated with the hnRNA of HeLa cells were also phosphorylated under in vivo labeling conditions.[162] The molecular weights of these proteins were 28,000, 30,000, 37,000, and 52,000. The same group of researchers also studied endogenous protein kinase activity and phosphoprotein phosphatase activity in nuclear hnRNP particles.[163,164] The endogenous activity phosphorylated predominantly the 28,000 and 37,000 mol wt polypeptides either in the presence or absence of cyclic AMP.[164] Schweiger and Schmidt[166] also noted that a 39,000 mol wt polypeptide, which had RNA binding activity, was labeled when 30S nuclear RNP particles were incubated with [γ-^{32}P]ATP. In a study by Prestayko et al.[38] at least two rat liver nuclear informofer proteins were phosphorylated.

More recently two of the polypeptides of the 40S RNP particles were shown to be phosphorylated.[167] These are the C_1 (42,000 mol wt) and C_2 (44,000 mol wt) polypeptides (Figure 5) and labeling may be achieved under in vivo or in vitro conditions. Figure 5 also

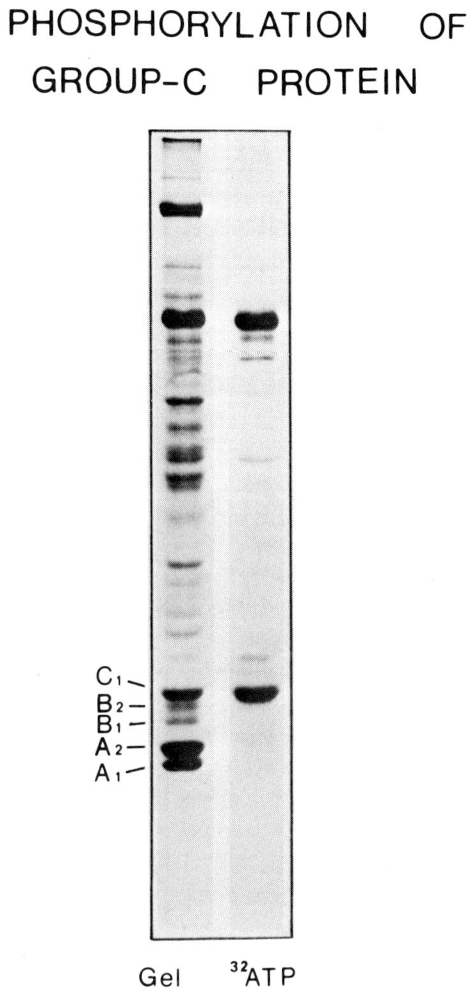

FIGURE 5. The phosphorylation of the group-C proteins. Monomer 40S particles were collected from a sucrose gradient and subjected to gel filtration chromatography through Bio-Gel® A-1.5m. All column fractions (of which this figure is representative) were incubated in gamma ^{32}P-ATP, resolved on gels, and subjected to autoradiographic analysis. Proteins C_1 and C_2 are the only major phosphorylated components of 40S hnRNP particles. If cells are labeled in vivo with ^{32}P orthophosphate the same results are obtained. The major phosphorylated high molecular weight band in the autoradiogram may be a dimer of C_1 as a protein of this size also coelutes with C_1 during ion exchange chromatography. (From Le-Stourgeon, W. M., Lothstein, L., Walker, B. W., and Beyer, A., *The Cell Nucleus,* Vol. 9, Part B, Busch, H., Ed., Academic Press, New York, 1980. With permission.)

indicates that another protein of higher molecular weight becomes phosphorylated, although it was suggested that this may be a dimer of the C group proteins. Of the six major proteins in the 40S particles, the C group proteins are the most tightly bound to RNA.

The various studies cited above all indicate that some proteins in the hnRNP complex are phosphorylated. However, there seems to be some disagreement as to how many and precisely which ones are phosphorylated. It is likely that the discrepancies stem from two problems. First, the molecular weights of the various proteins are not precisely defined and difficulty arises when reports from various laboratories are compared. Secondly, the particles larger than 40s contain more proteins. Thus, the higher number of phosphorylated proteins reported by Blanchard et al.[162] may be due to the fact that their particles cover a broad range of sizes. This was recently shown more clearly by Brunel and Lelay who analyzed the phosphoprotein content of hnRNP particles of various sizes.[168] The smallest particles (30 to 50S) had the least number of phosphoproteins, whereas the largest particles (>100S) contained the largest number of species of phosphorylated proteins. Brunel and Lelay also suggest that the phosphorylated proteins are unrelated to the major species of proteins in the hnRNP particles. These unresolved conflicts in data will have to be settled when each of the proteins is isolated and more fully characterized.

Small nuclear RNAs are also found complexed with proteins. For example, U_1 and U_2 RNAs are present in particles that contain ten polypeptides.[169] Two of these proteins, designated B16 and C5 are phosphorylated. Since the function of the small nuclear RNAs is poorly understood, the role of phosphorylattion of the associated proteins is also unknown.

G. Phosphorylation of Nuclear Enzymes

1. Phosphorylation of RNA Polymerase

The direct involvement of protein phosphorylation in the transcriptive process is suggested by the finding that RNA polymerases from various sources are phosphorylated. In yeast, all three RNA polymerases are phosphorylated as shown by in vivo studies.[170,172] Rat liver nucleolar RNA polymerase was also shown to be phosphorylated on serine as well as threonine residues when isolated nuclei were incubated with [γ-^{32}P]ATP.[173]

These studies have been extended to systems where isolated RNA polymerase is phosphorylated by exogenous protein kinase. Kranias et al.[174] demonstrated the in vitro phosphorylation of predominantly the 25,000-dalton subunit, and to a lesser extent the 180,000-dalton polypeptide of calf thymus RNA polymerase II. This was accomplished using a homologous nuclear cAMP-dependent protein kinase. However, essentially the same result could be obtained with a calf thymus nuclear cyclic AMP-independent protein kinase.[175,176] In either case, incubation of RNA polymerase II with the protein kinase led to a stimulation of activity as much as three-fold over the control. Dephosphorylation of the polymerase by *E. coli* alkaline phosphatase lowered activity concomitant with loss of phosphate.

Earlier studies also reported protein kinase-mediated activation of RNA polymerases in lesser defined systems. Calf ovary RNA polymerases I and II were activated by homologous cAMP-dependent protein kinase.[177] Rat liver RNA polymerase I was stimulated by homologous cAMP-dependent protein kinase.[178] Dahmus also observed a five- to seven-fold stimulation of RNA polymerase II by a protein factor with extensive protein kinase activity from Novikoff hepatoma ascites cells.[179] More recently, phosporylation of the 220,000-dalton subunit of wheat embryo RNA polymerase II was demonstrated.[180] This reaction occurred either in vitro when the polymerase was incubated with a homologous kinase preparation or in vivo at the onset of germination. The phosphorylation resulted in a moderate increase in RNA polymerase activity. The authors proposed that stored molecules in dormant embryo cells may be activated by phosphorylation of RNA polymerase at the onset of germination.

Thus, there is considerable evidence that RNA polymerases can incorporate phosphate and that the phosphorylation results in stimulation of activity. However, the physiological significance of this is unclear, as well as the question of whether the phosphorylation is an in vivo regulatory process.

2. Poly(A) Polymerase

Newly synthesized nuclear messenger RNA(mRNA) precursors contain 100 to 200 aden-

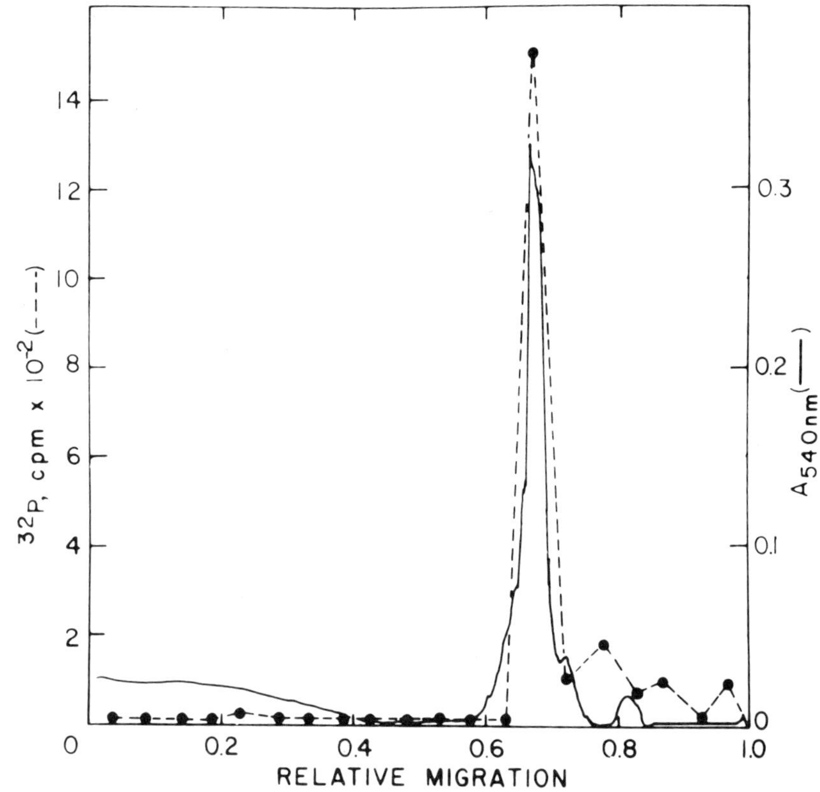

FIGURE 6. Polyacrylamide gel electrophoresis of radiolabeled hepatoma poly(A) polymerase. Nuclei isolated from hepatoma 3924A were incubated with [γ-^{32}]ATP (20 Ci/mmol). Poly(A) polymerase was extracted and purified essentially as described in Reference 183. After QAE-Sephadex® chromatography, enzyme was concentrated with polyethylene glycol, denatured, and subjected to polyacrylamide gel electrophoresis in the presence of sodium dodecylsulfate. Subsequent to electrophoresis, the gel was sliced, dissolved in protosol, and ^{32}P determined in a liquid scintillation counter. Purified hepatoma poly(A) polymerase (mol wt = 60,000), run on a parallel gel, was stained with Coomassie® brilliant blue and absorbance monitored at 540 nm on a Gilford® spectrophotometer with a linear transport attachment. The relative migrations of bovine serum albumin (mol wt = 67,000) and ovalbumin (mol wt = 44,000), run in parallel gels, were 0.6 and 0.75, respectively. (From Rose, K. M. and Jacob, S. T., *J. Biol. Chem.*, 254, 10256, 1979. With permission.)

ylate residues at their 3′ termini. The polyadenylation is catalyzed by a chromatin-associated poly(A) polymerse.[181,182] The precise function of the poly(A) tail is uncertain, although it has been suggested that it plays a role in the processing of the precursor to the smaller final product of messenger RNA. In any event, the activity of poly(A) polymerase appears to be regulated by phosphorylation. Rose and Jacob[183] found that nuclear poly(A) polymerase from rat liver and Morris hepatoma 3924A contained ^{32}P when it was isolated from nuclei that were previously incubated with [γ-^{32}P]ATP (Figure 6).[183] The hepatoma enzyme incorporated more than twice as much ^{32}P than the liver enzyme. Purified poly(A) polymerase also became phosphorylated when incubated with exogenous protein kinase. Predominantly serine, and to a lesser extent threonine, became labeled by the kinase. More importantly, the enzyme activity was enhanced as much as tenfold by the phosphorylation (Figure 7).

Rose and Jacob further studied the mechanism of enhancement of enzyme activity by phosphorylation. They found that the phosphorylated enzyme had an increased primer re-

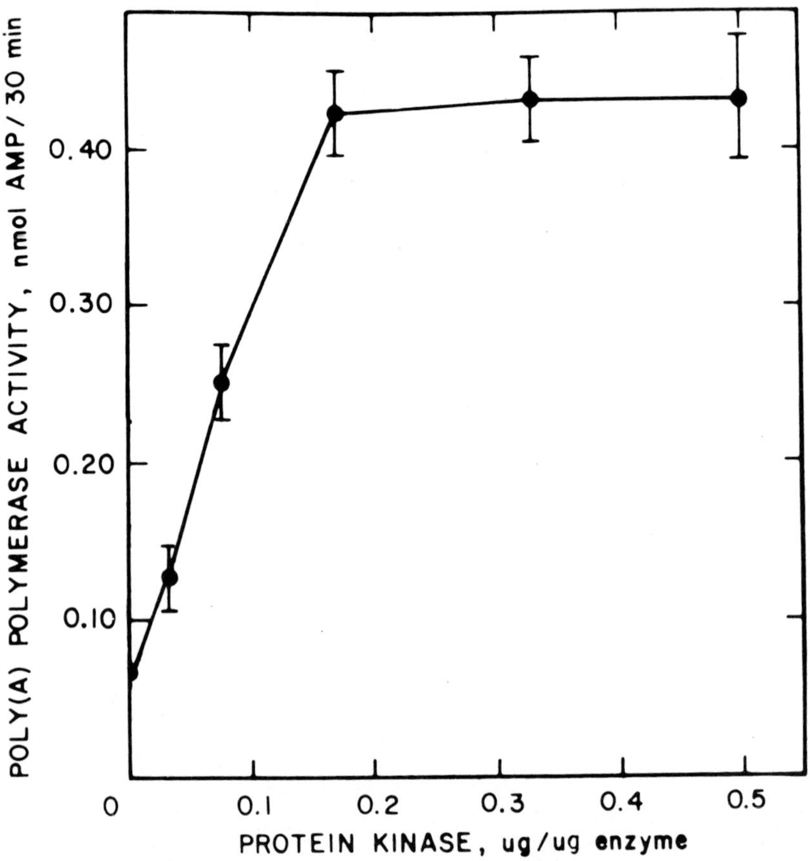

FIGURE 7. Effect of protein kinase on poly(A) synthesis. Protein kinase was obtained from hepatoma 3924A as outlined. Poly(A) polymerase was purified from liver nuclei by DEAE-Sephadex® chromatography and subjected to phosphocellulose chromatography using a shallow KCl gradient. Those poly(A) polymerase fractions eluting at 0.4 *M* KCl were further purified by hydroxylapatite and QAE-Sephadex® chromatography. Poly(A) synthesis was measured in assay mixtures containing 0.4 m*M* $MnCl_2$, 0.38 m*M* [^{3}H]ATP, 35 m*M* KCl, 0.5 m*M* dithiothreitol, 200 μg/mℓ of serum albumin, 52 m*M* Tris-HCl (pH 9), 375 μg/mℓ of poly(A), 1.7 μg of purified poly(A) polymerase, and protein kinase as indicated. Reactions were incubated at 37°C for 30 min. Assays were performed in quadruplicate and results expressed as the mean ± SE. (From Rose, K. M. and Jacob, S. T., *J. Biol. Chem.*, 254, 10256, 1979. With permission.)

quirement and synthesized a greater number of poly(A) chains.[184] However, the average poly(A) chain length did not increase. Thus, the rate, but not the extent of polyadenylation was increased by phosphorylation. Rose and Jacob suggest that the phosphorylated poly(A) polymerase has a greater affinity for poly A, thereby stabilizing the messenger RNA. This is consistent with their previous observation that the poly(A) polymerase is antigenically related and probably chemically identical to the poly(A) binding protein on polysomal messenger RNA.[185] These data suggest that poly(A) polymerase plays multiple roles and may be located in the cytoplasm as well as the nucleus. The phosphorylation of poly(A) polymerase represents one of the clearest examples of a cause and effect relationship between phosphorylation and activity.

3. *Histone Deacetylase*

Vidali et al.[186] isolated a protein fraction which specifically binds to histones 2A and 3 and deacetylates them. The protein has an acidic isoelectric point (pI 4.5), a molecular weight of 150,000 to 160,000, and contains 1.3% phosphate. The amino acid composition (Table 2) indicates that it, like many other phosphorylated nonhistone proteins, has a high content of potentially acidic amino acids. The relationship between its activity and its phosphate content has not been investigated.

H. Nucleoplasmic Phosphoproteins

1. Nucleoplasmin

The oocytes of the frog *Xenopus laevis* contain a protein which facilitates the assembly of nucleosomes in vitro from purified DNA and histones.[187] This protein is localized in the nucleoplasm of *Xenopus* oocytes, where it is the most abundant protein, accounting for 7.5 to 10% of total protein.[188-190] The same protein or a protein with immunologically similar determinants is also found in nuclei of a variety of other vertebrates.[189] Because of its abundance and its location in a soluble fraction of the nucleoplasm the protein has been named "nucleoplasmin".[191]

In addition to being an abundant protein in nuclei, nucleoplasmin has a number of unusual properties. It is an acidic protein with an isoelectric point near 5 and approximately 30% of its amino acid residues (Table 2) are potentially negatively charged.[188,189,192] Nucleoplasmin is phosphorylated probably at several sites, since multiple isoelectric variants have been observed (Figure 8).[188,189] Another curious property of nucleoplasmin is its unusual quarternary structure. The protein appears to be a pentamer composed of identical polypeptides of 29,000 daltons.[192]

Nucleoplasmin binds histones but does not bind nucleosomes or DNA.[192] Laskey and colleagues[187] propose that nucleoplasmin facilitiates nucleosome assembly by shielding the positive charges on histones from the DNA phosphates. This allows the histones to combine with the DNA in an orderly manner rather than form improperly placed salt bridges. The role of phosphorylation in this process or in maintaining the structure of nucleoplasmin is not known.

IV. SITES OF PHOSPHORYLATION

In recent years there has been increased interest in the structural nature of protein phosphorylation sites, following the demonstration that the amino acid residues surrounding these sites are involved in the specificity of protein kinases.[193,194] Sequences of amino acids near phosphorylation sites have been determined in most of the histones and in a large number of cytoplasmic proteins.[195] However, little is known about the sites of phosphorylation in the most highly phosphorylated proteins in the cell nucleus, the nonhistone proteins.

Recently, Mamrack et al.[143,144] found phosphorylation sites within clusters of acidic amino acids in two nonhistone proteins, C23 and B23. Table 4 indicates that three sites of phosphorylation are present in an acidic region of protein C23. The tryptic peptide (C23Ca) which contains these sites is present in three electrophoretic forms. Figure 9 illustrates that the electrophoretic mobility at pH 1.8 is directly proportional to the number of phosphoryl groups or the net negative charge of the molecule. At this pH the carboxyl groups are unionized so that the net charge of the molecule is determined only by the amino groups and phosphoryl groups. Thus, in peptides with multiple phosphorylation sites the electrophoretic mobility at low pH is a useful parameter in determining the level of phosphorylation.

Of the 42 residues in this segment of protein C23, 22 of them contain carboxyl groups. In all three sites of phosphorylation, an acidic amino acid lies either on the amino or carboxyl terminal side of the serine. The pattern of phosphorylation also appears to fit the pattern of phosphorylation of some sites in casein. Casein contains several serine residues which are phosphorylated when followed two positions in the sequence by an aspartic acid, glutamic

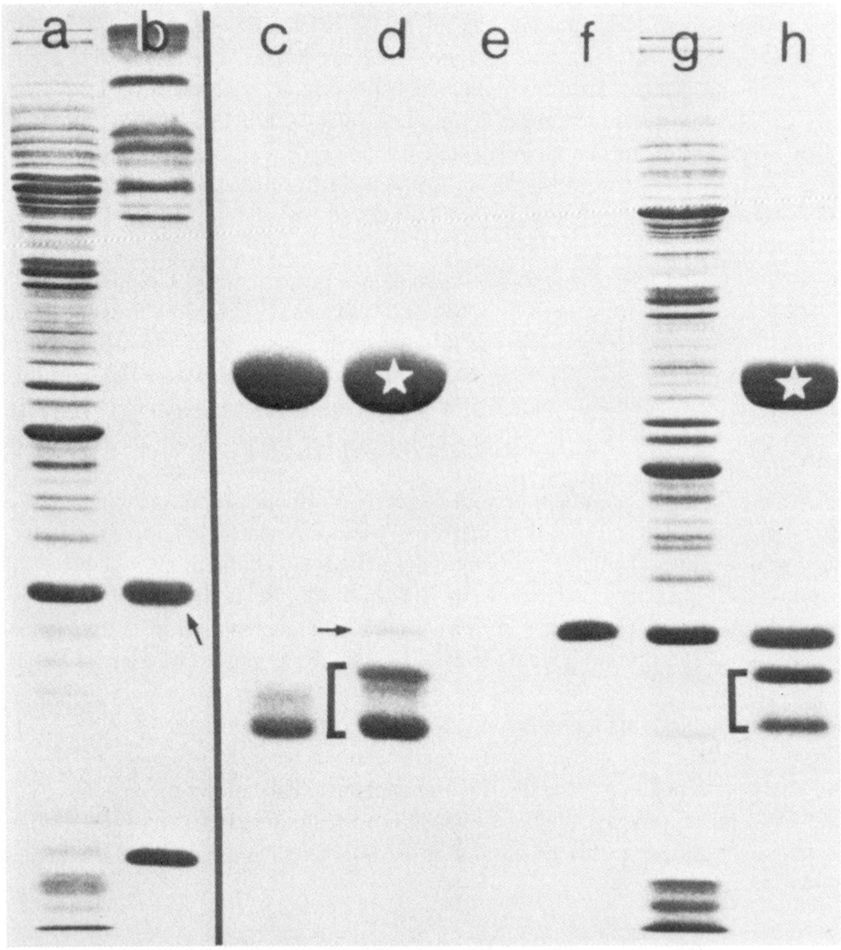

FIGURE 8. Gel electrophoresis of total nuclear proteins from 25 *Xenopus* oocytes labeled with [^{32}P]phosphate-labeled proteins obtained from nuclear material (80 nuclei) solubilized with NaDodSo$_4$/Triton® X-100-containing buffer by addition of antiserum and protein A-Sepharose® (lane d) is compared with the result of incubation with premium serum (lane c, stained with Coomassie® blue). The autoradiofluorogram of the same gel (lanes e and f) shows that the 30,000 mol wt protein (arrow) is the only labeled protein immunoprecipitated. Immunoprecipitate from nuclear material of 60 nuclei, treated with antiserum but not with protein A, and washed only with phosphate-buffer saline (lane h) is compared with the pattern of the proteins used for precipitation (lane g, from 25 nuclei). Stars, heavy chains of IgG; brackets, light chains of IgG. (From Krohne, G. and Franke, W. W., *Proc. Natl. Acad. Sci. U.S.A.*, 77, 1034, 1980. With permission.)

acid, or another phosphoserine.[193,196] All three sites of phosphorylation in peptide C23Ca follow this arrangement. If this is generally the pattern of sites of phosphorylation in nonhistone nuclear proteins, it may explain why casein has served as a good substrate for assaying cyclic nucleotide independent kinases found in the nucleus.

Nucleolar protein B23 contains a phosphorylation site which does not follow the casein arrangement (Table 4). This site is phosphorylated under in vivo but not in vitro labeling and is probably not phosphorylated by the same kinase that acts on the sites in C23-Ca.[197] However, the Thr-Pro-Ala-Lys sequence is identical to the C terminal sequence of peptide C23-Ca (Table 4). The phosphorylated threonine has an acidic amino acid at its N-terminal

Table 4
SITES PHOSPHORYLATED IN NUCLEOLAR NONHISTONE PROTEINS

Peptide C23-Ca, labeled in vitro[a]	Ala-Ala-Pro-Ala-Ala-Pro-Ala-PSER-Glu-Asp-Glu-Asp-Glu-Glu-Asp-Asp-Asp-Asp-Glu-Asp-Asp-Asp-Asp-Asp-PSER-Gln-Glu-PSER-Glu-Glu-Glu-Asp-Glu-Glu-Val-Met-Glu-Ile-Thr-Pro-Ala-Lys-
Peptide B23-T1A, unlabeled in vitro[b]	Asp-PTHR-Pro-Ala-Lys

[a] Data from Reference 144.
[b] Data from Reference 197.

side in contrast to the unphosphorylated threonine in C23Ca which is at the C-terminal side of an isoleucine.

In order to determine whether the phosphorylation sites of nonhistone proteins generally lie adjacent to acidic amino acids Jones et al.[198] surveyed 30 peptides from pronase digests of unfractionated nonhistone nuclear proteins. Each of the peptides contained 1 to 4 phosphorylated residues and ranged in chain length from 7 to 19 residues. A representative sample of the peptide is presented in Table 5. The amino acids that are most abundant in these sites are proline, glutamic acid, and glycine. In fact, all of the peptides contained these amino acids. Most of the peptides also contained multiple phosphoryl groups. The presence of acidic amino acids is consistent with the findings for the phosphopeptides from proteins C23 and B23. The prevalence of proline suggests that there is a requirement for nonhelical regions of secondary structure either in relation to kinase recognition or in relation to the function of these regions. The relative lack of hydrophobic amino acids may reflect the location of the sites of phosphorylation on the exterior of the protein molecules.

Jones and Olson also used another approach to the problem of the general nature of sites of phosphorylation.[199] The ^{32}P-labeled, unfractionated nonhistone nuclear proteins were subjected to partial acid hydrolysis and then fractionated by Dowex®-1 column chromatography followed by paper electrophoresis at pH 1.8. The latter system resolved the digest into inorganic phosphate phosphoserine, phosphothreonine which were negatively charged, and two products designated as A and B which were positively charged at pH 1.8 (Figure 10). Products A and B were identified as dipeptides containing phosphothreonine and phosphoserine, respectively. By subtractive Edman degradation they were shown to have the general structure, X-PSER or X-PTHR, where X is a nonphosphorylated amino acid. The two groups of dipeptides were further separated by two-dimensional electrophoresis into unique peptides (Figure 11) and subjected to amino acid analyses. The latter procedure provided a nearest neighbor analysis and revealed that a heterogeneous group of amino acids was on the amino terminal side of phosphoserine residues. These included valine, isoleucine, leucine, proline, alanine, glycine, and glutamic acid. In contrast, phosphothreonine residues were preceded predominantly by proline (Spot 2, Figure 11A). Kinetic studies on the hydrolysis of nonhistone proteins revealed that the dipeptides were released more rapidly than phosphoserine or phosphothreonine. Thus, they appear to be representative of sites of phosphorylation in nonhistone proteins.

The above study introduces a nearest-neighbor technique for the analyses of sites of phosphorylation. In addition, the finding of proline predominantly adjacent to phosphothreonine confirms the suggestion that nonhelical regions are abundant near phosphorylation sites.

From the limited information on sites of phosphorylation in nonhistone proteins, a few tentative conclusions may be drawn. Of the sites surveyed to date, the majority appear to be situated in nonhelical regions which have a high negative charge density. In those few

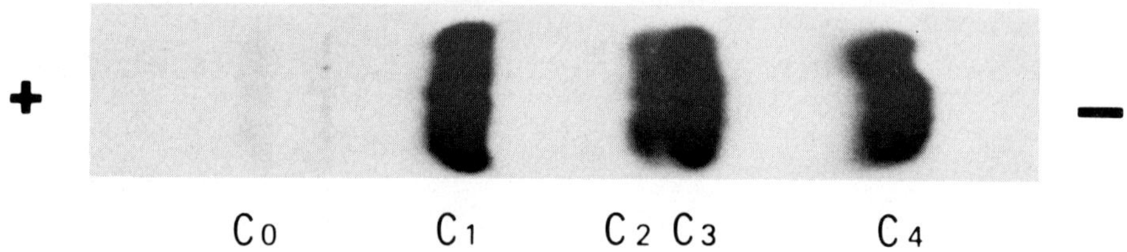

FIGURE 9a. Autoradiogram of tryptic fraction C from protein C23, separated by paper electrophoresis. The sample was desalted, placed on Whatman® 3MM paper, and electrophoresed at pH 1.8 for 1.5 hr at 3000 V; the paper was placed on Kodak® XR5 film for 24 hr. The major phosphopeptides are indicated. Peptide C_o was the only peptide that migrated toward the anode. (Reprinted with permission from Mamrack, M. D., Olson, M. O. J., and Busch, H., *Biochemistry,* 18, 3381, 1979. Copyright 1979 American Chemical Society.)

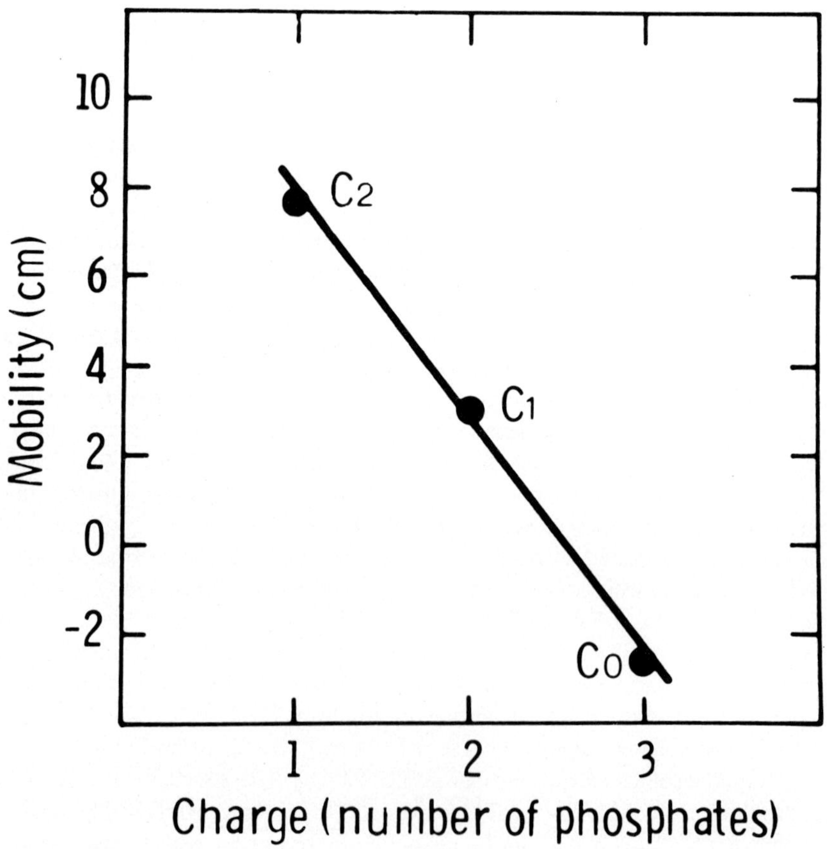

FIGURE 9b. Plot of relative mobility vs. presumed number of phosphate residues per peptide molecule. The data from Figure 9a were plotted to test the linearity of the relationship between electrophoretic mobility and integral multiples of the negative charge carried by each species of the Ca tryptic peptide. The peptides were assumed to have identical molecular weights as indicated by amino acid composition. (Reprinted with permission from Mamrack, M. D., Olson, M. O. J., and Busch, H., *Biochemistry,* 18, 3381, 1979. Copyright 1979 American Chemical Society.)

Table 5
AMINO ACID RATIOS OF ISOLATED PHOSPHOPEPTIDES[a]

	Peptide number									
Amino acid	**1**	**4**	**6**	**8**	**12**	**16**	**18**	**19**	**20**	**27**
Asx	0.7	0.9	2.2		0.9	0.9	0.6	0.9		1.1
Thr[b]	1.3		1.0	0.9		0.4		0.4	0.6	
Ser[b]	2.8	1.8	2.7	2.0	1.5	1.3	1.4	1.3	1.0	1.2
Glx	1.1	1.1	0.8	1.0	1.3	1.2	0.8	1.1	2.8	1.9
Pro	5.9	1.8	1.8	4.2	3.0	1.1	1.0	0.9	1.6	1.2
Gly	2.2	1.2	2.9	2.0	2.0	1.0	0.9	0.7	1.7	1.2
Ala	1.1		1.0	0.8	0.8	0.7	0.3	0.2		0.8
1/2 Cys						0.9				
Val								0.2		0.6
Tyr		0.8								
Lys	1.0			1.1		0.9		0.3		0.8
His			0.8							
Arg	1.1	1.2	1.0	0.8		2.0	0.9	1.8		0.7
PSer + PThr[c]	4	3	4	3	1	4	2	3	1	2
Approximate net charge (pH 7.5)[d]	−3	−5	−6	−2	−3	−3	−4	−4	−6	−4

[a] Tryptophan was not determined. Met, Ile, Leu, and Phe were not detected in these peptides.
[b] Threonine and serine values are uncorrected. The values represent the sum derived from phosphoamino acid breakdown (50 to 80% yield) plus any nonphosphorylated species.
[c] The sum of phosphoserine and phosphothreonine was calculated from the mobility at pH 1.8 and the number of basic residues.
[d] The net charge at pH 7.5 was calculated from the position of elution from a DEAE-Sephadex® column.

From Jones, C. E., Busch, H., and Olson, M. O. J., *Biochem. Biophys. Res. Commun.*, 90, 734, 1979.

sites for which sequences of surrounding amino acids are known[144,197] there is a high β turn potential.[200] This suggests that the sites of phosphorylation are in regions which loop out into space. This type of protein secondary structure also occurs at or near phosphorylation sites in proteins whose enzymatic activity is regulated by cyclic AMP-dependent protein kinase.[201] Thus, in nonhistone proteins regions of high negative charge density may be expected to loop out and be available for interaction with histones, or cations. The phosphorylation of these regions may regulate the level of charge density or the conformation of the region. The results of McCleary et al.[202] have already suggested that phosphorylation of nonhistone proteins is associated with binding to histones. Future research is required to determine the precise interactions of nonhistone proteins with other macromolecules and to ascertain whether phosphorylation regulates this interaction.

V. CONCLUSIONS

In the past decade, progress in the study of nonhistone phosphoproteins has been made in several areas:

1. Numerous phosphoproteins have been identified in various subcomponents of the nucleus, primarily by electrophoretic methods.
2. Changes in phosphorylation of nonhistone proteins have been correlated with functions of the nucleus, especially transcription.

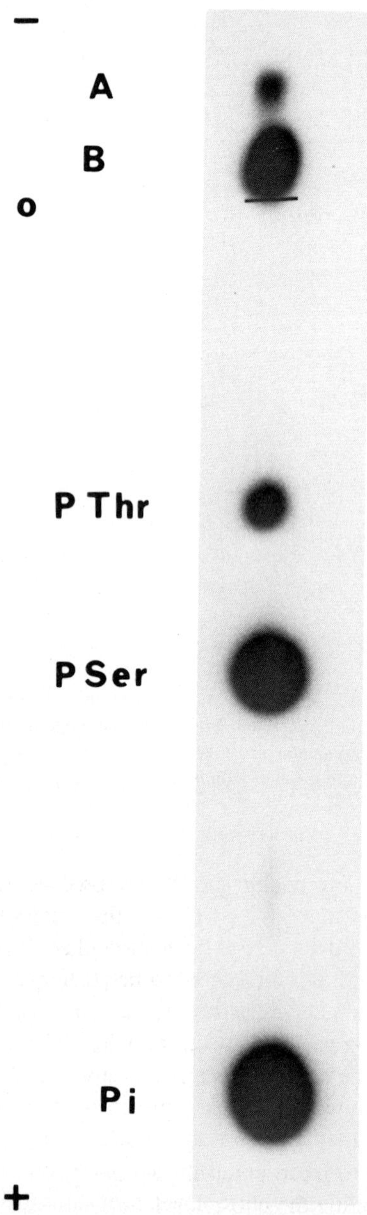

FIGURE 10. Autoradiograph of pH 1.8 paper electrophoretogram of partial acid hydrolysis (2 *N* HCl, 4 hr, 110°, *in vacuo*) products from acid-insoluble nonhistone chromatin proteins of Novikoff hepatoma. Electrophoresis was at 3000 V for 90 min in 2% formic acid, 8% acetic acid. (From Jones, C. E. and Olson, M. O. J., *Int. J. Peptide Protein Res.*, 6, 135, 1980. With permission.)

3. A dozen or more nuclear proteins have been isolated and partially characterized.
4. The sequences of amino acids surrounding the sites of phosphorylation have been determined in a few proteins.

In spite of the progress made, there is still no unified picture encompassing nonhistone

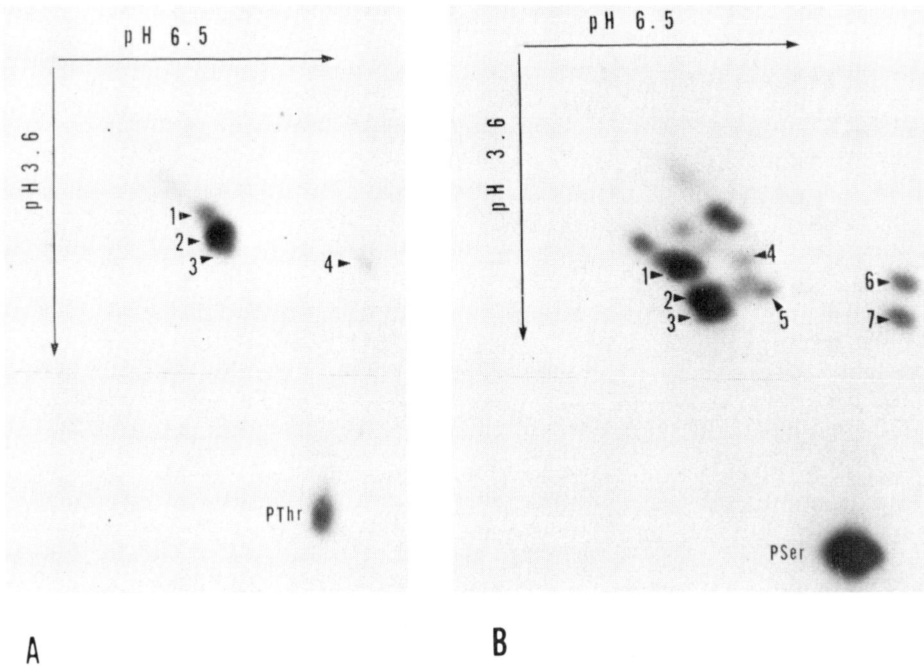

FIGURE 11. Autoradiograph of two-dimensional paper electrophoretograms of spot A (A) and spot B (B) isolated by preparative-level pH 1.8 paper electrophoresis of fraction I from a Dowex®-1 column. Radioactive phosphothreonine (A) or phosphoserine (B) were added as markers. (From Jones, C. E. and Olson, M. O. J., *Int. J. Peptide Protein Res.*, 6, 135, 1980. With permission.)

phosphoprotein function. In fact, the nuclear phosphoproteins are a diverse group of proteins which cover a broad range of functions. Although the evidence for involvement with transcription as proposed in the mid-1960s has been strengthened, additional roles have become apparent. In particular, phosphoproteins seem to be participants in the controlled assembly of various nuclear substructures such as nuclear membranes, nuclear and nucleolar RNP particles, and nucleosomes. In addition, several nuclear enzymes are phosphorylated: RNA polymerases, poly(A) polymerase, and histone deacetylase. Thus, the phosphoryl group is called upon to perform a multitude of tasks, some of them related to one another and others unrelated.

Future research will need to more precisely define the functions of individual phosphoproteins. The systems controlling the phosphorylation and dephosphorylation of the proteins also need to be elucidated. As the functions and the phosphorylation system are more clearly understood, structural information will become more important. Comparisons of sites of phosphorylation in different proteins should aid in understanding the specificity and complexity of the nuclear phosphoprotein kinases. Systems also need to be developed to study the effects of phosphorylation on the conformation of proteins and their interactions with other macromolecules.

The nucleus consists of a multitude of parts, like pieces in a giant three-dimensional puzzle. In the 1960s and 1970s a large number of the pieces, including numerous phosphoproteins, were found and classified. The task of the 1980s and the remainder of the century will be to determine how the pieces fit together or interact with each other. With continuing progress, the puzzle should be nearly assembled by the turn of the century, with a few pieces still missing.

ACKNOWLEDGMENTS

I thank Romie Brown for typing this manuscript. Part of the work reported in this chapter was supported by NIH grant GM28349.

REFERENCES

1. **Sanders, M. and Dixon, G. H.,** The biosynthesis of protamine in trout testis. IV. Sites of phosphorylation, *J. Biol. Chem.,* 247, 851, 1972.
2. **Daile, P., Carnegie, P. R., and Young, J. D.,** Synthetic substrate for cyclic AMP-dependent protein kinase, *Nature (London),* 257, 416, 1975.
3. **Olson, M. O. J.,** Metal ion effects on nuclear protein phosphorylation, in *Advances in Inorganic Biochemistry,* Vol. 3, Eichhorn, G. and Marzilli, L., Eds., Elsevier, New York, 1981, 167.
4. **Marshak, A. and Calvert, F.,** Specific activity of P^{32} in cell constituents of rabbit liver, *J. Cell. Comp. Physiol.,* 34, 451, 1949.
5. **Davidson, J. N., Frazer, S. C., and Hutchinson, W. C.,** Phosphorus compounds in the cell. 1. Protein-bound phosphorus fractions studies with the aid of radioactive phosphorus, *Biochem. J.,* 49, 311, 1951.
6. **Johnson, R. M. and Albert, S.,** Incorporation of P^{32} into ''phosphoprotein'' fraction of mammalian tissue, *J. Biol. Chem.,* 200, 335, 1953.
7. **Langan, T. A.,** A phosphoprotein preparation from liver nuclei and its effect on the inhibition of RNA synthesis by histones, in *Regulation of Nucleic Acid and Protein Synthesis,* Konigsberg, V. V. and Bosch, K., Eds., Elsevier, Amsterdam, 1967, 233.
8. **Kleinsmith, L. J., Allfrey, V. G., and Mirsky, A. E.,** Phosphoprotein metabolism in isolated lymphocyte nuclei, *Proc. Natl. Acad. Sci. U.S.A.,* 55, 1182, 1966.
9. **Kleinsmith, L. J. and Alfrey, U. G.,** Nuclear phosphoproteins. I. Isolation and characterization of a phosphoprotein fraction from calf thymus nuclei, *Biochim. Biophys. Acta,* 175, 123, 1969.
10. **LeStourgeon, W. M., Lothstein, L., Walker, B. W., and Beyer, A.,** The composition and topology of RNA and protein in monomer 40S ribonucleoprotein particles, in *The Cell Nucleus,* Vol. 9, Part B, Busch, H., Ed., Academic Press, New York, 1980.
11. **Olson, M. O. J., and Busch, H.,** Nucleolar proteins, in *Methods in Cell Biology,* Vol. 17, Stein, G., Stein, J., and Kleinsmith, L. J., Eds., Academic Press, New York, 1978, 163.
12. **Gerace, L. and Blobel, G.,** The nuclear envelope lamina is reversibly depolymerized during mitosis, *Cell,* 19, 277, 1980.
13. **Kleinsmith, L. J.,** Phosphorylation of nonhistone proteins, in *The Cell Nucleus,* Vol. 6, Busch, H., Ed., Academic Press, New York, 1978, 221.
14. **Jungman, R. A. and Kranias, E. G.,** Nuclear phosphoprotein kinases and regulation of gene transcription, *Int. J. Biochem.,* 8, 819, 1977.
15. **Ahmed, K. and Wilson, M. J.,** Chromatin controls in the prostate, in *The Cell Nucleus* Vol. 6, Busch, H., Ed., Academic Press, New York, 1978, 409.
16. **Bhorjee, J. S. and Pederson, T.,** Chromosomal proteins: tightly bound nucleic acid and its bearing on the measurement of nonhistone protein phosphorylation, *Anal. Biochem.,* 71, 393, 1976.
17. **Plimmer, R. H. A. and Bayliss, W. M.,** The separation of phosphorus from caseinogen by the action of enzymes and alkali, *J. Physiol. (London),* 33, 439, 1906.
18. **Schiltz, E. and Sekeris, C. E.,** Enzymatic phosphorylation of nuclear proteins by [γ-^{32}P]ATP in isolated rat liver nuclei, *H.-S. Z. Physiol. Chem.,* 350, 317, 1969.
19. **Chen, C-C., Bruegger, B. B., Halpern, R. M., Smith, R. A., and Smith, D. L.,** Occurrence and distribution of acid-labile histone phosphates in regenerating rat liver, *Biochemistry,* 13, 3785, 1974.
20. **Smith, D. L., Chen, C.-C., Bruegger, B. B., Noltz, S. L., Halpern, R. M., and Smith, R. A.,** Characterization of protein kinases forming acid-labile histone phosphates in Walker-256 carcinosarcoma cell nuclei, *Biochemistry,* 13, 3780, 1974.
21. **Olson, M. O. J., and Guetzow, K.,** Phosphoprotein phosphatase activity of Novikoff hepatoma nucleoli *Biochem. Biophys. Res. Commun.,* 70, 712, 1976.

22. **Pumo, D. E. and Kleinsmith, L. J.,** Methods for the assessment of nonhistone phosphorylation (acid-stable, alkali-labile linkages), in *Methods in Cell Biology,* Vol. 19, Stein, G., Stein, J., and Kleinsmith, L. J., Eds., Academic Press, New York, 1978, 119.
23. **Olson, M. O. J., Orrick, L. R., Jones, C. E., and Busch, H.,** Phosphorylation of acid-soluble nucleolar proteins of Novikoff hepatoma ascites cells *in vivo, J. Biol. Chem.,* 249, 2823, 1974.
24. **Schaffer, N. K.,** Analysis for O-phosphorylserine, *Methods Enzymol.,* 11, 702, 1967.
25. **Jones, C. E. and Olson, M. O. J.,** Phosphodipeptide analysis of nonhistone nuclear proteins from Novikoff hepatoma ascites cells, *Int. J. Peptide Protein Res.,* 16, 135, 1980.
26. **Hunter, T. and Sefton, B. M.,** The transforming gene product of Rous sarcoma virus phosphorylates tyrosine, *Proc. Natl. Acad. Sci. U.S.A.,* 77, 1311, 1980.
27. **Bylund, D. B. and Huang, T.-S.,** Decomposition of phosphoserine and phosphothreonine during acid hydrolysis, *Anal. Biochem.,* 73, 477, 1976.
28. **Richardson, W. S., Munksgaard, E. C., and Butler, W. T.,** Rat incisor phosphoprotein; the nature of the phosphate and quantitation of phosphoserine, *J. Biol. Chem.,* 253, 8042, 1978.
29. **Teng, C. S., Teng, C. T., and Allfrey, V. G.,** Studies of nuclear acidic proteins. Evidence for their phosphorylation, tissue specificity, selective binding to deoxyribonucleic acid and stimulatory effects on transcription, *J. Biol. Chem.,* 246, 3597, 1971.
30. **Platz, R. D., Kish, V. M., and Kleinsmith, L. J.,** Tissue specificity of non-histone chromatin phosphoproteins, *FEBS Lett.,* 12, 38, 1980.
31. **Rickwood, D., Riches, P. G., and MacGillivray, A. J.,** Studies of the *in vitro* phosphorylation of chromatin non-histone proteins in isolated nuclei, *Biochim. Biophys. Acta,* 299, 162, 1973.
32. **MacGillivray, A. J. and Rickwood, D.,** The heterogeneity of mouse-chromatin nonhistone proteins as evidenced by two-dimensional polyacrylamide gel electrophoresis and ion exchange chromatography, *Eur. J. Biochem.,* 41, 181, 1974.
33. **Platz, R. D. and Kleinsmith, L. J.,** Heterogeneity and specificity of non-histone nuclear phosphoproteins, *Comp. Biochem. Physiol.,* 55B, 9, 1976.
34. **LeStourgeon, W. M. and Rusch, H. P.,** Localization of nucleolar and chromatin residual acidic protein changes during differentiation in *Physarum polycephalum, Arch. Biochem. Biophys.,* 155, 144, 1973.
35. **Platz, R. D. and Hnilica, L. S.,** Phosphorylation of nonhistone chromatin proteins during sea urchin development, *Biochem. Biophys. Res Commun.,* 54, 222, 1973.
36. **Guguen-Guillouzo, C., Tichonicky, L., and Kruh, J.,** Hepatocyte chromosomal non-histone proteins in developing rat, *Eur. J. Biochem.,* 95, 235, 1979.
37. **Kleinsmith, L. J.,** Specific binding of phosphorylated non-histone chromatin proteins to deoxyribonucleic acid, *J. Biol. Chem.,* 248, 5648, 1973.
38. **Prestayko, A. W., Crane, P. M., and Busch, H.,** Phosphorylation and DNA binding of nuclear rat liver proteins soluble at low ionic strength, *Biochemistry,* 15, 414, 1976.
39. **Allfrey, V. G., Inoue, A., and Johnson, E. M.,** Use of DNA columns to separate and characterize nuclear nonhistone proteins, in *Chromosomal Proteins and Their Role in Regulation of Gene Expression,* Stein, G. S., Ed., Academic Press, New York, 1975, 265.
40. **Kostraba, N. C., Montagna, R. A., and Wang, T. Y.,** Study of the loosely bound nonhistone chromatin proteins. Stimulation of DNA templated by a specific DNA-binding fraction, *J. Biol. Chem.,* 250, 1548, 1975.
41. **Bluthmann, H.,** Specific binding of a nonhistone chromosomal protein from lymphocyte to DNA, *Eur. J. Biochem.,* 70, 233, 1976.
42. **Karn, J., Johnson, E. M., Vidali, G., and Allfrey, V. G.,** Differential phosphorylation and turnover of nuclear acidic proteins during the cell cycle of synchronized HeLa cells, *J. Biol. Chem.,* 249, 667, 1974.
43. **Platz, R. D., Stein, G. S., and Kleinsmith, L. J.,** Changes in the phosphorylation of non-histone chromatin proteins during the cell cycle of HeLa S_3 cells, *Biochem. Biophys. Res. Commun.,* 51, 735, 1973.
44. **Ezrailson, E. G., Olson, M. O. J., Guetzow, K. A., and Busch, H.,** Phosphorylation of non-histone chromatin proteins in normal and regenerating rat liver, Novikoff hepatoma and rat heart, *FEBS Lett.,* 62, 69, 1976.
45. **Thomson, J. A., Chiu, J.-F., Sakuma, K., and Hnilica, L. S.,** Nuclear proteins phosphokinases in normal and neoplastic tissues, *Cancer Res.,* 37, 3266, 1977.
46. **Brade, W. P., Thomson, J. A., Chiu, J.-F., and Hnilica, L. S.,** Chromatin-bound kinase activity and phosphorylation of chromatin non-histone proteins during early kidney regeneration after folic acid,. *Exp. Cell. Res.,* 84, 183, 1974.
47. **Liew, C. C. and Gornall, A. G.,** Covalent modification of nuclear proteins during aging, *Fed. Proc.,* 34, 186, 1975.
48. **Farron-Furstenthal, F.,** Protein kinases in hepatoma, and adult and fetal liver of the rat. I. Subcellular distribution, *Biochem. Biophys. Res. Commun.,* 67, 307, 1975.

49. **Lea, M. A. and Koch, M. R.,** Decreased incorporation of phosphate and amino-acids into low-solubility, nuclear proteins in rapidly growing hepatomas, *Int. J. Biochem.*, 8, 565, 1977.
50. **Thomson, J. A., Chiu, J.-F., and Hnilica, L. S.,** Nuclear phosphoprotein kinase activities in normal and neoplastic tissues, *Biochim. Biophys. Acta,* 407, 114, 1975.
51. **Ganpath, N., Prestayko, A. W., and Busch, H.,** Comparison of nuclear nonhistone phosphoproteins of rat liver and Novikoff hepatoma, *Cancer Res.*, 37, 1290, 1977.
52. **Ahmed, K. and Ishida, H.,** Effect of testosterone on nuclear phosphoproteins of rat ventral prostate, *Mol. Pharmacol.*, 7, 323, 1971.
53. **Ahmed, K. and Wilson, M. J.,** Chromatin-associated protein phosphokinases of rat ventral prostate. Characteristics and effects of androgenic status, *J. Biol. Chem.*, 250, 2370, 1975.
54. **Bottoms, G. D. and Jungmann, R. A.,** Effects of corticosterone on phosphorylation of rat liver nuclear proteins, *in vitro, Proc. Soc. Exp. Biol. Med.*, 144, 83, 1973.
55. **Allfrey, U. G., Johnson, E. M., Karn, J. and Vidali, G.,** Phosphorylation of nuclear proteins at times of gene activation, in *Protein Phosphorylation in Control Mechanisms,* Huijing, E. and Lee, E. Y. C., Eds., Academic Press,New York, 1973, 217.
56. **Liew, C. C., Suria, D., and Gornall, A. G.,** Effects of aldosterone on acetylation and phosphorylation of chromosomal proteins, *Endocrinology,* 93, 1025, 1973.
57. **Cohen, M. D. and Kleinsmith, L. J.,** Stimulation of uterine nonhistone protein phosphorylation and nuclear protein kinase activity by estradiol-17 beta, *Biochim. Biophys. Acta,* 435, 159, 1976.
58. **Keller, R. K., Chandra, T., Schrader, W. T., and O'Malley, B. W.,** Protein kinases of the chick oviduct: A study of the cytoplasmic and nuclear enzymes, *Biochemistry,* 15, 1958, 1976.
59. **Johnson, E. M. and Allfrey, V. G.,** Differential effects of cyclic adenosine-3,5′-monophosphate on phosphorylation of rat liver nuclear acidic proteins, *Arch. Biochem. Biophys.*, 152, 786, 1972.
60. **Johnson, E. M. and Hadden, J. W.,** Phosphorylation of lymphocyte nuclear acidic proteins: Regulation by cyclic nucleotides, *Science,* 187, 1198, 1975.
61. **Chiu, J.-F., Craddock, C., Getz, S., and Hnilica, L. S.,** Nonhistone chromatin protein phosphorylation during azo-dye carcinogenesis, *FEBS Lett.*, 33, 247, 1973.
62. **Chiu, J.-F., Brade, W. P., Thomson, J., Tsai, Y. H., and Hnilica, L. S.,** Non-histone protein phosphorylation in normal and neoplastic rat liver chromatin, *Exp. Cell. Res.*, 91, 200, 1975.
63. **Johnson, E. M., Karn, J., and Allfrey, V. G.,** Early nuclear events in the induction of lymphocytes proliferation by mitogens, *J. Biol. Chem.*, 249, 4990, 1974.
64. **Ishida, H. and Ahmed, K.,** Studies on phosphoproteins of submandibular gland nuclei isolated from isoproterenol-treated rats, *Exp. Cell Res.*, 78, 31, 1973.
65. **Ishida, H. and Ahmed, K.,** Studies on chromatin-associated protein phosphokinase of submandibular gland from isoproterenol-treated rats, *Exp. Cell Res.*, 84, 127, 1974.
66. **Brade, W. P., Chiu, J.-F., and Hnilica, L. S.,** Phosphorylation of rat liver nuclear acidic phosphoproteins after administration of α-1,2,3,4,5,6-hexachlorocyclohexane *in vivo, Mol. Pharmacol.*, 10, 398, 1974.
67. **Blankenship, J. and Bresnick, E.,** Effects of phenobarbital on the phosphorylation of acidic nuclear proteins of rat liver, *Biochim. Biophys. Acta,* 340, 218, 1974.
68. **Miki, K., Ishida, H., and Yamamoto, I.,** Effect of cytarabine and bleomycin on the phosphorylation of nuclear acidic protein, *FEBS Lett.*, 69, 249, 1976.
69. **Kuehn, G. D., Affolter, H.-U., Atmar, V. J., Seebeck, T., Gubler, U., and Braun, R.,** Polyamine-mediated phosphorylation of a nucleolar protein from *Physarum polycephalum* that stimulates rRNA synthesis, *Proc. Natl. Acad. Sci. U.S.A.*, 76, 2541, 1979.
70. **Bowen, B., Steinberg, J., Laemmli, U. K., and Weintraub, H.,** The detection of DNA-binding proteins by protein blotting, *Nucleic Acids Res.*, 8, 1, 1980.
71. **Towbin, H., Staehelin, T., and Gordon, J.,** Electrophoretic transfer of proteins from polyacrylamide gels to nitrocellulose sheets: Procedure and some applications, *Proc. Natl. Acad. Sci. U.S.A.*, 76, 4350, 1979.
72. **Farber, J., Stein, G., and Baserga, R.,** The regulation of RNA synthesis during mitosis, *Biochem. Biophys. Res. Commun.*, 47, 790, 1972.
73. **Pfeiffer, S. E. and Tolmack, L. J.,** RNA synthesis in synchronously growing populations of HeLa S_3 cells, *J. Cell. Physiol.*, 71, 77, 1968.
74. **Gallwitz, D. and Mueller, G. C.,** Histone synthesis *in vitro* on HeLa cell microsomes. The nature of the coupling to deoxyribonucleic acid synthesis, *J. Biol. Chem., 244,* 5947, 1969.
75. **Breindl, M. and Gallwitz, D.,** Identification of histone messenger RNA from HeLa cells. Appearance of histone mRNA in the cytoplasm and its translation in rabbit reticulocyte cell-free system, *Eur. J. Biochem.*, 32, 381, 1973.
76. **Robbins, E. and Borun, T. W.,** The cytoplasmic synthesis of histones in HeLa cells and its temporal relationship to DNA replication, *Proc. Natl. Acad. Sci. U.S.A.*, 57, 409, 1967.

77. **Klyzejko-Stefanowicz, L., Chiu, J. F., Tsai, P.-H., and Hnilica, L. S.,** Acceptor proteins in rat androgenic tissue chromatin, *Proc. Natl. Acad. Sci. U.S.A.,* 73, 1954, 1976.
78. **Nielsen, C. J., Sando, J. J., and Pratt, W. B.,** Evidence that dephosphorylation inactivates glucocorticoid receptors, *Proc. Natl. Acad. Sci. U.S.A.,* 74, 1398, 1977.
79. **Kamiyama, M., Dastugue, B., and Kruh, J.,** Action of phosphoproteins and protein kinase from rat liver chromatin on RNA synthesis, *Biochem. Biophys. Res. Commun.,* 49, 1345, 1971.
80. **Kamiyama, M., Dastugue, B., Defer, N., and Kruh, J.,** Liver chromatin non-histone proteins. Partial fractionation and mechanism of action on RNA synthesis, *Biochim. Biophys. Acta,* 277, 576, 1972.
81. **Shea, M. and Kleinsmith, L. J.,** Template-specific stimulation of RNA synthesis by phosphorylated non-histone chromatin proteins, *Biochem. Biophys. Res Commun.,* 50, 473, 1973.
82. **Kostraba, N. C., Montagna, R. A., and Wang, T. Y.,** Mode of action of non-histone proteins in the simulation of transcription from DNA, *Biochem. Biophys. Res. Commun.,* 72, 334, 1976.
83. **Kostraba, N. C. and Wang, T. Y.,** Transcriptional transformation of Walker tumor chromatin by non-histone proteins, *Cancer Res.,* 32, 2348, 1972.
84. **Thompson, J. A., Stein, J. L., Kleinsmith, L. J., and Stein, G. S.,** Activation of histone gene transcription by nonhistone chromsomal phosphoproteins, *Science,* 194, 428, 1976.
85. **Kleinsmith, L. J., Stein, J., and Stein, G.,** Dephosphorylation of nonhistone proteins specifically alters the pattern of gene transcription in reconstituted chromatin, *Proc. Natl. Acad. Sci. U.S.A.,* 73, 1174, 1976.
86. **Goodwin, G. H., Sanders, C., and Johns, E. W.,** A new group of chromatin-associated proteins with a high content of acidic and basic amino acids, *Eur. J. Biochem.,* 38, 14, 1973.
87. **Goodwin, G. H., Woodhead, L., and Johns, E. W.,** The presence of high mobility group non-histone chromatin proteins in isolated nucleosomes, *FEBS Lett.,* 73, 85, 1977.
88. **Levy, W. B., Connor, W., and Dixon, G. H.,** A subset of trout testis nucleosomes enriched in transcribed DNA sequences contains high mobility group proteins as major structural components, *J. Biol. Chem.,* 254, 609, 1979.
89. **Weisbrod, S. and Weintraub, H.,** Isolation of a subclass of nuclear proteins responsible for conferring a DNase I-sensitive structure on globin chromatin, *Proc. Natl. Acad. Sci. U.S.A.,* 76, 630, 1979.
90. **Weisbrod, S., Groudine, M., and Weintraub, H.,** Interaction of HMG 14 and 17 with actively transcribed genes, *Cell,* 19, 289, 1980.
91. **Goodwin, G. H., Walker, J. M., and Johns, E. W.** The high mobility group (HMG) non-histone chromosomal proteins, in *The Cell Nucleus,* Vol. 6, Busch, H., Ed., Academic Press, New York, 1978.
92. **Walker, J. M., Goodwin, G. H., Johns, E. W., Wietzes, P., and Gasstra, W.,** A comparison of the amino-terminal sequences of two calf-thymus chromatin non-histone proteins, *Int. J. Peptide Protein Res.,* 9, 220, 1977.
93. **Walker, J. M., Parker, B. P., and Johns, E. W.,** Isolation and partial sequence of the cyanogen bromide peptides from calf thymus non-histone chromosomal protein HMG-1, *Int. J. Peptide Protein Res.,* 12, 269, 1978.
94. **Walker, J. M., Hastings, J. R. B., and Johns, E. W.,** The primary structure of a non-histone chromosomal protein, *Eur. J. Biochem.,* 76, 461, 1977.
95. **Walker, J. M., Goodwin, G. H., and Johns, E. W.,** The primary structure of the nucleosome-associated chromosomal protein HMG-1, *FEBS Lett.,* 100, 394, 1979.
96. **Sterner, R., Vidali, G., Heinrikson, R. L., and Allfrey, V. G.,** Postsynthetic modification of high mobility group proteins. Evidence that high mobility group proteins are acetylated, *J. Biol. Chem.,* 253, 7601, 1978.
97. **Boffa, L. C., Sterner, R., Vidali, G., and Allfrey, V. G.,** Post-synthetic modifications of nuclear proteins. High mobility group proteins are methylated, *Biochem. Biophys. Res. Commun.,* 89, 1322, 1979.
98. **Saffer, J. D. and Glazer, R. I.,** The phosphorylation of high mobility group proteins 14 and 17 from Ehrlich ascites and L-1210 cells *in vitro, Biochem. Biophys. Res. Commun.,* 93, 1280, 1980.
99. **Arfmann, H.-A., Haase, E., and Schröter, J.,** High mobility group proteins from CHO cells. Their characterization and modifications during cell cycle, *Biochem. Biophys. Res. Commun.,* 101, 137, 1981.
100. **Inoue, A., Tei, Y., Hasuma, T., Yukioka, M., and Morsawa, S.,** Phosphorylation of HMG 17 by protein kinase NII from rat liver cell nuclei, *FEBS Lett.,* 117, 68, 1980.
101. **Arfmann, H.-A. and Baydoun, H.,** Preferential phosphorylation of HMG protein 17 *in vitro* by a nuclear protein kinase, *Z. Naturforsch.,* 36C, 319, 1981.
102. **Sun, Y.-C. S., Johnson, E. M., and Allfrey, V. G.,** Affinity purification of newly phosphorylated protein molecules, *J. Biol. Chem.,* 255, 742, 1980.
103. **Liew, C. C. and Chan, P. K.,** Identification of nonhistone chromatin proteins in chromatin subunits, *Proc. Natl. Acad. Sci. U.S.A.,* 73, 3458, 1976.
104. **Chan, P. K. and Liew, C. C.,** Identification of specific phosphoproteins in nuclease-digested chromatin subunits, *Can. J. Biochem.,* 55, 847, 1977.

105. **Chan, P.-K. and Liew, C.-C.,** Purification of a phosphoprotein from chromatin of rat liver, *Biochem. J.*, 183, 147, 1979.
106. **Bohm, J., Keil, G., and Knippers, R.,** Studies on protein-phosphorylation reactions in isolated chromatin, *Eur. J. Biochem.*, 78, 251, 1977.
107. **Defer, N., Kitzis, A., Levy, F., Tichonicky L., Sabatier, M.-M., and Kruh, J.,** Presence of non-histone proteins in nucleosomes, *Eur. J. Biochem.*, 88, 583, 1978.
108. **Kitzis, A., Tichonicky, L., Defer, N., and Kruh, J.,** Localization of phosphoproteins and of protein kinases in chromatin from hepatoma tissue-cultured cells, *Eur. J. Biochem.*, 111, 237, 1980.
109. **Saffer, J. D. and Coleman, J. E.,** Reversible phosphorylation of a nucleosome binding protein that stimulates transcription of nucleosome and deoxyribonucleic acid, *Biochemistry,* 19, 5874, 1980.
110. **MacGillivray, A. J., Johnston, C., MacFarlane, R., and Rickwood, D.,** The isolation and partial characterization of low-molecular-weight phosphorylated component of the non-histone proteins of mouse nuclei, *Biochem. J.*, 175, 35, 1978.
111. **Rickwood, D. and MacGillivray, A. J.,** Improved techniques for the fractionation of non-histone proteins of chromatin on hydroxyapatite, *Eur. J. Biochem.*, 51, 593, 1975.
112. **Grummt, I. and Grummt, F.,** Studies on the phosphorylation of nucleolar proteins, *FEBS Lett.*, 39, 129, 1974.
113. **Offenbacher, S. and Klein, E. S.,** Differential phosphorylation of rat liver nuclear non-histone proteins, *in vitro, Biochem. Biophys. Res. Commun.*, 66, 375, 1975.
114. **Pumo, D. E., Stein, G. S., and Kleinsmith, L. J.,** Stimulated phosphorylation of non-histone phosphoproteins in SV-40 transformed WI-38 human diploid fibroblasts, *Biochim. Biophys. Acta,* 402, 125, 1975.
115. **Kostraba, N. C. and Wang, T. Y.,** Inhibition of transcription *in vitro* by a non-histone protein isolated from Ehrlich ascites tumor chromatin, *J. Biol. Chem.*, 250, 8938, 1975.
116. **Kostraba, N. C., Loor, R. M., and Wang, T. Y.,** Tissue-specificity of the nonhistone protein that inhibits RNA synthesis, *in vitro, Biochem. Biophys. Res. Commun.*, 79, 347, 1977.
117. **Ohtsuki, K., Shiraishi, H., Yamada, E., Nakamura, M., and Ishida, N.,** A non-histone chromatin protein that is a specific phosphate acceptor of nuclear cAMP-independent protein kinase from mouse spleen cells, *J. Biol. Chem.*, 255, 2391, 1980.
118. **Goldknopf, I. L., Wilson, G., Ballal, N. R., and Busch, H.,** Chromatin conjugate protein A24 is cleaved and ubiquitin is lost during chicken erythropoiesis, *J. Biol. Chem.*, 255, 10555, 1980.
119. **Busch, H. and Smetana, K.,** *The Nucleolus,* Academic Press, New York, 1970.
120. **Kumar, A. and Warner, J. R.,** Characterization of ribosomal precursor particles from HeLa cell nucleoli, *J. Mol. Biol.*, 63, 233, 1972.
121. **Orrick, L. R., Olson, M. O. J., and Busch, H.,** Comparison of nucleolar proteins of normal rat liver and Novikoff hepatoma ascites cells by two-dimensional polyacrylamide gel electrophoresis, *Proc. Natl. Acad. Sci. U.S.A.*, 70, 1316, 1973.
122. **Kang, Y. J., Olson, M. O. J., and Busch, H.,** Phosphorylation of acid-soluble proteins in isolated nucleoli of Novikoff hepatoma ascites cells. Effects of divalent cations, *J. Biol. Chem.*, 249, 5580, 1974.
123. **Tsurugi, K., Morita, T., and Ogata, K.,** Identification and metabolic relationship between proteins of nucleolar 60-S particles and of ribosomal large subunits of rat liver by means of two-dimensional disc electrophoresis, *Eur. J. Biochem.*, 32, 555, 1973.
124. **Prestayko, A. W., Klomp, G. R., Schmoll, D. J., and Busch, H.,** Comparison of proteins of ribosomal subunits and nucleolar preribosomal particles from Novikoff hepatoma ascites cells by two-dimensional polyacrylamide gel electrophoresis, *Biochemistry,* 13, 1945, 1974.
125. **Olson, M. O. J., Prestayko, A. W., Jones, C. E., and Busch, H.,** Phosphorylation of proteins of ribosomes and nucleolar preribosomal particles from Novikoff hepatoma ascites cells, *J. Mol. Biol.*, 90, 161, 1974.
126. **Olson, M. O. J., Ezrailson, E. G., Guetzow, K., and Busch, H.,** Localization and phosphorylation of nuclear, nucleolar and extranucleolar nonhistone proteins of Novikoff hepatoma ascites cells, *J. Mol. Biol.*, 97, 611, 1975.
127. **Mamrack, M. D.,** Nucleolar Nonhistone Protein Phosphorylation: Sites of Phosphorylation within Highly Acidic Regions, Ph.D. thesis, Baylor College of Medicine, Houston, 1978.
128. **Kawashima, K. and Izawa, M.,** A nucleolus-specific phosphoprotein in mouse ascites tumor cells, *Mol. Biol. Rep.*, 3, 113, 1976.
129. **Tsutsui, K. and Oda, T.,** Isolation and characterization of a high molecular weight acid soluble nuclear protein from mouse ascites-sarcoma cells, *Eur. J. Biochem.*, 108, 497, 1980.
130. **Atmar, V. J., Daniels, G. R., and Kuehn, G. D.,** Polyamine stimulation of phosphorylation of nonhistone acidic protein in nuclei and nucleoli from *Physarum polycephalum, Eur. J. Biochem.*, 90, 29, 1978.
131. **Kang, Y.-J., Olson, M. O. J., Jones, C., and Busch, H.,** Nucleolar phosphoproteins of normal rat liver and Novikoff hepatoma ascites cells, *Cancer Res.*, 35, 1470, 1975.

132. **Ballal, N. R., Kang, Y.-J., Olson, M. O. J., and Busch, H.,** Changes in nucleolar proteins and their phosphorylation patterns during liver regeneration, *J. Biol. Chem.*, 250, 5921, 1975.
133. **Olson, M. O. J., Hatchett, S., Allan, R., Hawkins, T. C., and Busch, H.,** Nucleolus-associated phosphoprotein kinases of normal and regenerating liver and Novikoff hepatoma ascites cells, *Cancer Res.*, 38, 3421, 1978.
134. **Yu, F.-L.,** Increased levels of rat hepatic nuclear free and engaged RNA polymerase levels during liver regeneration, *Biochem. Biophys. Res. Commun.*, 64, 1107, 1975.
135. **Kawashima, K. and Izawa, M.,** Phosphorylation of a nucleolus-specific phosphoprotein, *in vitro, Biochem. Biophys. Res. Commun.*, 74, 265, 1977.
136. **Gamo, S. and Lindell, T. J.,** Presence of two protein kinases in highly purified rat liver nucleoli, *Life Sci.*, 15, 2179, 1975.
137. **Sakuma, K., Fujitani, H., Chium, J.-F., and Hnilica, L S.,** Nuclear and nucleolar protein phosphokinases in Novikoff hepatoma, *Int. J. Biochem.*, 9, 127, 1978.
138. **Grummt, I.,** Studies on the phosphorylation of nucleolar proteins, *FEBS Lett.*, 39, 125, 1974.
139. **Wilson, M. J. and Ahmed, K.,** Localization of protein phosphokinase activities in the nucleolus distinct from extranucleolar regions in rat ventral prostate nuclei, *Exp. Cell. Res.*, 93, 261, 1975.
140. **Daniels, G. R., Atmar, V. J., and Kuehn, G. D.,** A polyamine-activated protein kinase reaction from nuclei and nucleoli of *Physarum polycephalum* which phosphorylates a unique 70,000-Mr nonhistone protein, *Biochemistry*, 20, 2525, 1981.
141. **Olson, M. O. J. and Guetzow, K.,** Phosphoprotein phosphatase activity of Novikoff hepatoma nucleoli, *Biochem. Biophys. Res. Commun.*, 70, 717, 1976.
142. **Olson, M. O. J. and Guetzow, K. A.,** Nucleolar phosphoprotein phosphatase from Novikoff hepatoma and rat liver: characterization and partial purification, *Biochim. Biophys. Acta*, 526, 174, 1978.
143. **Mamrack, M. D., Olson, M. O. J., and Busch, H.,** Negatively charged phosphopeptides of nucleolar nonhistone proteins from Novikoff hepatoma, *Biochem. Biophys. Res. Commun.*, 76, 150, 1977.
144. **Mamrack, M. D., Olson, M. O. J., and Busch, H.,** Amino acid sequence and sites of phosphorylation in a highly acidic region of nucleolar nonhistone protein C23, *Biochemistry*, 18, 3381, 1979.
145. **Lischwe, M. A., Smetana, K., Olson, M. O. J., and Busch, H.,** Proteins C23 and B23 are the major nucleolar silver staining proteins, *Life Sci.*, 25, 701, 1979.
146. **Jones, C. E.,** Nuclear and Nucleolar Nonhistone Protein Phosphorylation Sites, Ph.D. thesis, Baylor College of Medicine, Houston, 1980.
147. **Olson, M. O. J., Guetzow, K., Allan, R., Mamrack, M., and Busch, H.,** Specificity of localization, degradation and distribution of acidic regions of nucleolar phosphoprotein C23, *Fed. Proc.*, 38, 493, 1979.
148. **Howell, W. M., Denton, T. E., and Diamond, J. R.,** Differential staining of the satellite regions of human acrocentric chromosomes, *Experientia*, 31, 260, 1975.
149. **Goodpasture, C. and Bloom, S. E.,** Visualization of nucleolar organizer regions in mammalian chromosomes using silver staining, *Chromosoma*, 53, 37, 1975.
150. **Dwyer, N. and Blobel, G.,** A modified procedure for isolation of a pore complex-lamina fraction from rat liver nuclei, *J. Cell Biol.*, 70, 581, 1976.
151. **Gerace, L., Blum, A., and Blobel, G.,** Immunocytochemical localization of the major polypeptides of the nuclear pore complex-lamina fraction. Interphase and mitotic distribution, *J. Cell Biol.*, 79, 546, 1978.
152. **Lam, K. S. and Kasper, C. B.,** Selective phosphorylation of a nuclear envelope polypeptide by an endogenous protein kinase, *Biochemistry*, 18, 307, 1979.
153. **Steer, R. C., Wilson, M. J., and Ahmed, K.,** Protein phosphokinase activity of rat liver nuclear membrane, *Exp. Cell. Res.*, 119, 403, 1979.
154. **Steer, R. C., Goueli, S. A., Wilson, M. J., and Ahmed, K.,** Cobalt-stimulated protein phosphokinase activity of the pore complex-lamina fraction from rat liver nuclear envelope, *Biochem. Biophys. Res. Commun.*, 92, 191, 1980.
155. **Berezney, R. and Coffey, D. S.,** The nuclear protein matrix: isolation, structure and functions, in *Advances in Enzyme Regulation*, Vol. 14, Weber, G., Ed., Pergamon Press, New York, 1977, 63.
156. **Pardoll, D. M., Vogelstein, B., and Coffey, D. S.,** A fixed site of DNA replication in eucaryotic cells, *Cell*, 19, 527, 1980.
157. **Allen, S. L., Berezney, R., and Coffey, D. S.,** Phosphorylation of nuclear matrix proteins in isolated regenerating rat liver nuclei, *Biochem. Biophys. Res. Commun.*, 75, 111, 1977.
158. **Gallinaro-Matringe, H., Stevenin, J., and Jacob, M.,** Salt dissociation of nuclear particles containing DNA-like RNA. Distribution of phosphorylated and nonphosphorylated species, *Biochemistry*, 14, 2547, 1975.
159. **Beyer, A. L., Chistenson, M. E., Walker, B. W., and Le Stourgeon, W. M.,** Identification and characterization of the packaging proteins of core 40S hnRNP particles, *Cell*, 11, 127, 1977.
160. **Suria, D. and Liew, C. C.,** Characterization of proteins associated with nuclear ribonucleoprotein particles to two-dimensional polyacrylamide gel electrophoresis, *Can. J. Biochem.*, 57, 32, 1979.

161. **Galinaro-Matringe, H. and Jacob, M.,** Nuclear particles from rat brain: complexity of the major proteins and their phosphorylation, *in vivo, FEBS Lett.*, 36, 105, 1973.
162. **Blanchard, J. M., Brunel, C., and Jeanteur, P.,** Phosphorylation *in vivo* of proteins associated with heterogeneous nuclear RNA in HeLa cell nuclei, *Eur. J. Biochem.*, 86, 301, 1978.
163. **Blanchard, J. M., Ducamp, C., and Jeanteur, P.,** Endogenous protein kinase activity in nuclear RNA particles from HeLa cells, *Nature (London)*, 253, 467, 1975.
164. **Blanchard, J. M., Brunel, C., and Jeanteur, P.,** Characterization of an endogeneous protein kinase activity in ribonucleoprotein structure containing heterogeneous nuclear RNA in HeLa cell nuclei, *Eur. J. Biochem.*, 79, 117, 1977.
165. **Periasamy, M., Brunel, C., Blanchard, J. M., and Jeanteur, P.,** Characterization of a phosphoprotein phosphatase activity in ribonucleoprotein particles from HeLa cell nuclei, *Biochem. Biophys. Res. Commun.*, 79, 1077, 1977.
166. **Schweiger, A. and Schmidt, D.,** Isolation of RNA-binding proteins from rat liver nuclear 30S particles, *FEBS Lett.*, 41, 17, 1974.
167. **Friedman, D.,** personal communication.
168. **Brunel, C. and Lelay, M.-N.,** Two dimensional analysis of proteins associated with heterogeneous nuclear RNA in various animal cell lines, *Eur. J. Biochem.*, 99, 273, 1979.
169. **Raj, N. B. K., Ro-Choi, T. S., and Busch, H.,** Nuclear ribonucleoprotein complexes containing U_1 and U_2 RNA, *Biochemistry*, 14, 4380, 1975.
170. **Bell, G. I., Valenzuela, P., and Rutter, W. J.,** Phosphorylation of yeast RNA polymerases, *Nature (London)*, 261, 429, 1976.
171. **Bell, G. I., Valenzuela, P., and Rutter, W.J.,** Phosphorylation of yeast DNA-dependent RNA polymerases *in vivo* and *in vitro, J. Biol. Chem.*, 252, 3082, 1977.
172. **Buhler, J. M., Iborra, F., Sentenac, A., and Fromageot, P.,** The presence of phosphorylated subunits in yeast RNA polymerases A and B, *FEBS Lett.*, 71, 37, 1976.
173. **Hirsch, J. and Martelo, O. J.,** Phosphorylation of rat liver ribonucleic acid polymerase I by nuclear protein kinases, *J. Biol. Chem.*, 251, 5408, 1976.
174. **Kranias, E. G., Schweppe, J., and Jungman, R. A.,** Phosphorylative and functional modifications of nucleoplasmic RNA polymerase II by homologous adenosine 3′,5′-monophosphate-dependent protein kinase from calf thymus and by heterologous phosphatase, *J. Biol. Chem.*, 252, 6750, 1977.
175. **Kranias, E. G. and Jungman, R. A.,** Purification and properties of a cyclic AMP-independent protein kinase from calf thymus nuclei, *Biochim. Biophys. Acta*, 517, 447, 1978.
176. **Kranias, E. G. and Jungman, R. A.,** Phosphorylation of calf thymus RNA polymerase II by nuclear cyclic 3′,5′-AMP-independent protein kinase, *Biochim. Biophys. Acta*, 517, 439, 1978.
177. **Jungmann, R. A., Hiestand, P. C., and Schweppe, J. S.,** Adenosine 3′,5′-monophosphate-dependent protein kinase and the stimulation of ovarian nuclear ribonucleic acid polymerase activities, *J. Biol. Chem.*, 249, 3444, 1974.
178. **Martelo, O. J. and Hirsch, J.,** Effect of nuclear protein kinases on mammalian RNA synthesis, *Biochem. Biophys. Res. Commun.*, 58, 1008, 1974.
179. **Dahmus, M. E.,** Stimulation of ascites tumor RNA polymerase II by protein kinase, *Biochemistry*, 15, 1821, 1976.
180. **Mazus, B., Szurmak, B., and Buchowicz, J.,** Phosphorylation *in vitro* and *in vivo* of the wheat embryo RNA polymerase II, *Acta Biochim. Polonica*, 27, 9, 1980.
181. **Rose, K. M., Bell, L. E., and Jacob, S. T.,** Specific inhibition of chromatin-associated poly(A) synthesis *in vitro* by cordycepin 5′-triphosphate, *Nature (London)*, 267, 178, 1977.
182. **Rose, K. M., Roe, F. J., and Jacob, S. T.** Two functional states of poly(adenylic acid) polymerase in isolated nuclei, *Biochim. Biophys. Acta*, 478, 180, 1977.
183. **Rose, K. M. and Jacob, S. T.,** Phosphorylation of nuclear poly(A) polymerase, *J. Biol. Chem.*, 254, 10256, 1979.
184. **Rose, K. M. and Jacob, S. T.,** Phosphorylation of nuclear poly (adenylic acid) polymerase by protein kinase: mechanism of enhanced poly (adenylic acid) synthesis, *Biochemistry*, 19, 1472, 1980.
185. **Rose, K. M., Jacob, S. T., and Kumar, A.,** Poly(A) polymerase and poly(A)-specific mRNA binding protein are antigenically related, *Nature (London)*, 279, 260, 1979.
186. **Vidali, G., Boffa, L. G., and Allfrey, V. G.,** Properties of an acidic histone-binding protein fraction from cell nuclei, *J. Biol. Chem.*, 247, 7365, 1972.
187. **Laskey, R. A., Honda, B. M., Mills, A. D., and Finch, J. T.,** Nucleosomes are assembled by an acidic protein which binds histones and transfers them to DNA, *Nature (London)*, 275, 416, 1978.
188. **Krohne, G. and Franke, W. W.,** Immunological identification and localization of the predominant nuclear protein of amphibian oocyte nucleus, *Proc. Natl. Acad. Sci. U.S.A.*, 77, 1034, 1980.
189. **Krohne, G. and Franke, W. W.,** A soluble acidic protein located in nuclei of diverse vertebrate species, *Exp. Cell. Res.*, 129, 167, 1980.

190. **Mills, A. D., Laskey, R. A., Black, P., and DeRobertis, E. M.,** An acidic protein which assembles nucleosomes *in vitro* is the most abundant protein in *Xenopus* oocyte nuclei, *J. Mol. Biol.,* 139, 561, 1980.
191. **Laskey, R. A. and Earnshaw, W. C.,** Nucleosome assembly, *Nature (London),* 286, 763, 1980.
192. **Earnshaw, W. C., Honda, B. M., Laskey, R. A., and Thomas, J. O.,** Assembly of nucleosomes: the reaction involving *X. laevis* nucleoplasmin, *Cell,* 21, 373, 1980.
193. **Mercier, J.-C., Grosclaude, F., and Ribadeau-Dumas, B.,** Structure primaire de la caseine α1-bovine-sequence complete, *Eur. J. Biochem.,* 23, 41, 1971.
194. **Matsuo, M., Huang, C.-H., and Huang, L. C.,** Evidence for an essential arginine recognition site on adenosine 3′-5′-cyclic monophosphate-dependent protein kinase of rabbit skeletal muscle, *Biochem. J.,* 173, 441, 1978.
195. **Williams, R. E.,** Phosphorylated sites in substrates of intracellular protein kinases: a common feature in amino acid sequences, *Sciences,* 192, 473, 1976.
196. **Brignon, G., Richadeau-Dumas, B., Mercier, J. C., Pelissier, J. P., and Das, B. C.,** Complete amino acid sequence of bovine alpha S2-casein, *FEBS Lett.,* 76, 274, 1977.
197. **Jones, C. E., Busch, H., and Olson, M. O. J.,** Sequence of a phosphorylation site in nucleolar protein B23, *Biochim. Biophys. Acta,* 667, 209, 1981.
198. **Jones, C. E., Busch, H., and Olson, M. O. J.,** A survey of peptides containing sites of phosphorylation in nonhistone nuclear proteins, *Biochem. Biophys. Res. Commun.,* 90, 734, 1979.
199. **Jones, C. E. and Olson, M. O. J.,** Phosphodipeptide analysis of nonhistone nuclear proteins from Novikoff hepatoma ascites cells, *Int. J. Peptide Protein Res.,* 135, 1980.
200. **Chou, P. Y. and Fasman, G. D.,** Beta-turns in proteins, *J. Mol. Biol.,* 115, 135, 1977.
201. **Small, D., Chou, P. Y., and Fasman, G. D.,** Occurrence of phosphorylated residues in predicted beta-turns: implications for beta-turn participation in control mechanisms, *Biochem. Biophys. Res. Commun.,* 79, 341, 1977.
202. **McCleary, A. P., Nooden, L. D., and Kleinsmith, L. J.,** Histone affinity chromatography as a tool for fractionating nonhistone chromatin proteins and studying histone-nonhistone protein interactions, *J. Biol. Chem.,* 253, 5199, 1978.

Chapter 4

NUCLEAR PROTEIN KINASES

Samuel J. Mitchell* and Lewis J. Kleinsmith

TABLE OF CONTENTS

* Samuel J. Mitchell passed away on November 17, 1980; this chapter, which was completed from his notes shortly afterward, is dedicated to his memory.

I. INTRODUCTION

Over the past decade protein phosphorylation has gradually come to be recognized as a widely occurring posttranslational mechanism for modifying the structure and activity of protein molecules. Among the large number of proteins now known to be subject to modification by phosphorylation-dephosphorylation reactions are enzymes (e.g., glycogen phosphorylase, phosphorylase kinase, glycogen synthase, pyruvate kinase, pyruvate dehydrogenase, and hormone-sensitive lipase to name but a few), ribosomal proteins, protein synthesis factors, membrane proteins, contractile proteins, microtubular proteins, mitochondrial proteins, viral proteins, and nuclear proteins (for reviews, see References 1 to 9). The functional significance of protein phosphorylation-dephosphorylation has been most clearly established for enzymes, where such modifications serve to either increase or decrease catalytic activity, depending on the enzyme in question. By analogy, it has been generally assumed that phosphorylation serves to regulate the activities of the other cellular proteins known to be phosphorylated, though it should be emphasized that such a regulatory role is yet to be directly demonstrated for most proteins subject to phosphorylation.

Attempts to define the physiological significance of a given protein phosphorylation reaction are complicated by the fact that proteins which do not appear to be phosphorylated in vivo can in some cases be made to serve as substrates for protein kinase in vitro.[10-13] Such phosphorylation induced under artificial in vitro conditions may cause changes in the conformation and functional activity of the protein in question, but the issue arises as to whether these changes are of any physiological relevance. In order to ward against the possibility of such erroneous conclusions concerning the regulatory role of protein phosphorylation, Krebs and Beavo[8] have established a set of criteria which must be met before one ascribes physiological significance to any particular protein phosphorylation reaction. Briefly, one must show that: (1) phosphorylation of the protein in question will occur in vitro using the appropriate protein kinase, (2) this phosphorylation causes changes in the function of the protein consistent with the role of the protein in vivo, (3) the levels of protein kinase in vivo are sufficient to induce the required level of phosphorylation, and (4) phosphorylation and dephosphorylation of the protein in vivo produce the same changes as in vitro. The last requirement is, of course, the most difficult one to fulfill, and yet is clearly essential because of the above-mentioned tendency to produce artifactual protein phosphorylations in vitro.

The highest concentration of phosphorylated proteins in the typical eukaryotic cell occurs in the nucleus, where the average number of phosphate groups per protein molecule is usually 20 to 30 times higher than in the cytoplasm.[14-16] Because of the potential significance of protein phosphorylation as a cellular regulatory mechanism, this finding of extensive phosphorylation of nuclear proteins is of considerable interest. The high concentration of protein-bound phosphate groups in the nucleus reflects both the presence of proteins with large numbers of bound phosphate groups, and the existence of many different types of phosphorylated proteins. In this chapter we will first describe the nature and possible physiological roles of the major substrates for nuclear protein kinases, and will then go on to discuss the properties and regulation of the kinases and phosphatases responsible for catalyzing the phosphorylation-dephosphorylation of these substrates.

II. SUBSTRATES FOR NUCLEAR PROTEIN KINASES

A. Nonhistones

The nonhistone protein fraction is highly enriched in phosphorylated proteins, with at least several dozen ^{32}P-labeled polypeptides being readily discernible by SDS-polyacrylamide gel electrophoresis.[16] Direct chemical analysis for protein-bound phosphate in the nonhistone protein fraction has independently indicated the presence of enough phosphorus to account for several phosphorylated amino acids per hundred residues, further confirming the presence of highly phosphorylated proteins among the nonhistones. Though the phosphorylation pattern of the nonhistone proteins has been shown to be tissue specific,[17-24] relatively little progress has been made in identifying and purifying the individual phosphorylated proteins in a native state in which their functional properties can be ascertained. Because the phosphorylation of nonhistone proteins has now been shown to undergo alterations in an impressive variety of developmental and physiological situations where changes in nuclear metabolism and gene activity occur (Table 1), there is much interest in isolating and identifying individual phosphorylated proteins so that their possible roles in nuclear functions can be investigated. In the process of investigating such proteins it has gradually become clear that a wide spectrum of different types of molecules are phosphorylated by protein kinases, and that the physiological roles of these different phosphorylations may therefore be quite diverse. The extent of this diversity will become increasingly apparent as we describe below some of the various types of nonhistone proteins known to serve as substrates for protein kinase-catalyzed phosphorylation.

1. Putative Transcriptional Regulatory Proteins

As a class, many of the nonhistone proteins which are substrates for protein kinase exhibit properties suggesting a possible role in the regulation of gene transcription. These properties of nonhistone phosphoproteins, which have been extensively discussed in a recent review,[7] include the facts that changes in their phosphorylation correlate with changes in gene activity, they are heterogeneous and tissue-specific, they bind specifically to DNA, and they alter the rate of RNA synthesis in cell-free systems. Though the above observations are all consistent with the idea that nonhistone phosphorylation is involved in the control of gene expression, such correlations are inherently indirect and in general do not provide direct evidence for a cause and effect relationship between protein phosphorylation and gene transcription. More rigorous proof would be provided, of course, by directly showing that nonhistone proteins in different states of phosphorylation have differing effects on RNA synthesis.

In an attempt to provide such evidence, Kleinsmith et al.[152] devised a procedure for enzymatically dephosphorylating nonhistone proteins prior to their addition to a cell-free transcriptional system. Transcription was found to be approximately 50% lower in the presence of dephosphorylated vs. native nonhistone proteins. Quantitation of the number of initiation sites confirmed that this reduction is due to a decrease in the number of available initiation sites for RNA synthesis, rather than an effect on RNA elongation, termination, or processing. Furthermore, use of a histone cDNA probe to monitor the transcription of a specific subset of gene sequences in this system provided evidence that the reduction in the total number of transcriptional initiation sites is not random throughout the DNA, but is a selective effect on certain genes.

Another experimental approach that has provided direct evidence for a link between protein phosphorylation and gene transcription involves the addition of purified protein kinase to isolated nuclei. Chuang et al.[112,113] have shown that the capacity of isolated adrenal medulla nuclei to phosphorylate nonhistone proteins and synthesize RNA is enhanced following

Table 1
SUMMARY OF EXPERIMENTALLY OBSERVED ALTERATIONS IN NONHISTONE PHOSPHORYLATION AND/OR KINASE ACTIVITY

System	Ref.
Growth and Development	
Lymphocyte activation	25—32
Avian erythrocytes	33, 34
Physarum polycephalum	19
HeLa cells	35—37
Reuben hepatoma cells	38
Yoshida sarcoma	39
Sea urchin embryos	40, 41
Trout embryos	42
Muscle differentiation	43—50
Neuronal and glial cells	51—55
Kidney regeneration	56
Barley germination	57, 58
Chinese hamster cells (K_{12})	59
WI-38 fibroblasts	60, 61
Chick embryo fibroblasts	62
Hamster kidney cells ($BHK_{21}C_{13}$)	63, 64
Testis and epididymis	65
Uterus	66
CHO cells (heat shock)	67
Liver (aging)	68
Liver (regeneration)	69—73
Liver (cell culture)	74
Landschutz tumor cells (amino acid starvation)	75
Hormonal Stimulation	
Testosterone (prostate)	76—80, 151
Antiandrogens (prostate)	81
Glucocorticoids (insect salivary glands)	82
Glucocorticoids (liver)	83—87
Glucocorticoids (lymphosarcoma)	88
Aldosterone (kidney)	89
Estradiol (uterus)	90
Estradiol (mammary carcinoma)	91—93
Estradiol (oviduct)	94
Estradiol (brain)	53, 54
Prolactin (mammary gland)	95
Chorionic gonadotropin (ovary)	96
Calcitonin (bone cells)	97
Triiodothyronine (liver)	98—104
Triiodothyronine (heart)	105, 106
Thyrotropin (thyroid)	107
Abscisic acid *(Lemna)*	57
Cyclic AMP (rat liver)	108
Cyclic AMP (rat heart)	109
Cyclic AMP (lymphocytes)	110
Cyclic AMP (mammary carcinoma)	91—93, 111
Cyclic AMP (adrenal medulla)	112, 113
Cyclic AMP (salivary gland)	114
Cyclic AMP (neurospora)	115
Cyclic AMP (pineal gland)	136

Table 1 (continued)
SUMMARY OF EXPERIMENTALLY OBSERVED ALTERATIONS IN NONHISTONE PHOSPHORYLATION AND/OR KINASE ACTIVITY

System	Ref.
Growth and Development	
Cyclic GMP (lymphocytes)	110
Cyclic GMP (salivary gland)	114
Malignant and Transformed Cells	
Mammary carcinoma	91—93, 116
Azo-dye carcinogenesis	117-119
Dimethylbenzanthracene carcinogenesis	111
Methylcholanthrene carcinogenesis	120
SV-40 transformed fibroblasts	121—125
Morris hepatomas	126—128
Novikoff hepatomas	71, 73, 129, 130
Friend erythroleukemia	131, 132
Ehrlich ascites	71, 129
Walker tumor	71, 129
Adenovirus transformed cells	120
Murine sarcoma virus transformed cells	120
Human leukemia	133
Chemical Agents	
Isoproterenol (salivary gland)	134, 135, 150
Isoproterenol (pineal gland)	136
Acetylcholine (lymphocytes)	110
Norepinephrine (glioma)	137
Carbachol (lymphocytes)	110
Phenobarbital (liver)	138
α-1,2,3,4,5,6-Hexachlorocyclohexane (liver)	139
Morphine (brain)	140, 141
Hemin (liver)	142
Vitamin D (kidney, liver)	143
Prostaglandin (lymphocytes)	110
2,4-Dichlorophenoxyacetic acid (soybeans)	144
Helanalin (Ehrlich ascites)	145
Cytarabine (salivary gland)	146
Bleomycin (salivary gland)	146
Chlorambucil (Yoshida sarcoma)	147
Melphalan (Yoshida sarcoma)	148
Cyclophosphamide (Yoshida sarcoma)	148
Eupahyssopin (Ehrlich ascites)	149

addition of protein kinase. If, however, the protein kinase is first incubated with the endogenous thermostable protein kinase inhibitor, the protein kinase is no longer capable of stimulating nuclear phosphorylation and RNA synthesis, strongly implying that it is the ability of the protein kinase to catalyze the phosphorylation of nuclear protein substrates which is responsible for the increase in RNA synthesis.

Experiments such as those cited above provide strong motivation for attempting to identify and isolate individual nuclear phosphoproteins which might function as transcriptional regulatory elements. Kostraba et al.[153-155] have isolated a nonhistone phosphoprotein of 11,000 mol wt which inhibits initiation of RNA transcription. This inhibitor is effective only with a eukaryotic RNA polymerase, acts preferentially with homologous native DNA, and interacts selectively with reiterated DNA sequences. However, the role that phosphorylation of this protein plays in controlling its activity, as well as the question of whether it acts selectively on particular genes or gene families, are critical issues which remain unanswered.

In addition to the above inhibitory protein, at least two other kinds of nonhistone phosphoproteins which may regulate transcription have been isolated, one from the nucleolus and the other from nervous tissue. Their properties will be briefly discussed in the following two sections.

2. Nucleolar Proteins

Over the years there have been numerous reports of the presence of both phosphorylated nonhistone proteins and protein kinase activity in isolated nucleoli.[70,73,80,101,102,156-172] Recently Kuehn and associates[170-172] have made a significant breakthrough in this area by isolating a dimeric phosphoprotein of 139,000 mol wt that appears to be involved in the control of ribosomal RNA synthesis. This protein, which has been purified without the use of strong dissociating agents from nuclei of the slime mold *Physarum polycephalum,* stimulates RNA polymerase I-catalyzed synthesis of ribosomal RNA using a homologous nucleolar deoxyribonucleoprotein complex as template. The phosphoprotein binds selectively to those restriction fragments of nucleolar DNA which contain the symmetry axis of the palindromic ribosomal DNA. After removal of the phosphate groups from the protein by treatment with alkaline phosphatase, both binding to ribosomal DNA and the ability to stimulate ribosomal RNA synthesis are inhibited, suggesting that the phosphorylation state of this phosphoprotein may be controlling ribosomal gene transcription.

3. S-100 Protein

A brain-specific protein, called S-100, is present in the cytoplasm of both neurons and glial cells. Small amounts of S-100 are also detectable in nuclei, and a divalent cation-dependent uptake of S-100 into the nucleus has been reported.[173] Although its nuclear function is yet to be clearly ascertained, exogenous S-100 has been shown to stimulate RNA polymerase I activity, but not RNA polymerase II activity, in isolated nuclei.[174] S-100 has also been reported to serve as a substrate for brain nuclear protein kinase,[175,176] raising the possibility that the increased RNA synthesis observed in the presence of S-100 is mediated by phosphorylation of S-100.

4. RNA Polymerase

The eukaryotic RNA polymerases are another class of proteins involved in gene transcription that are subject to phosphorylation. In yeast five subunits of polymerase I, two subunits of polymerase II, and three subunits of polymerase III are labeled after incubation of cells with ^{32}P-orthophosphate.[10,11,177] All three polymerases share a common subunit of 24,000 daltons which is phosphorylated. Phosphorylation of RNA polymerase has also been clearly established in vitro in yeast,[10,11] rat liver,[178,179] and calf thymus,[180,181] though in the case of yeast it is interesting to note that in addition to the polymerase subunits normally phosphorylated in vivo, an extra subunit of polymerase I and one from polymerase II are also phosphorylated in vitro.

Conflicting results have been obtained from experiments designed to ascertain whether phosphorylation regulates the catalytic activity of RNA polymerase. Negative results were obtained with yeast RNA polymerase,[10,11] but in rat liver,[178,182] calf thymus,[180,181] rat ovary,[183] and Novikoff ascites cells,[184] incubation with protein kinase was reported to stimulate RNA polymerase activity. Phosphorylation of RNA polymerase subunits was not unequivocally shown to be the explanation for the increased RNA polymerase activity in each of these instances;[185] however, in the case of calf thymus RNA polymerase II, a relatively strong case for this contention was made by showing that selective inhibition of protein kinase by the heat-stable protein kinase inhibitor causes a depression in phosphorylation of the 25,000-dalton polymerase subunit and a concomitant proportional decrease in polymerase II activity.[180]

It should also be pointed out in this section that phosphorylation of polypeptides which are not RNA polymerase subunits per se might also influence RNA polymerase activity. For example, Dokas et al.[186] have described several phosphorylated polypeptides which co-chromatograph with RNA polymerases I and II and whose phosphorylation state appears to influence the catalytic activity of the associated RNA polymerase, though these polypeptides are not the same size as any of the major polymerase subunits.

In addition to the above studies on eukaryotic RNA polymerases, bacterial RNA polymerase has also been reported to be subject to in vitro and in vivo phosphorylation.[187-189] Though supportive of a role of protein phosphorylation in regulating RNA polymerase activity, the relevance of studies carried out on bacteria for eukaryotic cells is always open to question.

5. *DNA-Binding Proteins*

It has been shown in several laboratories that some of the phosphorylated nonhistone proteins of eukaryotic cells bind directly to DNA, exhibiting a preference for homologous vs. heterologous DNA.[18,190,191] Though DNA-binding proteins may be involved in the regulation of transcription, as has been discussed above, they may also play a role in DNA replication and recombination. DNA-binding phosphoproteins with properties suggesting such a role have been isolated from mouse ascites cells, rat spermatocytes, meiotic cells of lilies, and animal virus-infected cells. Phosphorylation of the DNA-binding protein from mouse ascites cells reduces its binding to single-stranded but not duplex DNA, and abolishes the ability of the protein to stimulate DNA polymerase activity, suggesting a role for this "helix-destabilizing " protein in DNA replication.[192] Likewise, a DNA-binding protein from adenovirus[193] and the T-antigen of simian virus 40,[194-196] both of which are required for initiation of DNA synthesis, have been recently shown to be phosphoproteins. Independent support for the idea that protein phosphorylation may somehow be involved in the initiation of DNA replication comes from the report that DNA synthesis in cultured rat liver cells is stimulated by the addition of protein kinase, and that this stimulation can be prevented by the specific inhibitor of protein kinase catalytic subunits.[197]

Meiotic cells (rat spermatocytes[198] and lily microsporocytes[199]) also contain a helix-destabilizing protein whose properties are influenced by phosphorylation, but this protein is present mainly in the interval between S-phase of the meiotic cell cycle and the termination of chromosome pairing. Its virtual absence during DNA replication and its prominence during the period of chromosome pairing and crossing over suggest a possible function in genetic recombination.

6. *Poly(A) Polymerase*

In addition to the regulatory and enzymatic proteins associated with DNA transcription and replication, several other nuclear enzymes and proteins have been shown to be subject to phosphorylation by protein kinase. One of these is poly(A) polymerase, the enzyme responsible for the posttranscriptional addition of poly(A) tracts to newly synthesized RNA chains. Rose and Jacob[200,201] have recently reported that protein kinase-mediated phosphorylation of purified poly(A) polymerase is associated with an activation of the enzyme, the degree of activation being linearly dependent on the amount of protein kinase added to the incubation mixture. Phosphorylation of poly(A) polymerase appears to increase the affinity of the enzymes for its RNA substrate, thereby enhancing the rate of poly(A) synthesis; the net size of the poly(A) chains produced does not change, however. There is reason to believe that the above in vitro observations are of physiological significance, for the poly(A) polymerase of rapidly growing liver tumors has been reported to be more highly phosphorylated in vivo than the less active enzyme of normal liver.

7. Histone Deacetylase

The enzyme histone deacetylase, which catalyzes the hydrolysis of the ε-*N*-acetylysine linkage in histone molecules, has been extensively purified from calf thymus nuclei.[202] The isolated protein migrates in SDS-polyacrylamide gels as a polypeptide of 160,000 daltons and contains 1.3% phosphorus by weight, enough to account for several dozen phosphate groups! Though this number of phosphate groups seems extraordinarily high, multiple phosphorylation sites occur in the much smaller histone molecules,[5,13] so it is not inconceivable. Further purification and characterization of this interesting protein is needed to ascertain the significance of its high phosphate content.

8. Glucocorticoid Receptors

Gene expression in target cells responding to stimulation by steroid hormones is regulated at least in part by migration of steroid-receptor complexes into the nucleus and their subsequent binding to chromatin. Hence, steroid receptor protein must be among the constituents of the nonhistone protein fraction in such hormone target cells. Nielsen et al.[203] have reported that the glucocorticoid receptor protein from mouse fibroblasts and rat liver is inactivated by treatment with alkaline phosphatase, suggesting that the ability of this hormone receptor to bind its appropriate steroid may depend upon the protein being phosphorylated. Unfortunately, direct phosphorylation of the receptor has not yet been demonstrated because it has not been possible to purify the receptor.

9. Nucleosomal Proteins

Phosphorylated nonhistone proteins have been detected in chromatin fractions prepared by a wide variety of procedures. However, the relationship of these phosphoproteins to the particulate subunit of chromatin, the nucleosome, is yet to be clearly worked out. A specific subgroup of phosphorylated nonhistone proteins appears to be associated with the nucleosome,[132,204] and Chan and Liew have purified one of these proteins to homogeneity.[205] The HMG proteins, a group of relatively small nonhistone proteins characterized by high electrophoretic mobility in low pH polyacrylamide gels.[206] and thought to be intimately associated with chromatin subunits,[207] have also been reported to be phosphorylated.[88,208] Another small nonhistone phosphoprotein (10,000 mol wt), present in tiny amounts but detectable because it is highly phosphorylated, has been purified by MacGillivray and associates,[209] but its relationship to the other nucleosomal proteins is unclear.

Independent support for the idea that protein phosphorylation is intimately involved in nucleosome structure and function comes from several reports of the presence of protein kinase activity in isolated nucleosomes.[210-212] It has also been shown that the nucleosomal substructure of chromatin influences the ability of nonhistone proteins to be phosphorylated, for a large spectrum of nonhistones can be phosphorylated by protein kinase in chromatin whose structure has been disintegrated by nuclease treatment than in chromatin which is intact.[213]

10. Proteins Associated with Heterogeneous Nuclear RNA

The heterogeneous nuclear RNA molecules (HnRNA), which are the precursors of messenger RNA in eukaryotic cells, are associated with a mixture of nonhistone proteins to form ribonucleoprotein particles. Several phosphoproteins, as well as protein kinase activity, have been detected in this nonhistone protein fraction.[214-223] The role played by phosphorylation of these proteins is yet to be ascertained, though an involvement in RNA processing and/or transport into the cytoplasm are obvious possibilities. One of these HnRNA-associated phosphoproteins is probably the poly(A) polymerase discussed earlier. It is worth noting that the phosphorylated proteins bind more tightly to RNA than do nonphosphorylated species of similar molecular weight, raising the possiblity that phosphorylation controls protein-RNA interactions. In the case of poly(A) polymerase, this could explain why the phosphorylated form of the enzyme is more active, i.e., because it binds its substrate with higher affinity.

11. Nuclear Envelope Proteins

Like other biological membranes, the nuclear envelope has been reported to contain enzymes catalyzing the phosphorylation and dephosphorylation of endogenous proteins.[224-227] The major species subject to phosphorylation is a nuclear membrane polypeptide of 68,000 mol wt that appears to be concentrated in the region of the nuclear pore complexes, suggesting the possible involvement of this protein phosphorylation reaction in the exchange of materials between nucleus and cytoplasm.

12. Nuclear Matrix Proteins

After extraction of chromatin from the nucleus, a residual protein network termed the nuclear matrix remains.[228] In addition to a proposed role as a structural framework for maintaining the shape of the nucleus, the nuclear matrix has been claimed to be the site of DNA replication.[229] The nuclear matrix consists of a small number of polypeptides, several of which are subject to phosphorylation. It is interesting to note that in regenerating liver, a maximal level of phosphorylation of these proteins occurs at 12 hr after partial hepatectomy, a few hours before the onset of DNA replication.[72]

B. Histones

When the ability of isolated nuclei to phosphorylate endogenous proteins was first discovered,[230] it became immediately apparent that histones as well as nonhistone proteins are involved as substrates. Subsequent investigations have revealed that all major classes of histones, as well as the closely related protamines, are subject to phosphorylation (see[1,5,13,231-235] for reviews). In particular, phosphorylation of these proteins has been correlated with cell proliferation and hormonal stimulation in a variety of systems. The behavior of the individual types of histones and protamines as substrates for protein kinases in such systems will be briefly summarized in the following sections.

1. Histone H1

The most extensively studied substrate for nuclear protein kinase is clearly histone H1, the largest and most variable of the histones. This molecule exists as several distinct subfractions within any given tissue,[236] such microheterogeneity presumably being permitted because H1 is not an integral component of the nucleosomal octomer.[237,238] The phosphorylation pattern of histone H1 is exceptionally complex, involving multiple protein kinases and half a dozen or more phosphorylation sites. The occurrence of multiple phosphorylation sites not only implies a great deal of complexity in the regulation of histone H1 function, but also makes it difficult to experimentally ascertain the physiological significance of histone phosphorylation, since investigators often report data on the overall phosphorylation of H1 molecules without investigating which sites are involved.

At present, it is possible to subdivide H1 phosphorylation into several major categories. The first such category involves phosphorylation of serine 37,* which appears to be subject

* Since amino acid insertions and deletions occur in various H1 histones, numerical assignments used in this section will not apply precisely to all sources of H1, but will usually be accurate within a few residues.

to hormonal control via activation of cyclic AMP-dependent protein kinase. Langan[239,240] was the first to report this type of histone phosphorylation, observing that injection of glucagon into rats causes a marked increase in the phosphorylation of this particular serine in histone H1. Phosphorylation of this serine could also be induced with injections of dibutyryl cyclic AMP, implicating activation of cyclic AMP-dependent protein kinase as the mechanism by which glucagon stimulates H1 phosphorylation. This conclusion is supported by the fact that cyclic AMP-dependent protein kinase catalyzes phosphorylation of the same serine in vitro as is phosphorylated in response to glucagon in vivo.[241] One of the unusual features of hormone-induced H1 phosphorylation is that in vivo, only about 1% of the H1 histone molecules become phosphorylated. This fact has been cited to support the idea that H1 phosphorylation is involved in regulating transcription of a small number of genes whose encoded enzymes are produced in response to glucagon, though there is no direct evidence for this idea and no satisfactory model indicating how phosphorylated H1 molecules could recognize the specific genes that need to be activated. There is evidence indicating that H1 phosphorylation influences histone binding to DNA in vitro,[242,243] but the physiological significance of this effect is unclear.

In addition to glucagon, the hormones thyrotropin[244,245] and thyroxin,[246] as well as nerve growth factor[247] and epidermal growth factor,[248] have all been reported to enhance histone H1 phosphorylation. The precise site of phosphorylation has not been ascertained in these cases, but at least for those agents which increase cyclic AMP levels (thyrotropin, epidermal growth factor, nerve growth factor), it might be expected that the same serine 37 is involved, for this is the only site in histone H1 known to be phosphorylated by cyclic AMP-dependent protein kinase. One problem associated with studies involving hormonal stimulation is that the physiological effects of hormones are usually pleiotropic, involving induction of enzyme activities as well as a stimulation of cell proliferation. Since cell proliferation is associated with a type of H1 phosphorylation different from the cyclic AMP-dependent phosphorylation of serine 37, studies on hormone stimulation are difficult to interpret in the absence of direct analysis of the sites being phosphorylated.

In addition to the hormonally induced, cAMP-dependent phosphorylation of serine 37, the other major category of H1 phosphorylation is that associated with cell proliferation. The earliest observations suggesting a functional relationship between H1 phosphorylation and cell proliferation involved X-irradiation of dividing cells, which was found to inhibit both cell division and H1 phosphorylation.[249-251] Shortly thereafter it was reported that extensive histone H1 phosphorylation is a general occurrence in rapidly dividing cells, and that a positive correlation exists between the phosphate content of H1 histone and cell replication rate.[252-255] Furthermore, when rapidly growing cells cease dividing, they lose essentially all traces of phosphorylated H1, suggesting that H1 phosphorylation plays a direct role in the proliferation process.

Such observations prompted investigations on the relationship between histone phosphorylation and the various stages of the cell cycle as a logical approach for ascertaining the physiological relevance of histone phosphorylation to cell growth and division. Bradbury and co-workers[256,257] investigated the cell cycle of the slime mold *Physarum polycephalum,* whose naturally synchronous mitotic cycle obviates the need for inducing synchrony by artificial means which may in turn introduce artifacts of their own. Utilizing long-term double-labeling experiments with ^{3}H-lysine and ^{32}P-orthophosphate to determine the phosphate content of H1 throughout the cell cycle, these workers were able to show that a dramatic increase in the phosphorylation of H1 occurs during late G2, just prior to chromosome condensation. H1-bound phosphate levels begin to drop during mitosis, returning to basal levels by the beginning of G1. Furthermore, when nuclei isolated from various stages of the cell cycle were assayed for protein kinase activity, a burst in activity was seen in mid-G2, just preceding the increase in H1 phosphorylation. This temporal sequence of increased nuclear protein kinase followed by enhanced H1 phosphorylation, chromosome condensation, and finally histone dephosphorylation led Bradbury et al.[258] to propose that H1 phosphorylation is the trigger for chromosome condensation. As one means of testing this idea, experiments were carried out in which exogenous histone kinase was added to intact cells in early or mid-G2; under these conditions chromosome condensation and mitosis began up to 40 min earlier than in untreated cells, supporting the idea that H1 phosphorylation is involved in initiating chromosome condensation and mitosis.[258,259]

Investigations of H1 phosphorylation during the cell cycle of mammalian cells[234,260-271] have revealed a pattern which is generally similar to that occurring in *Physarum,* though the details are different enough to raise some questions about the validity of the proposed role of H1 phosphorylation in chromosome condensation. In CHO cells, for example, extensive phosphorylation of H1 does not occur until the beginning of prophase, after the condensation of chromatin has alrerady begun.[234] It has also been shown that the dephosphorylation of H1 which occurs at the end of mitosis is not required for uncoiling and dispersing the condensed chromosomes, for addition of phosphatase inhibitors to dividing cells does not alter the timing of chromosome dispersion, even though the hyperphosphorylated state of H1 persists into G1 under such conditions.[272] Nonetheless, that H1 phosphorylation is somehow related to chromosome condensation is strongly suggested by the recent isolation of a mammalian temperature-sensitive cell-cycle mutant in which H1 phosphorylation is depressed and chromosome condensation is abnormal at the nonpermissive temperature.[273] Unfortunately, the situation is further complicated by the fact that H3 phosphorylation is also depressed in this mutant. In any case, a cause and effect relationship between histone phosphorylation and chromosome condensation cannot be proven in such mutants until the primary defect is localized to histone kinase, if in fact that is where it occurs.

In addition to the extensive phosphorylation of histone H1 which occurs as growing cells enter mitosis, this histone is also phosphorylated during interphase. Using tryptic digestion to characterize the phosphorylation sites involved, Gurley et al.[234,268] have concluded that interphase phosphorylation can be divided into at least two types, both of which can be distinguished from the phosphorylation associated with the beginning of mitosis. First, phosphorylation of a single serine in the carboxyl terminal portion of H1 is initiated at least 2 hr before any cells enter cells enter S phase, indicating that old histones from previous cell cycles are being phosphorylated. This type of phosphorylation is cumulative, occurring on increasing numbers of H1 molecules as the cell progresses, through G1, S, and G2. It has been postulated by Gurley et al. that this phosphorylation is related to a submicroscopic change in chromatin structure which occurs during and subsequent to DNA replication. There is some biochemical evidence that DNA-histone interactions progressively change during such periods,[274] though there is no direct evidence that H1 phosphorylation is involved.

As cells enter S phase, a second type of phosphorylation is superimposed on H1, involving up to two additional serines in the carboxyl-terminal portion of the molecule. This phosphorylation is not cumulative, involving no more than 10% of the H1 molecules at any given time. For this reason it has been proposed that H1 phosphorylation of this particular type may be associated with DNA replication forks or with the deposition of histone on the newly synthesized DNA.

Tryptic digestion of H1 molecules has also been utilized to characterize the previously discussed phosphorylation which occurs as cells enter mitosis. At this time as many as six serine and threonine residues, localized in both the amino and carboxyl halves of H1, are found to be phosphorylated, including the sites phosphorylated during interphase. This phosphorylation of multiple sites involves all molecules of H1, and has therefore been referred to as "superphosphorylation" to distinguish it from the less extensive interphase phosphorylation.

Histone H1 phosphorylation is thus seen to be extraordinarily complex, involving both the cyclic AMP-dependent phosphorylation of a single serine, and the growth associated, cyclic AMP-independent phosphorylation of a half dozen or more serine and threonine residues at varying points in the cell cycle. If this complexity isn't enough, at least two other types of H1 phosphorylation have been reported. Langan[13] has described the in vitro phosphorylation of serine 106 by a cyclic AMP-independent protein kinase, but no in vivo phosphorylation of this residue has been detected and its significance is therefore unclear. And Smith et al.[275-277] have described the presence of phosphorylated lysine in histone H1, an unusual phosphorylation reaction which has not been extensively characterized.

2. Histone H2A

The phosphorylation of histone H2A has been reported in both proliferating and nonproliferating tissues[264,265,278-287] and thus differs from the growth-associated phosphorylation of H1. In growing cells H2A phosphorylation appears to be independent of cell cycle position occurring during G1, S, G2, and M.[263,264] H2A phosphorylation is thus independent of histone synthesis and H1 phosphorylation. Gurley et al.[288] have compared H2A phosphorylation in cultured lines of two strains of deer mice which differ greatly in heterochromatin content. Though H1 phosphorylation was found to be identical in these two cell lines, H2A phosphorylation was enhanced in the more heterochromatic cell type, suggesting an involvement of H2A phosphorylation in heterochromatin structure.

As in the case of H1, more than one phosphorylation site has been identified in histone H2A. The N-terminal serine is subject to in vivo phosphorylation on its hydroxyl group (as well as acetylation of the amino group),[232,278] while serine 19 has been shown to be selectively phosphorylated by cyclic AMP-dependent protein kinase in vitro.[289] The extensive ^{32}P-labeling of H2A observed in cell cultures suggests that other phosphorylation sites may be identified in the future.

3. Histone H2B

Histone H2B phosphorylation has been reported in several differentiating cell types, including developing trout testis[231,232] and avian erythrocytes.[283,284] In trout testis phosphorylation of serine 6 of H2B occurs during periods of DNA and histone synthesis in spermatogonia and primary spermatocytes.[231,232] Though suggesting that H2B phosphorylation may be growth associated, Gurley et al.[234,290] have failed to detect H2B phosphorylation in dividing CHO cells. In trout testis only about 5% of the H2B molecules are involved, however, so the physiological significance of this phosphorylation and its relationship to cell growth and division remain in question.

In vitro, H2B is readily phosphorylated by cyclic AMP-dependent protein kinase.[289,291,292] Serine residues 14, 32, and 36 have all been reported to be phosphorylated under such conditions, but phosphorylation of these sites in vivo has not as yet been reported.

4. Histone H3

Histone H3, like histone H1, is phosphorylated predominantly in proliferating cells.[232,234,267,268,270,283,284] At least two phosphorylation sites have been identified in H3 molecules labeled with ^{32}P in vivo,[232,293] serines 10 and 28. Gurley et al.[234,267,268,270] have observed that in CHO cells, H3 phosphorylation occurs only at the time of mitosis, coincident with superphosphorylation of histone H1. As is the case for H1, H3 is rapidly dephosphorylated as cells enter G1. The parallel behavior between phosphorylation of H3 and superphosphorylation of H1 suggests the possibility that both may serve similar functions in the initiation or regulation of chromosome condensation. The recent observation that phosphorylation of H3 alters the stability of the H3-H4 complex,[294] an integral component of the nucleosome, provides at least indirect support for the idea that phosphorylation of this histone influences chromatin structure.

5. Histone H4

Phosphorylation of histone H4 has been reported in trout testis,[231,232,278-280,295] avian erythroid cells,[233,282,284] regenerating rat liver,[276,277] and to a limited extent in cultured CHO cells.[234] In trout testis and avian erythroid cells, the phosphorylation site has been identified as the hydroxyl group of the N-terminal acetyl-serine residue. Phosphorylation of this residue in trout testis occurs with an appreciable lag following histone synthesis, and has therefore been suggested to play a role in modifying the gross structure or activity of chromatin after the correct binding of this histone to DNA has occurred.[231] In avian erythrocytes, however, phosphorylation of the same site occurs coincident with histone H4 synthesis in the cytoplasm, and so is a cytoplasmic rather than a nuclear event. In this case the phosphate group is removed shortly after the H4 molecule enters the nucleus, suggesting that this type of histone H4 phosphorylation may be involved in mediating the early steps of histone-DNA binding and chromatin assembly, rather than influencing chromatin structure of activity per se.[233]

In addition to phosphorylation of the hydroxyl group the N-terminal serine, phosphorylation of histidine has also been reported to occur in H4 histone molecules in vivo.[276,277] In regenerating rat liver preexisting H4 histone molecules in vivo.[276,277] In regenerating rat liver preexisting H4 molecules are phosphorylated in this way at the time of DNA synthesis, suggesting a possible involvement of this histone modification in DNA replication, e.g., by loosening the interaction between histones so that DNA replication may proceed.[277]

In vitro, protein kinase-mediated phosphorylation of serine 47 has been reported[339] in addition to the two types of in vivo phosphorylations mentioned above. The physiological significance of the phosphorylation of serine 47 is uncertain because it has thus far not been measured in studies involving intact cells.

6. Histone H5

Histone H5 is an unusual lysine-rich, serine-rich histone found only in the nuclei of nucleated avian erythrocytes. This histone replaces histone H1 in developing avian erythrocytes as the nucleus becomes progressively more compact and metabolically inert.[296] Histone H5 gradually becomes more and more phosphorylated during this period; eventually at least nine sites are phosphorylated,[297] including serines 3 and 7, and a series of sites in the C-terminal region similar to those occurring in histone H1.[298] At the terminal stage of erythrocyte development H5 becomes completely dephosphorylated, suggesting that phosphorylation and dephosphorylation of this molecule may be involved in mediating the dramatic changes in chromatin condensation which occur during maturation of this cell type.

7. Protamines

The protamines are a group of highly basic polypeptides which replace the histones during differentiation of sperm cells in certain vertebrates. Phosphorylation of protamines occurs rapidly after their synthesis and entry into the nucleus, followed by a slow period of dephosphorylation lasting over a period of 5 to 10 days.[232,299] During this time period the histones are displaced from the chromatin and the chromatin gradually becomes more and more condensed, but the exact role of protamine phosphorylation and dephosphorylation in these events remains obscure. At least four serines, three of which are adjacent to each other, have been identified as sites of phosphorylation in protamines.[232,300,301]

III. ISOLATION AND CHARACTERIZATION OF NUCLEAR PROTEIN KINASES

A. Types of Nuclear Protein Kinases

We have now seen that a large number of different nuclear substrates for protein kinases exist, including at least a dozen types of nonhistone proteins and all the major classes of

Table 2
RELATIVE ACTIVITIES OF VARIOUS SUBSTRATES AS PHOSPHATE DONORS FOR IN VITRO PHOSPHORYLATION OF NONHISTONE PROTEINS[15]

Substrate	Relative activity
ATP	1.0
GPT	0.57
ITP	0.38
CTP	0.22
UTP	0.06
dATP	1.0
ADP	0.02
PP	<0.01

histones and protamines. A multitude of nuclear protein kinases appear to be involved in catalyzing the phosphorylation of these various substrates, though the exact number is uncertain at present because of the difficulty in purifying these enzymes to homogeneity. Most of the nuclear protein kinases which have been studied catalyze phosphorylation of the hydroxyl groups of serine and threonine residues, forming *O*-phosphoryl linkages which can be identified by their stability in hot acid and lability in alkali.[302] Although ATP is the classic phosphate donor for such reactions, GTP and other nucleoside- and deoxynucleoside-triphosphates are also capable of donating their terminal phosphate groups to nuclear proteins in vitro (Table 2).[15] The relative contributions of these various phosphate donors for in vivo nuclear phosphorylation is not clear, nor has it been possible to determine whether these various donors exhibit any selectivity in the protein sites they phosphorylate.

In addition to phosphorylating serine and threonine residues, nuclear protein kinases which catalyze the phosphorylation of histidine and lysine residues have also been detected.[275-277,303] Because the *N*-phosphoryl linkage is extremely labile in acid, most procedures for monitoring chromosomal protein phosphorylation or nuclear protein kinase activity preclude detection of *N*-phosphorylated amino acids.[304] When precautions are taken to prevent hydrolysis of the acid-labile N-P linkage, however, this type of phosphorylation can be shown to be roughly equivalent in qualitative significance to formation of acid-stable (O-P) phosphoryl linkages. Very few investigations have been made of the nuclear kinases responsible for N-P phosphorylation, but it has been reported that at least two cyclic nucleotide-independent histone kinases of this type exist, one catalyzing the phosphorylation of lysine residues in histone H1, and the other catalyzing phosphorylation of histidine residues in histone H4.[275,303]

Protein kinases have been detected in every nuclear subfraction in which they have been looked for. The presence of protein kinase activity bound to chromatin has been observed in numerous tissues and cell types, including thymus[15,305-308] liver,[71,99,126,129,142,309-319] lymphocytes,[28,320] oviduct,[94,321] ovary,[322] kidney,[56,64] cardiac muscle,[46,323,324] salivary gland,[135] prostate,[79] thyroid,[325] brain,[140,141,326] mammary gland,[327,328] *Physarum,*[329] and a variety of tumors and cell culture lines.[37,61,62,71,125,126,129,318,330-332] In a few instances, this chromatin-associated protein kinase activity has been localized to the subunit structure of chromatin, the nucleosome.[210-212,333] Protein kinase activity has also been detected in nucleoli,[73,80,102,157,159-161,164,165,167-170,334] RNP particles,[217,218,220,223,325] the nuclear envelope,[224-226] and even the nuclear sap,[306,325,330,335,336] though few investigations involve isolation techniques which would make this latter localization apparent.

The exact number of different protein kinases present in the nucleus is difficult to ascertain. Kish and Kleinsmith[315] were the first to report extensive heterogeneity of chromatin-bound protein kinases, detecting at least 12 peaks of kinase activity in beef liver nonhistone phosphoproteins fractioned by ion-exchange chromatography. Extensive multiplicity of nuclear protein kinases was subsequently observed by several other laboratories, which reported the presence of five to ten chromatin-bound kinases in Novikoff hepatoma,[71,129,320] Walker tumor cells,[71] HeLa cells,[37,332] normal and transformed human fibroblasts,[61,320] lymphocytes,[320] cardiac muscle,[324] kidney cells,[64,320] and ovarian cells.[331] A similar degree of heterogeneity has also been detected in the protein kinases isolated from purified nucleoli.[73] Though one cannot completely rule out aggregation with other proteins as an explanation for the widely observed multiplicity of nuclear protein kinases, the fact that these different peaks of protein kinase activity behave differently in phosphorylating exogenously added substrates suggests there is some physiological significance to the observed heterogeneity. A more precise answer to the question of how many nuclear protein kinases actually exist requires purification and identification of the individual enzymes, a task which is beginning to yield definitive results.

For example, at least three protein kinases involved in catalyzing the phosphorylation of histone H1 have been clearly distinguished from each other.[337] Histone kinase I, a cyclic AMP-dependent protein kinase which catalyzes the phosphorylation of serine 37 in histone H1, is the same enzyme as the cyclic AMP-dependent protein kinase found in the cytoplasm of most eukaryotic cells. In spite of the term "histone kinase" this enzyme is not specific for histones; among the other proteins whose phosphorylation it catalyzes are phosphorylase kinase, glycogen synthetase, pyruvate kinase, fructose 1,6-diphosphatase, hormone-sensitive lipase, and troponin.[1,3-6,8,9] The amino acid sequences surrounding the site phosphorylated by this enzyme share common features, especially the presence of basic amino acids N-terminal to the site of phosphorylation. Cyclic AMP-dependent protein kinase can be subfractionated by DEAE ion-exchange chromatography into two subclasses, designated Types I and II, which differ from each other predominantly in their regulatory subunits. Though both forms of this protein kinase have been detected in nuclei,[338] in some tissues the Type I enzyme appears to be predominantly localized in the nucleus, while the Type II enzyme is restricted to the cytoplasm.[339]

Histone kinase II, like histone kinase I (cyclic AMP-dependent protein kinase), is widely distributed in animal tissues.[240] It catalyzes phosphorylation preferentially on serine 106 of histone Hl in vitro, though its physiological significance is unclear since no in vivo phosphorylation of this residue has thus far been detected.[13] Histone kinase II is cyclic AMP-independent, and can be distinguished from the catalytic subunit of cyclic AMP-dependent protein kinase by virtue of the fact that it catalyzes phosporylation of a separate site and is unaffected by the heat-stable inhibitor of cyclic AMP-dependent protein kinase.[337] This latter point is especially important, since the mere finding that cyclic AMP has no effect on an isolated protein kinase does not in itself distinguish whether the enzyme is a true cyclic AMP-independent protein kinase, or is simply the isolated catalytic subunit of cyclic AMP-dependent protein kinase.[340] The heat-stable inhibitor of cyclic AMP-dependent protein kinase is often used to distinguish between these alternatives, since it specifically inhibits the enzymatic activity of the isolated catalytic subunit of cyclic AMP-dependent protein kinase.[341] Caution is needed in interpreting such experiments carried out on chromatin-bound kinases, however, since the catalytic subunit of cyclic AMP-dependent protein kinase apparently loses its susceptibility to inhibition by the heat-stable inhibitor when it is tightly bound to chromatin.[113]

The third enzyme involved in phosphorylating histone H1 is called the "growth-associated histone kinase" or "histone kinase G" because it is present only in growing cells.[261,329,337] This enzyme, which is also cyclic AMP-independent, phosphorylates at least four serine and threonine residues, mainly in the C-terminal region of histone H1, though one phosphorylation site in the N-terminal region has also been identified.[13] Kinase G is activated during the G2 phase of the cell cycle, just prior to H1 superphosphorylation.[342] In *Physarum* this kinase activity has been resolved by ion-exchange chromatography into two apparently different components with slightly different times of appearance during the G2 phase of the cell cycle, suggesting the possibility that these two forms of the enzyme sequentially phosphorylate different sites in the H1 histone.[329]

In addition to the three types of histone kinase described above, several other protein kinases which phosphorylate specific histones have been identified. A histone kinase stimulated by cyclic AMP and acting preferentially on histones 2B and 1 has been identified in the nuclear sap of isolated rat liver nuclei.[336] This enzyme differs from histone kinase I (cycli AMP-dependent protein kinase) with respect to bivalent cation requirements and response to the heat-stable protein kinase inhibitor. A similar cyclic AMP-activated histone kinase distinct from histone kinase I has also been identified in the nuclear sap of thyroid cells.[325]

A histone kinase which preferentially phosphorylates a single site in histone H3 has been isolated from thymoctye chromatin;[308] this enzyme is not affected by either cyclic AMP or the heat-stable inhibitor of cyclic AMP-dependent protein kinase.

At least three protein kinases specific for histone H4 have been identified. The first, designated H4-kinase (I), preferentially catalyzes phosphorylation of serine 47, while H4-kinase (II) phosphorylates the N-terminal serine of H4 (as well as the analogous N-terminal serine occurring in histone H2A).[343] Neither of these H4 kinases is cyclic AMP-dependent, nor do either behave like the isolated catalytic subunit of cyclic AMP-dependent protein kinase. A third type of protein kinase preferentially phosphorylating histone H4 is present in nucleoli; though the phosphorylation site(s) have not yet been identified in this case, the nucleolar kinase differs from the other two H4-specific kinases in being stimulated by cyclic GMP.[334]

Less progress has been made in identifying protein kinases which phosphorylate nonhistone proteins, but here too, multiple enzymes appear to be involved. Two major forms of nonhistone protein kinase, which are cyclic nucleotide independent, have recently been purified to homogeneity.[307,344-347] These enzymes, designated NI and NII, prefer acidic proteins as substrates and are only marginally inhibited by the heat-stable inhibitor, suggesting they are not the free catalytic subunit of cyclic AMP-dependent protein kinase. Both NI and NII exhibit an unusual activation by Fe^{++}. A similar cyclic nucleotide-independent nuclear protein kinase which preferentially phosphorylates acidic proteins and is optimally activated by Co^{++} has been purified from rat mammary gland.[328]

In addition to the above three cyclic nucleotide-independent enzymes phosphorylating nonhistone proteins, another cyclic nucleotide-independent protein kinase identified in nuclei turns out to be identical to the catalytic subunit of cytoplasmic cyclic AMP-dependent protein kinase. This isolated subunit, which has been reported to be translocated into the nucleus in cells stimulated by hormones, cyclic AMP, or neurotransmitters,[322,340,348-361] is apparently capable of phosphorylating nonhistones[113,358] as well as histones.

A nuclear protein kinase activity stimulated by Ca^{++}-calmodulin has recently been identified in rat liver;[338] though this enzyme was detected using basic proteins (histones or protamine) as substrates, its substrate specificity and possible relationships to the other more extensively characterized protein kinases described above is unclear at present. A nonhistone kinase activated by Mn^{++} has also been identified in tumor cell nuclei.[129] Though not yet purified to homogeneity, its apparent association with malignant cells makes it an enzyme of considerable interest. A cyclic GMP-activated protein kinase which preferentially phosphorylates a specific nonhistone nucleolar protein has been identified in nucleoli,[167] while cyclic AMP-activated protein kinase fractions which prefer nonhistone proteins as substrates have been identified in chromatin.[315] Though none of the enzymatic activities mentioned in this paragraph have been purified to homogeneity, their differential responses to such potential regulators as Ca^{++}-calmodulin, Mn^{++}, cyclic GMP, and cyclic AMP suggest that they are different enzymes. Even if this differential behavior were to be caused by the association of a variety of regulatory subunits with the same common protein kinase, it would be of considerable physiological interest and relevance.

The variety of different nuclear protein kinases which have now been isolated and characterized provides considerable support for the idea that a large number of such enzymes exist. We have seen for histones alone that four different kinases phosphorylating histone H1 have already been identified (kinases I, II, G, and the kinase that phosphorylates lysine), plus four kinases for histone H4 (H4-kinases I and II, the kinase that phosphorylates histidine, and the cyclic GMP-activated protein kinase), and one more kinase each for histones H2B and H3. For nonhistone proteins there are kinases NI and NII, the isolated catalytic subunit of cyclic AMP-dependent protein kinase, and perhaps enzymes stimulated by Co^{++}, Ca^{++}-calmodulin, Mn^{++}, cyclic GMP, and cyclic AMP. Thus in this brief listing alone we can account for at least 18 different types of enzymes, and more seem certain to be identified. The reason for so many different forms of the same enzyme isn't known for certain, but one obvious possibility is that it allows the phosphorylation of different substrates to be regulated independently. Some of the various types of regulation thought to be involved will be discussed below.

B. Regulation of Nuclear Protein Kinases

Many factors have been implicated in the regulation of nuclear protein kinase activity, including ions, polyamines, cyclic nucleotides, nuclear translocation of cytoplasmic protein kinase, protein kinase autophosphorylation, interaction of protein kinase with other macromolecules, synthesis of new protein kinase, and finally, protein kinase inhibitors. These various possibilities will be elaborated upon in the following sections.

1. Ions

When studied in vitro, nuclear protein kinases require a divalent cation for optimal activity. Mg^{++} is the divalent cation most commonly employed for this purpose, but other ions can be substituted when working with particular kinases. For example, the nonhistone kinases NI and NII are roughly five- to ten-fold more active in the presence of 10 m*M* Fe^{++} than they are with 10 m*M* Mg^{++}.[347] Likewise, a nonhistone kinase requiring Co^{++} for optimal activity in vitro has also been described.[328] Though the physiological relevance of such observations is difficult to ascertain, at least in the case of Fe^{++} it is known that significant amounts of nonheme iron occur in nuclei.[362] Such observations suggest the possibility that alterations in the relative concentrations of specific divalent cations present in the nucleus might influence the spectrum of nuclear proteins being phosphorylated in vivo. Indirect support for this idea has come from the demonstration that different profiles of phosphorylated protein are obtained when isolated nuclei are incubated in the presence of either Mg^{++}, Co^{++}, Zn^{++}, or Mn^{++}.[157]

In general, calcium ions are not capable of substituting for Mg^{++} when assaying for nuclear protein kinase activity in vitro.[15,157,347] However, it has been reported that incubation of brain slices in a medium containing Ca^{++} stimulates phosphorylation of nonhistone proteins and inhibits the phosphorylation of histones.[52,55] Presumably this effect of calcium ions is an indirect one, perhaps mediated by the calcium-binding protein, calmodulin. This interpretation is supported by the observation that the Ca^{++}-calmodulin complex stimulates the activity of at least one nuclear protein kinase in vitro, while Ca^{++} or calmodulin by itself has no effect.[338]

2. Polyamines

Polyamines such as spermine, spermidine, and putrescine have been reported to stimulate nonhistone protein phosphorylation in intact cells,[363] in isolated nuclei[170,364] and nucleoli,[170] and in partially purified nuclear protein kinase preparations.[365,366] Spermidine appears to be the most effective polyamine, but it is worth noting that a mixture of spermidine and spermine seems to be much more effective than either polyamine alone.[170] In most experiments the polyamine-induced stimulation observed is no more than several-fold, but in isolated nuclei and nucleoli of *Physarum,* stimulations in the range of 6- to 30-fold have been noted.[170] An explanation for this apparent discrepancy may be found in the discovery that polyamines preferentially stimulate the phosphorylation of particular nonhistone proteins.[170,362] Hence, the overall degree of stimulation observed will depend on what percentage of the proteins making up the particular substrate being studied are polyamine sensitive. The ability of polyamines to stimulate nuclear protein phosphorylation is not due to substitution of polyamines for Mg^{++}, since the polyamine effect is detectable even in the presence of optimal concentrations of magnesium ion.[170,364]

The potential physiological significance of the ability of polyamines to stimulate nuclear protein phosphorylation is illustrated by events occurring in the androgen-stimulated prostate gland. Administration of testosterone to castrated rats enhances the rate of nonhistone protein phosphorylation in prostate nuclei.[76,77] Polyamine content in this tissue is also elevated in response to androgen, though the kinetics are more sluggish that the increase in nonhistone protein phosphorylation.[367] However, this discrepancy does not preclude a role for polyamines in the observed stimulation of nuclear protein phosphorylation, since a rapid increase in a cytoplasmic polyamine-binding protein has also been observed to occur in prostate cells in response to testosterone.[368] If this binding protein serves to transport polyamines into the nucleus, nuclear polyamine concentrations could be elevated prior to the major burst in new polyamine synthesis. It is interesting to note that the polyamine-binding protein is itself phosphorylated, and its activity in binding polyamines can be modulated by phosphorylation-dephosphorylation. Hence, one possible scenario is that androgens act to stimulate phosphorylation of this polyamine-binding protein (perhaps by elevating Type I cyclic AMP-dependent protein kinase[369]), followed by transport of the protein-bound polyamine into the nucleus where it can stimulate phosphorylation of specific nonhistone proteins.

3. Cyclic Nucleotides

The physiological role of cyclic AMP in regulating the activity of cytoplasmic cyclic AMP-dependent protein kinase is well established, but the situation regarding nuclear protein kinases is less clear. Cyclic AMP has been reported to alter the patterns of nuclear protein phosphorylation in several different cell types when administered to intact animals, organs, or tissue slices.[91-93,108,111,114,115,136,241] Although such experiments provide convincing evidence for the physiological occurrence of cyclic AMP-mediated control of nuclear protein phosphorylation, they do not prove that the cyclic AMP is acting directly on nuclear protein kinases. An alternative possibility, to be discussed in detail in the following section, is that cyclic AMP is binding to the cytoplasmic cyclic AMP-dependent protein kinase, causing it to dissociate into its regulatory and catalytic subunits, followed by translocation of the catalytic subunit into the nucleus.

Though such a translocation mechanism may be responsible for some of the effects of cyclic AMP on nuclear protein phosphorylation, there are several reasons for believing that cyclic AMP also acts directly at the nuclear level. First of all, cyclic AMP has been shown to stimulate protein phosphorylation in isolated nuclei[108,109] and chromatin.[115] Second, some partially purified fractions of nuclear protein kinase have been reported to be either stimulated[46,66,79,165,180,186,305,315,322,324,325,330,335,336,338,339,370] or inhibited[79,165,186,315] by cyclic

AMP, even though unfractionated nuclear protein kinase preparations are generally insensitive to cyclic nucleotides. At least one of the cyclic AMP-dependent nuclear protein kinases is identical to the cytoplasmic enzyme, and is probably not a result of cytoplasmic contamination because it is detectable in nuclei isolated by a nonaqueous procedure preventing exchange of water-soluble components between nucleus and cytoplasm.[322] Third, cyclic AMP-binding proteins have been detected in nuclei using radioactive cyclic AMP,[92,94,305,322,335,370-378] as well as the photoreactive analog 8-azido cyclic AMP.[378,379] Among the cyclic AMP-binding proteins detected in this way are the regulatory subunits of the two isozymic forms of cyclic AMP-dependent protein kinase.[379] Fourth, cyclic AMP itself,[380] as well as adenylate cyclase,[381-384] have been detected in nuclei, though the possibility of cytoplasmic contamination is difficult to rule out in such experiments. Finally, cyclic AMP has been shown to stimulate both RNA synthesis[178,183,186,187,385-389] and RNA release[390] when added to isolated nuclei or subnuclear fractions, indicating that this nucleotide does not need to interact with cytoplasmic protein kinase in order to exert effects on the nucleus.

Taken together, the above observations strongly suggest that cyclic AMP acts directly at the nuclear level. Though the most straightforward interpretation is that cyclic AMP works by activating nuclear protein kinases, the possibility that cyclic AMP binds to other regulatory proteins whose actions have nothing to do with protein phosphorylation should not be overlooked, especially since such cyclic AMP-binding proteins are known to interact with DNA in regulating gene transcription in bacteria.[391]

As mentioned in the above paragraph, partially purified preparations of nuclear protein kinases have been reported to be either stimulated or inhibited by cyclic AMP, depending on the enzyme fraction and substrate being studied. Stimulation of protein phosphorylation by cyclic AMP is consistent with what one normally observes with cytoplasmic cyclic AMP-dependent protein kinase, but the question arises as to how one can explain the apparent inhibition of nuclear protein phosphorylation induced by cyclic AMP in some cases. Though this is not the normally expected effect, there is some precedent for cyclic AMP-mediated inhibition of protein phosphorylation in several systems other than nuclei.[392-404] The mechanism by which cyclic AMP depresses protein phosphorylation has not been established in all such situations, but in the few cases where the issue has been extensively explored, it turns out to be due to a cyclic AMP-induced increase in the activity of a protein phosphatase rather than to a direct action of cyclic AMP on protein kinase.[186,401,403,404]

The possible role of cyclic GMP in controlling nuclear protein phosphorylation has not been investigated as extensively as the role of cyclic AMP, but here, too, some intriguing observations have been made. Cyclic GMP, and agents which elevate intracellular levels of cyclic GMP, have been shown to stimulate the phosphorylation of selected nonhistone proteins in intact lymphocytes.[110] Cyclic GMP-activated protein kinases capable of catalyzing the phosphorylation of histones as well as nonhistones have been detected in nuclei,[167,334] as have cyclic GMP-binding proteins,[373] guanylate cyclase,[405]and cyclic GMP itself.[406-408] Most interesting is the observation that cyclic GMP (but not cyclic AMP) is preferentially localized at genetically active sites on polytene chromosomes.[408] It is worth noting in this context that cyclic GMP has been claimed to alter the rate of RNA synthesis in isolated nuclei,[409] as well as in RNA polymerase fractions,[388,410] though there is a conflict as to whether the effect on RNA polymerase II is inhibitory or stimulatory. The association of cyclic GMP with active chromosomal loci is consistent with the idea that this cyclic nucleotide is involved in controlling RNA synthesis through effects on the phosphorylation of nonhistone proteins, though as in the case of cyclic AMP, the possible interaction of cyclic GMP with regulatory molecules that have nothing to do with protein phosphorylation should not be overlooked.

4. Nuclear Translocation of Cytoplasmic Protein Kinase

Evidence has now been accumulated from studies on at least five different cell types indicating that elevation of cytoplasmic cyclic AMP levels can induce migration into the nucleus of cytoplasmic cyclic AMP-dependent protein kinase, either as the holoenzyme or as the free catalytic subunit. Such nuclear translocation has been reported in ovarian cells stimulated by gonadotropins or cyclic AMP,[322,340,348-350] in liver cells stimulated by glucagon or cyclic AMP,[351,352] in the transynaptic induction of tyrosine hydroxylase in adrenal medulla,[353,357] and cyclic AMP-stimulated neuroblastoma[358] and mammary carcinoma cells.[359-361] Because the experimental demonstration of protein kinase translocation required measurements of the relative amounts of protein kinase associated with nuclear and soluble fractions after subcellular fractions, the results are plagued by potential pitfalls such as enzyme redistribution and artifactual adsorption to subcellular organelles. Though artifacts of this type clearly occur in some cases,[411] they can be minimized by precautions such as homogenization in buffers containing 0.1 to 0.15 *M* salt or isolating nuclei in nonaqueous media. Both approaches have been taken in confirming the reality of nuclear translocation of cytoplasmic protein kinase. Additional support for the reality of protein kinase translocation comes from the changes in the phosphorylation of endogenous nuclear proteins which are observed subsequent to the detection of kinase translocation in liver,[239,240] ovary,[96] adrenal medulla,[357] neuroblastoma,[358] and mammary carcinoma.[91,92,111] In liver it is H1 histone which is phosphorylated subsequent to protein kinase translocation, but in the other cell types studied, nonhistone proteins appear to be the major substrate for the translocated kinase.

5. Protein Kinase Autophosphorylation

While much of the nuclear protein kinase activity can be separated from the bulk of its nuclear protein substrates, virtually all nuclear protein kinase preparations retain endogenous substrate that can be phosphorylated upon the addition of ATP and Mg^{++}. The possibility that some of this endogenous substrate is the protein kinase itself seems a distinct possibility in view of reports that highly purified cytoplasmic protein kinases catalyze their own autophosphorylation. In the case of type II cyclic AMP-dependent protein kinase, the regulatory subunit is phosphorylated by an intramolecular reaction involving the intact holoenzyme.[412-419] The result of this type of phosphorylation is to decrease the affinity of the regulatory subunit for the catalytic subunit, which in essence results in an increased sensitivity of the enzyme to activation by cyclic AMP. Studies on the autophosphorylation of Type I cyclic AMP-dependent protein kinase have led to the conclusion that the catalytic subunit, rather than the regulatory subunit, becomes phosphorylated during autophosphorylation.[420] Though analogous modifications of nuclear protein kinases are yet to be unequivocally demonstrated, autophosphorylation is yet another type of potential regulation which should not be overlooked.

6. Interaction of Protein Kinases with Other Macromolecules

Protein kinases can be either stimulated or inhibited by interacting with a variety of different modulator proteins. For example, cyclic AMP-dependent protein kinase is selectively inhibited by a thermostable protein of about 24,000 mol wt which binds to and inactivates the catalytic subunit of the enzyme.[341,421] A second heat-stable inhibitor has been described which is slightly smaller (15,000 mol wt) and inhibits cyclic AMP-dependent and cyclic GMP-dependent, as well as cyclic-independent, protein kinases.[422] A protein which selectively stimulates the ability of cyclic GMP-dependent protein kinase to phosphorylate histones, but not other substrates, has also been described.[423-426] This stimulatory modulator is very acidic, exhibiting an isoelectric point near 4; the resulting polyanionic nature of the modulator may have something to do with its mode of action, since other polyanions, such as polydeoxyribonucleotides, have also been reported to stimulate cyclic GMP-dependent protein kinase.[427]

Because the above protein kinase modulators have been purified from whole tissue extracts, it is not clear whether they play a physiological role in regulating nuclear protein phosphorylation. The presence of protein kinase inhibitors endogenous to the nucleus has been clearly demonstrated,[428-430] but the two nuclear inhibitors which have been characterized thus far turn out to be different from the modulators described above. One is a large (150,000 mol wt) heat-labile protein, while the other is a family of two small (<2000 mol wt) oligonucleotides.[429] Both these inhibitors are specific for nuclear cyclic AMP-independent protein kinases, being inert toward cytosol protein kinases either in the presence or absence of cyclic AMP. The effects of the two nuclear inhibitors are additive, and are exerted equally on the major isozymic forms of nuclear protein kinase. The physiological relevance of these inhibitors is not clear, but it is apparent that their presence in nuclei keeps protein kinase activity depressed significantly below what it would be if the inhibitors were not present.

At least one in vivo situation has been described in which there is evidence for the production of a macromolecule which modulates nuclear protein kinase activity, and that involves lymphocytes stimulated by antibodies to surface immunoglobulins (Ig). The increased rate of phosphorylation of nonhistone proteins which occurs in such cells can be induced by incubating control nuclei with cell-free extracts from anti-Ig stimulated cells.[30] The active substance in the extracts is not protein kinase itself, but behaves like a nondialyzable protein factor.

Another category of proteins known to influence nuclear protein kinase activity is the histones. In the presence of histones, the rate of phosphorylation of nonhistone proteins is dramatically increased.[431-433] This stimulation is a specific effect which cannot be elicited by other small basic proteins, such as cytochrome c. The idea that it is due to some type of selective interaction between histones and the nonhistone phosphoprotein substrates has been reinforced by affinity chromatography experiments which demonstrate that certain nonhistone phosphoproteins selectively bind to histones.[434]

Though histones have been generally observed to stimulate nonhistone protein phosphorylation, a specific nonhistone protein of 13,000 mol wt (especially histone 2A), and stimulated by double-stranded DNA.[435] The phosphorylation of this protein is catalyzed by a nuclear cyclic AMP-independent protein kinase, but the functional significance of the protein itself is as yet unknown.

7. *Synthesis of New Protein Kinase*

Without a purified, well-characterized protein kinase to employ either as a standard or as a means to make antibodies, unambiguous differentiation between *de novo* enzyme synthesis and enzyme activation is difficult to attain. One possibility is to label newly synthesized proteins with deuterated amino acids, allowing them to be distinguished from preexisting proteins on a metrizamide density gradient. When this approach was utilized to investigate the mechanism underlying the increase in histone kinase activity which occurs during the G2 phase of the cell cycle, it was concluded that activation of preexisting enzyme rather than synthesis of new enzyme is largely responsible.[342]

There is at least one situation, however, in which cells clearly synthesize new protein kinase, and that is during malignant transformation by certain tumor viruses. In this case of the Rous sarcoma virus, which has been the most extensively studied in this regard, the transforming gene codes for a new protein kinase[436-439] which exhibits the unusual property of phosphorylating tyrosine residues.[440] This protein kinase, which closely resembles a normally occurring cellular protein kinase,[441-445] phosphorylates several cellular proteins, but the subcellular localization and functional significance of these proteins are unknown at present.[446] Nuclear substrates for the Rous-encoded protein kinase, if they exist, are yet to be identified, but several nonhistone proteins have been shown to be phosphorylated by the protein kinase encoded by another transforming virus, namely SV40.[447] The close association between malignant transformation and the synthesis of new protein kinases opens up exciting possibilities for studying the role of nuclear protein phosphorylation in regulating cellular activities.

8. Other Protein Kinase Inhibitors

In addition to the macromolecule inhibitors of protein kinases described in Section III.B.6, several low molecular weight substances are known to inhibit protein kinase activity. For example hemin, a factor known to regulate protein synthesis by its inhibitory effects on a cytoplasmic protein kinase,[448-450] also inhibits the phosphorylation of nonhistones proteins by chromatin-associated protein kinase.[142] Though the physiological significance of this finding is not clear, it is interesting to note that as hemin accumulates in developing red blood cells, gene transcription gradually slows down.

Adenosine analogs such as 3′-deoxyadenosine (cordycepin) have also been shown to be potent inhibitors of nuclear protein kinases,[451-454.] This observation is particularly noteworthy because cordycepin is often employed experimentally as a tool for selectively inhibiting poly(A) formation. It should be realized that the ability of cordycepin to depress nuclear protein phosphorylation, which is not widely appreciated, may in fact be responsible for some of the results attributed to decreased poly(A) synthesis in such experiments.

IV. NUCLEAR PROTEIN PHOSPHATASES

A discussion of the enzymes involved in catalyzing nuclear protein phosphorylation would not be complete without a brief mention of nuclear phosphatases, since the phosphorylation state of any given protein reflects the relative balance between its rates of phosphorylation and dephosphorylation. It was first reported in the late 1960s that nuclei contain an enzymatic activity capable of removing phosphate groups from nuclear proteins and releasing them in the form of inorganic phosphate.[455,456] A variety of different phosphatase activities have since been described in nuclei, including alkaline phosphatase,[457-459] acid phosphatase,[459,460] histone phosphatase,[456,461-464] and nonhistone phosphatase.[186,465-468] Such enzymes have been detected in the nuclear sap,[468] nuclear RNP particles,[221] nucleoli,[460,465,466] nuclear envelope,[227] and nuclear pore complex.[458] Relatively little phosphatase activity is present in chromatin or the nonhistone protein fraction insolated from chromatin, however.[15,468]

The relative importance of the various types of phosphatases in dephosphorylating nuclear proteins is yet to be elucidated, primarily because only a few nuclear phosphatases have been purified and characterized. The acid and alkaline phosphatases exhibit relatively low efficiency with nuclear proteins as substrates, and are more likely involved in the metabolism of low molecular weight phosphate esters. A histone phosphatase has been extensively purified from liver nuclei which dephosphorylates histones and protamines, but not nonhistones or low molecular weight phosphate esters.[456,461-463] This protein phosphatase has a subunit molecular weight of 34,000 and exhibits optimal enzymatic activity at a pH of about 7. It appears in multiple forms which are believed to be due to association of the same catalytic subunit with different association of the same catalytic subunit with different proteins or regulatory factors. The enzyme does not require divalent cations or cyclic nucleotides, but is 90% inhibited by 1 m*M* Zn^{++}. This inhibition by zinc ions is in marked contrast to the behavior of alkaline phosphatase, which is a zinc-requiring metalloenzyme. The purified histone phosphatase is similar in many ways to a cytoplasmic protein phosphatase which dephosphorylates phosphorylase, protein kinase, and other cytoplasmic phosphoproteins.

A protein phosphatase which catalyzes dephosphorylation of nucleolar nonhistone proteins as well as H1 histone has been extensively purified from isolated nucleoli.[465,466] This nucleolar phosphatase differs from histone phosphatase in that it requires very low ionic strength for optimal activity. Ion-exchange chromatography resolves the nucleolar phosphoprotein phosphatase into three peaks of activity, but their instability has thus far precluded any further analysis.

Multiple peaks of nonhistone phosphatase activity have also been detected in nuclear sap extracts fractionated by ion-exchange chromatography.[468] One nonhistone phosphtase isolated in this way has been useful as an experimental tool for dephosphorylating nonhistone proteins prior to their addition back to cell-free transcription systems.[152] A phosphatase which dephosphorylates a specific nonhistone protein and which is activated by cyclic AMP has also been identified in soluble extracts of rat liver nuclei.[186] Though the mechanism of the cyclic AMP effect is not known in this case, it has been shown that the cyclic AMP-dependence of cytoplasmic protein phosphatase depends on the substrate being dephosphorylated; for example, the dephosphorylation of cyclic AMP-dependent protein kinase is activated by cyclic AMP because the phosphatase works better on the dissociated regulatory subunit than on the intact holoenzyme, and the cyclic AMP acts to promote this dissociation.[469]

Since so few studies on nuclear protein phosphatases have been carried out, it is uncertain what role, if any, these enzymes play in regulating the phosphorylation state of nuclear proteins. When this issue was investigated in the androgen-stimulated prostate, alterations in histone protein phosphorylation were not found to be accompanied by significant changes in histone phosphatase activity.[464] On the other hand, inhibition of nuclear protein phosphatases by divalent cations and their activation by cyclic AMP provide potential mechanisms for regulating protein phosphorylation through effects on nuclear phosphatases. More studies will clearly be needed in order to ascertain whether such regulation in fact occurs.

V. CONCLUDING REMARKS

This review has attempted to summarize what is currently known about the various nuclear enzymes and substrates involved in protein phosphorylation-dephosphorylation reactions. When phosphorylated proteins were first discovered as components of the cell nucleus in the mid-1960s, it was hypothesized that they play a role in the regulation of gene expression. The data collected in subsequent years were by a large consistent with this hypothesis, and even led to the formulation of a general model which attempted to explain how the phosphorylation of specific nonhistone proteins might lead to displacement of histones from specific genes and their ensuing transcription. Since this model has recently been described in detail elsewhere,[7] it need not be repeated here.

What is now becoming increasingly apparent, however, is that the multiplicity of enzymes and substrates involved in nuclear protein phosphorylation precludes any single model from being a complete explanation for the significance of nuclear protein phosphorylation. Thus, protein phosphorylation may be a mechanism for controlling nuclear enzymes (e.g., RNA polymerase, histone deacetylase, poly(A) polymerase, etc.), RNA-packaging proteins, chromatin structural proteins, hormone-receptors, gene-specific transcriptional regulators, and so forth. The challenge for the future is clearly to work out the details for each of these potential types of regulation, a task which will require the identification and purification of all the individual components involved. Recent progress in purifying individual nuclear protein kinases and nonhistone proteins, taken in conjunction with the availability of recombinant DNA techniques for isolating specific regions of the genome, provide reasons for believing that such a goal can now be realistically pursued.

REFERENCES

1. **Langan, T. A.** Protein kinases and protein kinase substrates, *Adv. Cyclic Nucleotide Res.*, 3, 99, 1973.
2. **Taborsky, G.** Phosphoproteins, *Adv. Protein Chem.*, 28, 1, 1974.
3. **Rubin, C. S. and Rosen, O. M.**, Protein phosphorylation, *Annu. Rev. Biochem.*, 44, 831, 1975.
4. **Trewavas, A.**, Post-translational modification of proteins by phosphorylation, *Annu. Rev. Plant Physiol.*, 27, 349, 1976.
5. **Johnson, E. M.**, Cyclic AMP-dependent protein kinase and its nuclear substrate proteins, *Adv. Cyclic Nucleotide Res.*, 8, 267, 1977.
6. **Nimmo, H. G. and Cohen, P.**, Hormonal control of protein phosphorylation, *Adv. Cyclic Nucleotide Res.*, 8, 145, 1977.
7. **Kleinsmith, L. J.**, Phosphorylation of nonhistone proteins, in *The Cell Nucleus*, Vol. 6, Busch, H., Ed., Academic Press, New York, 1978, 221.
8. **Krebs, E. G. and Beavo, J. A.**, Phosphorylation-dephosphorylation of enzymes, *Annu. Rev. Biochem.*, 48, 923, 1979.
9. **Glass, D. B. and Krebs, E. G.**, Protein phosphorylation catalyzed by cyclic AMP-dependent and cyclic GMP-dependent protein kinases, *Annu. Rev. Pharmacol. Toxicol.*, 20, 363, 1980.
10. **Bell, G. I., Valenzuela, P., and Rutter, W. J.**, Phosphorylation of yeast RNA polymerases, *Nature (London)*, 261, 429, 1976.
11. **Bell, G. I., Valanzuela, P., and Rutter, W. J.**, Phosphorylation of yeast DNA-dependent RNA polymerases *in vivo* and *in vitro*. Isolation of enzymes and identification of phosphorylated subunits, *J. Biol. Chem.*, 252, 3082, 1977.
12. **Thomson, J. A., Brade, W. P., Chiu, J.-F., Hnilica, L. S., and McGill, M.**, *In vivo* and *in vitro* phosphorylation of nuclear proteins in rat liver, *Can. J. Biochem.*, 54, 86, 1976.
13. **Langan, T. A.**, Methods for the assessment of site-specific histone phosphorylation, *Methods Cell Biol.*, 19, 127, 1978.
14. **Langan, T. A.**, A phosphoprotein preparation from liver nuclei and its effect on the inhibition of RNA synthesis by histones, in *Regulation of Nucleic Acid and Protein Biosynthesis*, Koningsberger, V. V. and Bosch, L., Eds., Elsevier, Amsterdam, 1967, 233.
15. **Kleinsmith, L. J. and Allfrey, V. G.**, Nuclear phosphoproteins. I. Isolation and characterization of a phosphoprotein fraction from calf thymus nuclei, *Biochim. Biophys. Acta*, 175, 123, 1969.
16. **Kleinsmith, L. J.**, Acidic nuclear proteins, in *Acidic Proteins of the Nucleus*, Cameron, I. L. and Jeter, J. R., Jr., Eds., Academic, New York, 1974, 103.
17. **Platz, R. D., Kish, V. M., and Kleinsmith, L. J.**, Tissue specificity of non-histone chromatin phosphoproteins, *FEBS Lett.*, 12, 38, 1970.
18. **Teng, C. S., Teng, C. T., and Allfrey, V. G.**, Studies of nuclear acidic proteins. Evidence for their phosphorylation, tissue specificity, selective binding to deoxyribonucleic acid, and stimulatory effects on transcription, *J. Biol. Chem.*, 246, 3597, 1971.
19. **LeStourgeon, W. M. and Rusch, H. P.**, Nuclear acidic protein changes during differentiation in *Physarum polycephalum*, *Science*, 174, 1233, 1971.
20. **LeStourgeon, W. M. and Rusch, H. P.**, Localization of nucleolar and chromatin residual acidic protein changes during differentiation in *Physarum polycephalum*, *Arch. Biochem. Biophys.*, 155, 144, 1973.
21. **Vidali, G., Boffa, L. C., Littau, V. C., Allfrey, K. M., and Allfrey, V. G.**, Changes in nuclear acidic protein complement of red blood cells during embryonic development, *J. Biol. Chem.*, 248, 4065, 1973.
22. **MacGillivray, A. J. and Rickwood, D.**, The heterogeneity of mouse-chromatin nonhistone proteins as evidenced by two-dimensional polyacrylamide-gel electrophoresis and ion-exchange chromatography, *Eur. J. Biochem.*, 41, 181, 1974.
23. **Platz, R. D. Kleinsmith, L. J.**, Heterogeneity and specificity of non-histone nuclear phosphoproteins, *Comp. Biochem. Physiol.*, 55B, 9, 1976.
24. **O'Hara, I. and Yanagihara, T.**, Nuclear chromatin proteins from rabbit cerebrum, cerebellum and liver: synthesis and phosphorylation, *J. Neurochem.*, 29, 1065, 1977.
25. **Kleinsmith, L. J., Allfrey, V. G., and Mirsky, A. E.**, Phosphorylation of nuclear protein early in the course of gene activation in lymphocytes, *Science*, 154, 780, 1966.
26. **Pogo, B. G. T. and Katz, J. R.**, Early events in lymphocyte transformation by phytohaemagglutinin. II. Synthesis and phosphorylation of nuclear proteins, *Differentiation*, 2, 119, 1974.
27. **Johnson, E. M., Karn, J., and Allfrey, V. G.**, Early nuclear events in the induction of lymphocyte proliferation by mitogens. Effects of concanavalin A on the phosphorylation and distribution of non-histone chromatin proteins, *J. Biol. Chem.*, 249, 4990, 1974.
28. **Sons, W., Unsöld, H. J., and Knippers, R.**, Increase of chromatin-bound protein kinase after stimulation of lymphocytes by concanavalin A, *Eur. J. Biochem.*, 65, 263, 1976.

29. **Kishimoto, T., Nishizawa, Y., Kikutani, H., and Yamamura, Y.,** Biphasic effects of cyclic AMP on IgG production and on the changes of non-histone nuclear proteins induced with anti-immunoglobulin and enhancing soluble factor, *J. Immunol.*, 118, 2027, 1977.
30. **Nishizawa, Y., Kishimoto, T., Kikutani, H., and Yamamura, Y.,** Induction and properties of cytoplasmic factor(s) which enhance nuclear nonhistone protein phosphorylation in lymphocytes stimulated by anti-Ig, *J. Exp. Med.*, 146, 653, 1977.
31. **Stott, D. I. and Williamson, A. R.,** Non-histone chromatin proteins of B lymphocytes stimulated by lipopolysaccharide. II. Phosphorylation, *Biochim. Biophys. Acta*, 521, 739, 1978.
32. **Facchini, A., Astaldi, G. C. B., Cocco, L., Wijermans, P., Manzoli, F. A., and Astaldi, A.,** Early events in thymocyte activation. II. Changes in nonhistone chromatin proteins induced by a thymus-dependent human serum factor, *J. Immunol.*, 123, 1577, 1979.
33. **Gershey, E. L. and Kleinsmith, L. J.,** Phosphorylation of nuclear proteins in avian erythrocytes, *Biochim. Biophys. Acta*, 194, 519, 1969.
34. **Shelton, K. R., Seligy, V. L., and Neelin, J. M.,** Phosphate incorporation into ''nuclear'' residual proteins of goose erythrocytes, *Arch. Biochem. Biophys.*, 153, 375, 1972.
35. **Platz, R. D., Stein, G. S., and Kleinsmith, L. J.,** Changes in the phosphorylation of non-histone chromatin proteins during the cell cycle of HeLa S_3 cells, *Biochem. Biophys. Res. Comm.*, 51, 735, 1973.
36. **Karn, J., Johnson, E. M., Vidali, G., and Allfrey, V. G.,** Differential phosphorylation and turnover of nuclear acidic proteins during the cell cycle of synchronized HeLa cells, *J. Biol. Chem.*, 249, 667, 1974.
37. **Phillips, I. R., Shephard, E. A., Stein, J. L., Kleinsmith, L. J., and Stein, G. S.,** Nuclear protein kinase activities during the cell cycle of HeLa S_3 cells, *Biochim. Biophys. Acta*, 565, 326, 1979.
38. **Defer, N., Tichonicky, L., Kruh, J., Zoutewelle, G., and Van Wijk, R.,** Variations in some molecular events during the early phases of the Reuber H 35 cell cycle. II. Chromatin protein phosphorylation and protein kinases, *Biochimie*, 61, 855, 1979.
39. **Riches, P. G., Harrap, K. R., and Sellwood, S. M.,** Phosphorylation of nuclear proteins in rodent tissues, *Biochem. Soc. Trans.*, 1, 684, 1973.
40. **Platz, R. D. and Hnilica, L. S.,** Phosphorylation of nonhistone chromatin proteins during sea urchin development, *Biochem. Biophys. Res. Commun.*, 54, 222, 1973.
41. **Ševaljević, L., Popović, Ž., and Konstantinović, M.** Investigation on stage-related changes of sea urchin chromatin proteins by hydroxyapatite chromatography, *Int. J. Biochem.*, 6, 903, 1975.
42. **Trevithick, J. R.,** Patterns of incorporation of (^{3}H)leucine and (^{32}P)orthophosphate into nonhistone proteins of Rainbow Trout during early development, *Can. J. Biochem.*, 52, 408, 1974.
43. **Mân, N. T., Morris, G. E., and Cole, R. J.,** Phosphorylation of nuclear proteins during muscle differentiation in vitro, *FEBS Lett.*, 42, 257, 1974.
44. **Mân, N. T., Morris, G. E., and Cole, R. J.,** Gene activation during muscle differentiation and the role of nonhistone chromosomal protein phosphorylation, *Dev. Biol.*, 47, 81, 1975.
45. **Piras, M. M. and Piras, R.,** Nuclear protein phosphokinase activities and phosphorylation of chromosomal proteins in embryonic and adult muscles of the chick, *Mol. Cell. Biochem.*, 16, 119, 1977.
46. **Limas, C. J.,** Myocardial nuclear protein kinases during postnatal development, *Am. J. Physiol.*, 235, H338, 1978.
47. **Leibovitch, M. P., Tichonicky, L., and Kruh, J.,** Chromatin protein kinases and phosphoproteins during myoblast growth and differentiation, *Biochem. Biophys. Res. Commun.*, 81, 623, 1978.
48. **Liew, C.-C. and Sole, M. J.,** Nuclear proteins in the heart of the cardiomyopathic Syrian hamster. Phosphorylation of phenol-soluble nonhistone proteins, *Circ. Res.*, 42, 637, 1978.
49. **Lough, J. and Ingram, V. M.,** Change in nuclear phosphoprotein during in vitro myogenesis, *Exp. Cell Res.*, 114, 349, 1978.
50. **Mân, N. T., Morris, G. E., and Cole, R. J.,** Two-dimensional gel analysis of nuclear proteins during muscle differentiation *in vitro*. II. Changes in protein phosphorylation, *Exp. Cell Res.*, 126, 383, 1980.
51. **Fleischer-Lambropoulos, H., Sarkander, H.-I., Brade, W. P.,** Phosphorylation of nonhistone chromatin proteins from neuronal and glial nuclei-enriched fractions of rat brain, *FEBS Lett.*, 45, 329, 1974.
52. **Kanungo, M. S. and Thakur, M. K.,** Phosphorylation of chromosomal proteins as a function of age and its modulation by calcium, *Biochem. Biophys. Res. Commun.*, 79, 1031, 1977.
53. **Lee, N. M. and Loh, H. H.,** Phosphorylation and methylation of chromatin proteins from mouse brain nuclei, *J. Neurochem.*, 29, 547, 1977.
54. **Kanungo, M. S. and Thakur, M. K.,** Effects of oestradiol on covalent modifications of chromosomal proteins and transcription of chromatin of the brain of rats of various ages, *J. Steroid Biochem.*, 11, 879, 1979.
55. **Kanungo, M. S. and Thakur, M. K.,** Phosphorylation and acetylation of nonhistone chromosomal proteins of the brain of rats of various ages and their modulation by calcium and estradiol., *Biochem. Biophys. Res. Commun.*, 86, 14, 1979.

56. **Brade, W. P., Thomson, J. A., Chiu, J.-F., and Hnilica, L. S.,** Chromatin-bound kinase activity and phosphorylation of chromatin non-histone proteins during early kidney regeneration after folic acid, *Exp. Cell Res.*, 84, 183, 1974.
57. **Chapman, K. S. R., Trewavas, A., and van Loon, L. C.,** Regulation of the phosphorylation of chromatin-associated proteins in *Lemna* and *Hordeum, Plant Physiol.*, 55, 293, 1975.
58. **Trewavas, A.,** Changes in phosphorylated and unphosphorylated nuclear proteins during germination in barley, *Phytochemistry,* 15, 363, 1976.
59. **Rieber, M. and Bacalao, J.,** Alterations in nuclear phosphoproteins of a temperature-sensitive Chinese hamster cell line exposed to non-permissive conditions, *Exp. Cell Res.*, 85, 334, 1974.
60. **Pumo, D. E., Stein, G. S., and Kleinsmith, L. J.,** Phosphorylation of nonhistone chromosomal proteins early during the prereplicative phase of the cell cycle of WI-38 diploid fibroblasts, *Cell Differ.*, 5, 45, 1976.
61. **Bombik, B. M. and Baserga, R.,** Effect of cyclic AMP on chromatin-bound protein kinases in WI-38 fibroblasts stimulated to proliferate, *Biochim. Biophys. Acta,* 443, 343, 1976.
62. **Kruh, J., Courtois, Y., and Tichonicky, L.,** Phosphorylation and protein kinase activities of chromosomal non histone proteins from chick embryo fibroblasts, *Biochimie,* 57, 1323, 1975.
63. **Marty de Morales, M., Blat, C., and Harel, L.,** Changes in the phosphorylation of non-histone chromosomal proteins in relationship to DNA and RNA synthesis in BHK_{21} C_{13} cells, *Exp. Cell Res.*, 86, 111, 1974.
64. **Blat, C., Marty de Morales, M., and Harel, L.,** Kinase activities in the non-histone chromosomal proteins of resting and proliferating BHK21 C13 cells, *Exp. Cell Res.*, 98, 104, 1976.
65. **Kadohama, N. and Turkington, R. W.,** Changes in acidic chromatin proteins during the hormone-dependent development of rat testis and epididymis, *J. Biol. Chem.*, 249, 6225, 1974.
66. **Eppenberger, U., Roos, W., Fabbro, D., Sury, A., Weber, J., Bechtel, E., Huber, P., and Jungmann, R. A.,** Ontogeny of the adenosine-3′:5′-phosphate-dependent protein-kinase system during early uterine development, *Eur. J. Biochem.*, 98, 253, 1979.
67. **Caizergues-Ferrer, M., Bouche, G., and Amalric, F.,** Phosphorylated proteins involved in the regulation of rRNA synthesis in CHO cells recovering from heat shock, *FEBS Lett.*, 116, 261, 1980.
68. **Liew, C. C. and Gornall, A. G.,** Covalent modification of nuclear proteins during aging, *Feb. Proc.*, 34, 186, 1975.
69. **Ezrailson, E. G.,Olson, M. O. J., Guetzow, K. A., and Busch, H.,** Phosphorylation of non-histone chromatin proteins in normal and regenerating rat liver, Novikoff hepatoma and rat heart, *FEBS Lett.*, 62, 69, 1976.
70. **Ballal, N. R., Kang, Y.-J., Olson, M. O. J., and Busch, H.,** Changes in nucleolar proteins and their phosphorylation patterns during liver regeneration, *J. Biol. Chem.*, 250, 5921, 1975.
71. **Thomson, J. A., Chiu, J.-F., Sakuma, K., and Hnilica, L. S.,** Nuclear protein phosphokinases in normal and neoplastic tissues, *Cancer Res.*, 37, 3266, 1977.
72. **Allen, S. L., Berezney, R., and Coffey, D. S.,** Phosphorylation of nuclear matrix proteins in isolated regenerating rat liver nuclei, *Biochem. Biophys. Res Commun.*, 75, 111, 1977.
73. **Olson, M. O. J., Hatchett, S., Allan, R., Hawkins, T. C., and Busch, H.,** Nucleolus-associated phosphoprotein kinases of normal and regenerating liver and Novikoff hepatoma ascites cells, *Cancer Res.*, 38, 3421, 1978.
74. **Guguen-Guillouzo, C., Tichonicky, L., Szajnert, M. F., and Kruh, J.,** Changes in some chromatin and cytoplasmic enzymes of perinatal rat hepatocytes during culture, *In Vitro,* 16, 1, 1980.
75. **Jolicoeur, P., Lemay, A., Labrie, F., and Steiner, A. L.,** Phosphorylation of chromosomal and ribosomal proteins and intracellular levels of cyclic 3′-5′-adenosine monophosphate and cyclic 3′-5′-guanosine monophosphate during amino acid starvation in Landschutz tumour cells, *Exp. Cell Res.*, 89, 231, 1974.
76. **Ahmed, K. and Ishida, H.,** Effect of testosterone on nuclear phosphoproteins of rat ventral prostate, *Mol. Pharmacol.*, 7, 323, 1971.
77. **Schauder, P., Starman, B. J., and Williams, R. H.,** Effect of testosterone on phenol-soluble nuclear acidic proteins of rat ventral prostate, *Proc. Soc. Exp. Biol. Med.*, 145, 331, 1974.
78. **Kadohama, N. and Anderson, K. M.,** Nuclear non-histone proteins from rat ventral prostate cells undergoing hypertrophy or hyperplasia, *Exp. Cell Res.*, 99, 135, 1976.
79. **Ahmed, K. and Wilson, M. J.,** Chromatin-associated protein phosphokinases of rat ventral prostate. Characteristics and effects of adrogenic status, *J. Biol. Chem.*, 250, 2370, 1975.
80. **Wilson, M. J. and Ahmed, K.,** The differential response of prostatic nucleolar and extra-nucleolar protein phosphokinase activities following androgen deprivation, *Endocrinol. Res. Commun.*, 3, 63, 1976.
81. **Wilson, M. J., Davis, A. T., and Ahmed, K.,** Effects of antiandrogens (cyproterone acetate and flutamide) on the activity of nuclear protein phosphokinases and phosphatases of rat ventral prostate, *Mol. Pharmacol.*, 17, 212, 1980.

82. **Schmidt, E. C. and Goodman, R. M.,** Hormone-induced alterations in nonhistone protein. Methylation and phosphorylation in *Sciara coprophila, Differentiation,* 6, 177, 1976.
83. **Allfrey, V. G., Johnson, E. M., Karn, J., and Vidali, G.,** Phosphorylation of nuclear proteins at times of gene activation, in *Protein Phosphorylation in Control Mechanisms, Huijing, F. and Lee, E. Y. C., Eds., Academic Press, New York,* 1973, 217.
84. **Bottoms, G. D. and Jungmann, R. A.,** Effects of corticosterone on phosphorylation of rat liver nuclear proteins *in vitro, Proc. Soc. Exp. Biol. Med.,* 144, 83, 1973.
85. **Schauder, P., Starman, B., and Williams, R. H.,** Effect of cortisol on phosphorylation of acidic proteins in liver nuclei from adrenalectomized rats, *Experientia,* 30, 1277, 1974.
86. **Trajković, D., Ribarac-Stepić, and Kanazir D.,** The effect of cortisol on the phosphorylation of rat liver nuclear acidic proteins and the role of these proteins in biosynthesis of nuclear RNA, *Arch. Int. Physiol. Biochim.,* 82, 211, 1974.
87. **Kanazir, D. T., Trajković, D. P., Ribarac-Stepić, N., Popić, S. D., and Metlaš, R.,** Cortisol dependent acute metabolic responses in rat liver cells, *J. Steroid Biochem.,* 9, 467, 1978.
88. **Milholland, R. J., Ip, M. M., and Rosen, F.,** The effect of hydrocortisone treatment on the in vivo phosphorylation of a subgroup of non-histone nuclear proteins in the mouse lymphosarcoma P1798, *Biochem. Biophys. Res. Commun.,* 88, 993, 1979.
89. **Liew, C. C., Suria, D., and Gornall, A. G.,** Effects of aldosterone on acetylation and phosphorylation of chromosomal proteins, *Endocrinology,* 93, 1025, 1973.
90. **Cohen, M. E. and Kleinsmith, L. J.,** Stimulation of uterine nonhistone protein phosphorylation and nuclear protein kinase activity by estradiol-17β, *Biochim. Biophys. Acta,* 435, 159, 1976.
91. **Cho-Chung, Y. S. and Redler, B. H.,** Dibutyryl cyclic AMP mimics ovariectomy: nuclear protein phosphorylation in mammary tumor regression, *Science,* 197, 272, 1977.
92. **Cho-Chung, Y. S.,** Antagonistic action between cyclic adenosine 3′:5′-monophosphate and estrogen in rat mammary tumor growth control, *Cancer Res.,* 38, 4071, 1978.
93. **Cho-Chung, Y. S. and Doud, F. J.,** Antagonistic action between cyclic AMP and estrogen in phosphorylation of mammary tumor nuclear proteins, *Cancer Lett.,* 5, 219, 1978.
94. **Keller, R. K., Chandra, T., Schrader, W. T., and O'Malley, B. W.,** Protein kinases of the chick oviduct: a study of the cytoplasmic and nuclear enzymes, *Biochemistry,* 15, 1958, 1976.
95. **Turkington, R. W. and Riddle, M.,** Hormone-dependent phosphorylation of nuclear proteins during mammary gland differentiation *in vitro, J. Biol. Chem.,* 244, 6040, 1969.
96. **Jungmann, R. A. and Schweppe, J. S.,** Mechanism of action of gonadotropin. I. Evidence for gonadotropin-induced modifications of ovarian nuclear basic and acidic protein biosynthesis, phosphorylation, and acetylation, *J. Biol. Chem.,* 247, 5535, 1972.
97. **Bustos, S. E., Schuster, G. S., Dirksen, T. R., McKinney, R. V., and Lai, Y.-L.,** Effect of calcitonin on the phospohorylation of non-histone from cultured bone cells, *J. Cell. Physiol.,* 92, 43, 1977.
98. **Figassa, E., Taningher, M., Gallo, G., and Orunesu, M.,** Stimulation of nuclear protein kinase activity in rat liver induced by the administration of 3,5′-triiodo-l-thyronine (T_3), *Experientia,* 31, 522, 1975.
99. **Kruh, J. and Tichonicky, L.,** Effect of triiodothyronine on rat liver chromatin protein kinase, *Eur. J. Biochem.,* 62, 109, 1976.
100. **Taningher, M., Molinari, M. P., and Cesarone, C. F.,** Early increase in the phosphorylation of liver chromatin non-histone proteins from thyroidectomized rats treated with triiodothyronine, *Endocrinology,* 101, 1221, 1977.
101. **Fugassa, E., Gallo, G., and Pertica, M.,** Increased in vitro phosphorylation of rat liver nucleolar proteins following triiodothyronine administration, *Experientia,* 32, 1379, 1976.
102. **Fugassa, E., Gallo, G., Pertica, M., Voci, A., and Orunesu, M.,** Increased activity of rat liver nucleolar protein kinase following triiodothyronine administration, *Biochim. Biophys. Acta.,* 485, 350, 1977.
103. **Roques, M., Tirard, A., and DeGroot, J.,** Liver protein kinase activity and triiodothyronine, *Endocrinology,* 100, 967, 1977.
104. **Coleoni, A. H. and DeGroot, L. J.,** Liver nuclear protein phosphorylation *in vivo* and the effect of triiodothyronine, *Endocrinology,* 106, 1103, 1980.
105. **Gibson, K., Tichonicky, L., and Kruh, J.,** The effect of triiodothyronine on myocardial protein kinases, *Mol. Cell. Biochem,.* 9, 79, 1975.
106. **Limas, C. J. and Chan-Stier, C.,** Myocardial chromatin activation in experimental hyperthyroidism in rats. Role of nuclear non-histone proteins, *Circ. Res.,* 42, 311, 1978.
107. **See, Y. P. and Burrow, G. N.,** Effect of thyrotropin on the phosphorylation of thyroid chromosomal proteins, *Can. J. Biochem.,* 57, 523, 1979.
108. **Johnson, E. M. and Allfrey, V. G.,** Differential effects of cyclic adenosine-3′,5′-monophosphate on phosporylation of rat liver nuclear proteins, *Arch. Biochem. Biophys.,* 152, 786, 1972.
109. **Akhtar, R. A. and Itzhaki, S.,** Studies *in vitro* of the effects of adenosine 3′:5′-cyclic monophosphate on the phosphorylation of nuclear proteins in isolated rat heart nuclei, *Biochem. J.,* 161, 487, 1977.

110. **Johnson, E. M. and Hadden, J. W.**, Phosphorylation of lymphocyte nuclear acidic proteins: regulation by cyclic nucleotides, *Science*, 187, 1198, 1975.
111. **Cho-Chung, Y. S., Redler, B. H., and Lewallen, R. P.**, Nuclear protein phosphorylation and hormone-dependent mammary tumor regression following dibutyryl cyclic adenosine 3′:5′-monophosphate treatment or ovariectomy, *Cancer Res.*, 38, 3405, 1978.
112. **Chuang, D. M., Hollenbeck, R., and Costa, E.**, Enhanced template activity in chromatin from adrenal medulla after phosphorylation of chromosomal proteins, *Science*, 193, 60, 1976.
113. **Chuang, D.-M., Hollenbeck, R. A., and Costa, E.**, Protein phosphorylation in nuclei of adrenal medulla incubated with cyclic adenosine 3′:5′-monophosphate-dependent protein kinase, *J. Biol. Chem.*, 252, 8365, 1977.
114. **Itzhaki, S. and Capps, M. J.**, The phosphorylation of salivary gland chromatin proteins following treatment of rats with dibutyryl cyclic AMP and dibutyryl cyclic GMP, *Gen. Pharmacol.*, 9, 355, 1978.
115. **Judewicz, N. D. and Torres, H. N.**, Control of chromatin transcription by a cyclic adenosine 3′,5′-monophosphate-stimulated phosphorylation in *Neurospora crassa*, *FEBS Lett.*, 107, 160, 1979.
116. **Kadohama, N. and Turkington, R. W.**, Altered populations of acidic chromatin proteins in breast cancer cells, *Cancer Res.*, 33, 1194, 1973.
117. **Chiu, J.-F., Craddock, C., Getz, S., and Hnilica, L. S.**, Nonhistone chromatin protein phosphorylation during azo-dye carcinogenesis, *FEBS Lett.*, 33, 247, 1973.
118. **Ahmed, K.**, Increased phosphorylation of nuclear phosphoproteins in precancerous liver, *Res. Commun. Chem. Pathol. Pharmacol.*, 9, 771, 1974.
119. **Chiu, J.-F., Brade, W. P., Thomson, J., Tsai, Y.-H., and Hnilica, L. S.**, Non-histone protein phosphorylation in normal and neoplastic rat liver chromatin, *Exp. Cell Res.*, 91, 200, 1975.
120. **Segawa, K., Oda, K., Yuasa, Y., Shiroki, K., and Shimojo, H.**, Phosphorylation of chromatin- and ribosome-associated proteins in cells transformed by adenovirus, murine sarcoma virus, and methylcholanthrene, *J. Virol.*, 27, 800, 1978.
121. **Krause, M. O., Kleinsmith, L. J., and Stein, G. S.**, Properties of the genome in normal and SV40 transformed W138 human diploid fibroblasts. I. Composition and metabolism of nonhistone chromosomal proteins, *Exp. Cell Res.*, 92, 164, 1975.
122. **Krause, M. O., Kleinsmith, L. J., and Stein, G. S.**, Properties of the genome in normal and SV-40 transformed WI-38 human diploid fibroblasts. III. Turnover of nonhistone chromosomal proteins and their phosphate groups, *Life Sci.*, 16, 1047, 1975.
123. **Krause, M. O., Noonan, K. D., Kleinsmith, L. J., and Stein, G. S.**, The effect of SV40 transformation on the chromosomal proteins of 3T3 mouse embryo fibroblasts, *Cell Differ.*, 5, 83, 1976.
124. **Pumo, D. E., Stein, G. S., and Kleinsmith, L. J.**, Stimulated phosphorylation of non-histone phosphoproteins in SV-40 transformed WI-38 human diploid fibroblasts, *Biochim. Biophys. Acta*, 402, 125, 1975.
125. **Segawa, K. and Oda, K.**, Protein kinases and their protein substrates associated with chromatin and ribosomes in SV40-transformed rat cells, *Biochim. Biophys. Acta*, 521, 374, 1978.
126. **Farron-Furstenthal, F.**, Protein kinases in hepatome, and adult and fetal liver of the rat. I. Subcellular distribution, *Biochem. Biophys. Res. Commun.*, 67, 307, 1975.
127. **Lea, M. A. and Koch, M. R.**, Decreased incorporation of phosphate and amino acids into low-solubility, nuclear proteins in rapidly growing hepatomas, *Int. J. Biochem.*, 8, 565, 1977.
128. **Dayal, V. K., Saha, A., and Lea, M. A.**, Partial characterization of nonhistone nuclear proteins which are decreased in hepatomas of the rat, *Int. J. Biochem.*, 10, 759, 1979.
129. **Thomas, J. A., Chiu, J.-F., and Hnilica, L. S.**, Nuclear phosphoprotein kinase activities in normal and neoplastic tissues, *Biochim. Biophys. Acta*, 407, 114, 1975.
130. **Ganpath, N., Prestayko, A. W., and Busch, H.**, Comparison of nuclear nonhistone phosphoproteins of rat liver and Novikoff hepatoma, *Cancer Rds.*, 37, 1290, 1977.
131. **Houkom, E. C., Neumann, J. R., and Ingram, V. M.**, Experimental approaches to embryonic erythropoiesis, *Ann. Immunol.*, 128C, 709, 1977.
132. **Neumann, J., Whittaker, R., Blanchard, B., and Ingram, V.**, Nucleosome-associated proteins and phosphoproteins of differentiating Friend erythroleukemia cells, *Nucleic Acids Res.*, 5, 1675, 1978.
133. **Seeber, S., Meshkov, T., Brucksch, K. P., Käding, J., Schmidt, C. G., and Busch, H.**, Comparative studies on phenol-soluble nonhistone chromatin proteins in normal and leukaemic human leukocytes, *Klin. Wochenschr.*, 57, 257, 1979.
134. **Ishida, H. and Ahmed, K.**, Studies on phosphoproteins of submandibular gland nuclei isolated from isoproterenol-treated rats, *Exp. Cell Res.*, 78, 31, 1973.
135. **Ishida, H. and Ahmed, K.**, Studies on chromatin-associated protein phosphokinase of submandibular gland from isoproterenol-treated rats, *Exp. Cell Res.*, 84, 127, 1974.
136. **Winters, K. E., Morrissey, J. J., Loos, P. J., and Lovenberg, W.**, Pineal protein phosphorylation during serotonin N-acetyltranserase induction, *Proc. Natl. Acad. Sci. U.S.A.*, 74, 1928, 1977.

137. **Peruman, A. S. and Murthi, V. K.,** Effect of norepinephrine on chromatin protein phosphorylation in C-6 glioma cells, *Life Sci.*, 26, 447, 1980.
138. **Blankenship, J. and Bresnick, E.,** Effects of phenobarbital on the phosphorylation of acidic nuclear proteins of rat liver, *Biochim. Biophys. Acta,* 340, 218, 1974.
139. **Brade, W. P., Chiu, J.-F., and Hnilica, L. S.,** Phosphorylation of rat liver nuclear acidic phosphoproteins after administration of α-1,2,3,4,5,6-hexachlorocyclohexane *in vivo, Mol. Pharmacol.*, 10, 398, 1974.
140. **Oguri, K., Lee, N. M., and Loh, H. H.,** Apparent protein kinase activity in oligondendroglial chromatin after chronic morphine treatment, *Biochem. Pharmacol.*, 25, 2371, 1976.
141. **Hook, V. Y. H., Lee, N. M., and Loh, H. H.** Partial purification of nuclear protein kinase from small dense nuclei of mouse brain and the effect of chronic morphine treatment, *J. Neurochem.*, 34, 1274, 1980.
142. **Loeb, J. E. and Creuzet, C.,** Action of hemin on chromatin protein kinases of rat liver, *FEBS Lett.*, 81, 423, 1977.
143. **Hirschmann, P. N. and Itzhaki, S.** Differential effects of 25-hydroxy-cholecalciferol on non-histone protein phosphorylation in the liver and kidney of the vitamin D-deficient rat, *Biochem. Soc. Trans.*, 4, 339, 1976.
144. **Murray, M. G. and Key, J. L.,** 2,4-Dichlorophenoxyacetic acid-enhanced phosphorylation of soybean nuclear proteins, *Plant Physiol.*, 61, 190, 1978.
145. **Hall, I. H., Lee, K.-H., Mar, E. C., Starnes, C. O., and Waddell, T. G.,** Antitumor agents. XXI. A proposed mechanism for inhibition of cancer growth by tenulin and helenalin and related cyclopentenones, *J. Med. Chem.*, 20, 333, 1977.
146. **Miki, K., Ishida, H., and Yamamoto, I.,** Effect of cytarabine and bleomycin on the phosphorylation of nuclear acidic protein, *FEBS Lett.*, 69, 249, 1976.
147. **Riches, P. G., Sellwood, S. M., and Harrap, K. R.,** Some effects of chlorambucil on nuclear protein phosphorylation in the Yoshida ascites sarcoma, *Chem.-Biol. Interactions,* 18, 11, 1977.
148. **Wilkinson, R., Birbeck, M. S. C., and Harrap, K. R.,** Enhancement of the nuclear reactivity of alkylating agents by prednisolone, *Cancer Res.*, 39, 4256, 1979.
149. **Hall, I. H., Lee, K.-H., and Eigebaly, S. A.,** Antitumor agents. XXIX. Effects of eupahyssopin on nucleic acid, protein, anaerobic, and aerobic metabolism of Ehrlich ascites tumor cells, *J. Pharm. Sci.*, 67, 1232, 1978.
150. **Itzhaki, S. and Capps, M. J.,** Biochemical studies in the nuclei of rat salivary glands stimulated with isoproterenol: phosphorylation of chromatin proteins, *Gen. Pharmacol.*, 9, 347, 1978.
151. **Kadohama, N. and Anderson, K. M.,** Acetylation and phosphorylation of nuclear proteins from growing or dividing rat ventral prostate cells, *Can. J. Biochem.*, 55, 513, 1977.
152. **Kleinsmith, L. J., Stein, J., and Stein, G.,** Dephosphorylation of nonhistone proteins specifically alters the pattern of gene transcription in reconstituted chromatin, *Proc. Natl. Acad. Sci. U.S.A.*, 73, 1174, 1976.
153. **Kostraba, N. C. and Wang, T. Y.** Inhibition of transcription *in vitro* by a non-histone protein isolated from Ehrlich ascites tumor chromatin, *J. Biol. Chem.*, 250, 8938, 1975.
154. **Kostraba, N. C., Newman, R. S., and Wang, T. Y.,** Selective inhibition of transcription by a nonhistone protein isolated from Ehrlich ascites tumor chromatin, *Arch. Biochem. Biophys.*, 179, 100, 1977.
155. **Kostraba, N. C., Loor, R. M., and Wang, T. Y.,** Tissue-specificity of the nonhistone protein that inhibits RNA synthesis, *in vitro, Biochem. Biophys. Res. Commun.*, 79, 347, 1977.
156. **Olson, M. O. J., Orrick, L. R., Jones, C., and Busch, H.,** Phosphorylation of acid-soluble nucleolar proteins of Novikoff hepatoma ascites cells *in vivo, J. Biol. Chem.*, 249, 2823, 1974.
157. **Kang, Y.-J., Olson, M. O. J., and Busch, H.,** Phosphorylation of acid-soluble proteins in isolated nucleoli of Novikoff hepatoma ascites cells. Effects of divalent cations, *J. Biol. Chem.*, 249, 5580, 1974.
158. **Olson, M. O. J., Prestayko, A. W., Jones, C. E., and Busch, H.,** Phosphorylation of proteins of ribosomes and nucleolar preribosomal particles from Novikoff hepatoma ascites cells, *J. Mol. Biol.*, 90, 161, 1974.
159. **Gamo, S. and Lindell, T. J.,** Presence of two protein kinases in highly purified rat liver nucleoli, *Life Sci.*, 15, 2179, 1974.
160. **Grummt, I.,** Studies on the phosphorylation of nucleolar proteins. Identification of a nucleolus-associated protein kinase activity, *FEBS Lett.*, 39, 125, 1974.
161. **Grummt, I. and Grummt, F.,** Studies on the phosphorylation of nucleolar proteins. Comparison of the phosphorylation patterns of nuclear and nucleolar proteins labeled in vivo and in vitro, *FEBS Lett.*, 39, 129, 1974.
162. **Prestayko, A. W., Olson, M. O. J., and Busch, H.,** Phosphorylation of proteins of ribosomes and nucleolar preribosomal particles in vivo in Novikoff hepatoma asictes cells, *FEBS Lett.*, 44, 131, 1974.
163. **Olson, M. O. J., Ezrailson, E. G., Guetzow, K., and Busch, H.,** Localization and phosphorylation of nuclear, nucleolar and extranucleolar non-histone proteins of Novikoff hepatoma ascites cells, *J. Mol. Biol.*, 97, 611, 1975.

164. **Kang, Y.-J., Olson, M. O. J., Jones, C., and Busch, H.,** Nucleolar phosphoproteins of normal rat liver and Novikoff hepatoma ascites cells, *Cancer Res.*, 35, 1470, 1975.
165. **Wilson, M. J. and Ahmed, K.,** Localization of protein phosphokinase activities in the nucleolus distinct from extra-nucleolar regions in rat ventral prostate nuclei, *Exp. Cell Res.*, 93, 261, 1975.
166. **Kawashima, K. and Izawa, M.,** A nucleolus-specific phosphoprotein in mouse ascites tumor cells, *Mol. Biol. Rep.*, 3, 113, 1976.
167. **Kawashima, K. and Izawa, M.,** Phosphorylation of a nucleolus-specific phosphoprotein, *in vitro, Biochem. Biophys. Res. Commun.*, 74, 265, 1977.
168. **Sakuma, K., Fujitani, H., Chiu, J.-F., and Hnilica, L. S.,** Nuclear and nucleolar protein phosphokinases in Novikoff hepatoma, *Int. J. Biochem.*, 9, 127, 1978.
169. **Mamrack, M. D., Olson, M. O. J., and Busch, H.,** Amino acid sequence and sites of phosphorylation in a highly acidic region of nucleolar nonhistone protein C23, *Biochemistry*, 18, 3381, 1979.
170. **Atmar, V. J., Daniels, G. R., and Kuehn, G. D.,** Polyamine stimulation of phosphorylation of nonhistone acidic protein in nuclei and nucleoli from *Physarum polycephalum, Eur. J. Biochem.*, 90, 29, 1978.
171. **Kuehn, G. D., Affolter, H.-U., Atmar, V. J., Seebeck, T., Gubler, U., and Braun, R.,** Polyamine-mediated phosphorylation of a nucleolar protein from *Physarum polycephalum* that stimulates rRNA synthesis, *Proc. Natl. Acad. Sci. U.S.A.*, 76, 2541, 1979.
172. **Atmar, V. J., Daniels, G. R., Kuehn, G. D., and Braun, R.,** Opposing kinetic effects of an acidic nucleolar phosphoprotein from *Physarum polycephalum* on homologous and heterologous transcription systems, *FEBS Lett.*, 114, 205, 1980.
173. **Michetti, F., Miani, N., DeRenzis, G., Caniglia, A., and Correr, S.,** Nuclear localization of S-100 protein, *J. Neurochem.*, 22, 239, 1974.
174. **Miani, N., Michetti, F., DeRenzis, G., and Caniglia, A.** Effect of a brain-specific protein (S-100 protein) on the nucleolar RNA polymerase activity in isolated brain nuclei, *Experientia*, 29, 1499, 1973.
175. **Perumal, A. S.,** Nuclear protein kinase of brain: effect of S-100 protein, *Res. Commun. Chem. Pathol. Pharmacol.*, 13, 489, 1976.
176. **Perumal, A. S. and Rapport, M. M.,** *In vitro* phosphorylation of S-100 protein by brain nuclear protein kinases, *Life Sci.*, 22, 803, 1978.
177. **Buhler, J. M., Iborra, F., Sentenac, A., and Fromageot, P.,** The presence of phosphorylated subunits in yeast RNA polymerases A and B, *FEBS Lett.*, 71, 37, 1976.
178. **Hirsch, J. and Martelo, O. J.,** Phosphorylation of rat liver ribonucleic acid polymerase I by nuclear protein kinases, *J. Biol. Chem.*, 251, 5408, 1976.
179. **Hirsch, J. and Martelo, O. J.,** Purification and properties of a nuclear protein kinase associated with ribonucleic acid polymerase, *Biochem. J.*, 169, 355, 1978.
180. **Kranias, E. G., Schweppe, J. S., and Jungmann, R. A.,** Phosphorylative and functional modifications of nucleoplasmic RNA polymerase II by homologous adenosine 3′:5′-monophosphate-dependent protein kinase from calf thymus and by heterologous phosphatase, *J. Biol. Chem.*, 252, 6750, 1977.
181. **Kranias, E. G. and Jungmann, R. A.,** Phosphorylation of calf thymus RNA polymerase II by nuclear cyclic 3′,5′-AMP-dependent protein kinase, *Biochim. Biophys. Acta*, 517, 439, 1978.
182. **Martelo, O. J. and Hirsch, J.,** Effect of nuclear protein kinases on mammalian RNA synthesis, *Biochem. Biophys. Res. Commun.*, 58, 1008, 1974.
183. **Jungmann, R. A., Hiestand, P. C., and Schweppe, J. S.,** Adenosine 3′:5′-monophosphate-dependent protein kinase and the stimulation of ovarian nuclear ribonucleic acid polymerase activity, *J. Biol. Chem.*, 249, 5444, 1974.
184. **Dahmus, M. E.,** Stimulation of ascites tumor RNA polymerase II by protein kinase, *Biochemistry*, 15, 1821, 1976.
185. **Kuroiwa, A., Sekimizu, K., Ueno, K., Mizuno, D., and Natori, S.,** Separation of stimulatory factor of RNA polymerase II from protein kinase activity of Ehrlich ascites tumor cells, *FEBS Lett.*, 75, 183, 1977.
186. **Dokas, L., Rittschof, D., and Kleinsmith, L. J.,** Effects of cyclic adenosine 3′:5′-monophosphate on phosphoprotein kinase and phosphatase fractions prepared from rat liver nuclei, *Arch. Biochem. Biophs.*, 191, 578, 1978.
187. **Martelo, O. J., Woo, S. L. C., Reimann, E. M., and Davie, E. W.,** Effect of protein kinase on ribonucleic acid polymerase, *Biochemistry*, 9, 4807, 1970.
188. **Martelo, O. J., Woo, S. L. C., and Davie, E. W.,** Phosphorylation of *Escherichia coli* RNA polymerase by rabbit skeletal muscle protein kinase, *J. Mol. Biol.*, 87, 685, 1974.
189. **Zillig, W., Fujiki, H., Blum, W., Janeković, D., Schweiger, M., Rahmsdorf, H.-J., Ponta, H., and Hirsch-Kauffmann, M.,** *In vivo* and *in vitro* phosphorylation of DNA-dependent RNA polymerase of *Escherichia coli* by bacteriophage-T_7-induced protein kinase, *Proc. Natl. Acad. Sci. U.S.A.*, 72, 2506, 1975.
190. **Kleinsmith, L. J.,** Specific binding of phosphorylated non-histone chromatin proteins to deoxyribonucleic acid, *J. Biol. Chem.*, 248, 5648, 1973.

191. **Kostraba, N. C., Montagna, R. A., and Wang, T. Y.,** Study of loosely bound non-histone chromatin proteins. Stimulation of deoxyribonucleic acid-templated ribonucleic acid synthesis by a specific deoxyribonucleic acid-binding phosphoprotein fraction, *J. Biol. Chem.,* 250, 1548, 1975.
192. **Otto, B., Baynes, M., and Knippers, R.,** A single-strand specific DNA-binding protein from mouse cells that stimulates DNA polymerase. Its modification by phosphorylation, *Eur. J. Biochem.,* 73, 17, 1977.
193. **Linné, T., Jörnvall, H., and Philipson, L.,** Purification and characterization of the phosphorylated DNA-binding protein from adenovirus-Type-2-infected cells, *Eur. J. Biochem.,* 76, 481, 1977.
194. **Tegtmeyer, P., Rundell, K., and Collins, J. K.,** Modification of simian virus 40 protein A, *J. Virol.,* 21, 647, 1977.
195. **Edwards, C. A. F., Khoury, G., and Martin, R. G.,** Phosphorylation of T-antigen and control of T-antigen expression in cells transformed by wild-type and tsA mutants of simian virus 40, *J. Virol.,* 29, 753, 1979.
196. **Montenarh, M. and Henning, R.,** Simian virus 40 T-antigen phosphorylation is variable, *FEBS Lett.,* 114, 107, 1980.
197. **Boynton, A. L. and Whitfield, J. F.,** A possible involvement of type H cAMP-dependent protein kinase in the initiation of DNA synthesis by rat liver cells, *Exp. Cell Res.,* 126, 477, 1980.
198. **Mather, J. and Hotta, Y.,** A phosphorylatable DNA-binding protein associated with a lipoprotein fraction from rat spermatocyte nuclei, *Exp. Cell Res.,* 109, 181, 1977.
199. **Hotta, Y. and Stern, H.,** The effect of dephosphorylation on the properties of a helix-destabilizing protein from meiotic cells and its partial reversal by a protein kinase, *Eur. J. Biochem.,* 95, 31, 1979.
200. **Rose, K. M. and Jacob, S. T.,** Phosphorylation of nuclear poly(A) polymerase. Comparison of liver and hepatoma enzymes, *J. Biol. Chem.,* 254, 10256, 1979.
201. **Rose, K. M. and Jacob, S. T.,** Phosphorylation of nuclear poly(adenylic acid) polymerase by protein kinase: mechanism of enhanced poly(adenylic acid) synthesis, *Biochemistry,* 19, 1472, 1980.
202. **Vidali, G., Boffa, L. C., and Allfrey, V. G.,** Properties of an acidic histone-binding protein fraction from cell nuclei. Selective precipitation and deacetylation of histones F2A1 and F3, *J. Biol. Chem.,* 247, 7365, 1972.
203. **Nielsen, C. J., Sando, J. J., and Pratt, W. B.,** Evidence that dephosphorylation inactivates glucocorticoid receptors, *Proc. Natl. Acad. Sci. U.S.A.,* 74, 1398, 1977.
204. **Chan, P. K. and Liew, C. C.,** Identification of specific phosphoproteins in nuclease-digested chromatin subunits, *Can. J. Biochem.,* 55, 847, 1977.
205. **Chan, P.-K. and Liew, C.-C.,** Purification of a phosphoprotein from chromatin of rat liver, *Biochem. J.,* 183, 147, 1979.
206. **Goodwin, G. H., Sanders, C., and Johns, E. W.,** A new group of chromatin-associated proteins with a high content of acidic and basic amino acids, *Eur. J. Biochem.,* 38, 14, 1973,
207. **Weisbrod, S. and Weintraub, H.,** Isolation of a sublcass of nuclear proteins responsible for conferring a DNase I-sensitive structure on globin chromatin, *Proc. Natl. Acad. Sci. U.S.A.,* 76, 630, 1979.
208. **Sun, I. Y.-C., Johnson, E. M., and Allfrey, V. G.,** Affinity purification of newly phosphorylated protein molecules. Thiophosphorylation and recovery of histones H1, H2B, and H3 and the high mobility group protein HMG-1 using adenosine 5′-0-(3-thiotriphosphate) and cyclic AMP-dependent protein kinase, *J. Biol. Chem.,* 255, 742, 1980.
209. **MacGillivray, A. J., Johnston, C., MacFarlane, R., and Rickwood, D.,** The isolation and partial characterization of low-molecular-weight phosphorylated component of the non-histone proteins of mouse nuclei, *Biochem. J.,* 175, 35, 1978.
210. **Defer, N., Kitzis, A., Levy, F., Tichonicky, L., Sabatier, M., and Kruh, J.,** Presence of non-histone proteins in nucleosomes, *Eur. J. Biochem.,* 88, 583, 1978.
211. **Neumann, J. R., Owens, B. B., and Ingram, V. M.,** Nucleosome-associated protein kinases in murine erythroleukemia cells, *Arch. Biochem. Biophys.,* 197, 447, 1979.
212. **Phillips, I. R., Shephard, E. A., Tatcher, W. B., Stein, J. L., and Stein, G. S.,** Evidence for nonhistone chromosomal protein kinase activity associated with nucleosomes isolated from HeLa S_3 cells, *FEBS Lett.,* 106, 56, 1979.
213. **Böhm, J., Keil, G., and Knippers, R.,** Studies on protein-phosphorylation reactions in isolated chromatin, *Eur. J. Biochem.,* 78, 251, 1977.
214. **Gallinaro-Matringe, H. and Jacob, M.,** Nuclear particles from rat brain: complexity of the major proteins and their phosphorylation *in vivo, FEBS Lett.,* 36, 105, 1973.
215. **Schweiger, A. and Schmidt, D.,** Isolation of RNA-binding proteins from rat liver nuclear 30 S-particles, *FEBS Lett.,* 41, 17, 1974.
216. **Gallinaro-Matringe, H., Stevenin, J., and Jacob, M.,** Salt dissociation of nuclear particles containing DNA-like RNA. Distribution of phosphorylated and nonphosphorylated species, *Biochemistry,* 14, 2547, 1975.

217. **Blanchard, J. M., Ducamp, Ch., Jeanteur, Ph.,** Endogenous protein kinase activity in nuclear RNP particles from HeLa cells, *Nature (London),* 253, 467, 1975.
218. **Blanchard, J.-M., Brunel, C., and Jeanteur, P.,** Characterisation of an endogenous protein kinase activity in ribonucleoprotein structures containing heterogenous nuclear RNA in HeLa cell nuclei, *Eur. J. Biochem.,* 79, 117, 1977.
219. **Karn, J., Vidali, G., Boffa, L. C., and Allfrey, V. G.,** Characterization of the non-histone nuclear proteins associated with rapidly labeled heterogeneous nuclear RNA, *J. Biol Chem.,* 252, 7307, 1977.
220. **Blanchard, J.-M., Brunel, C., and Jeanteur, P.,** Phosphorylation *in vivo* and *in vitro* of proteins from HeLa cells: heterogeneous nuclear ribonucleoprotein particles, *Biochem. Soc. Trans.,* 5, 670, 1977.
221. **Periasamy, M., Brunel, C., Blanchard, J.-M ., and Jeanteur, P.,** Characterization of a phosphoprotein phosphatase activity in ribonucleoprotein particles from HeLa cell nuclei, *Biochem. Biophys. Res. Commun.,* 79, 1077, 1977.
222. **Blanchard, J.-M., Brunel, C., and Jeanteur, P.,** Phosphorylation *in vivo* of proteins associated with heterogeneous nuclear RNA in HeLa cell nuclei, *Eur. J. Biochem.,* 86, 301, 1978.
223. **Lelay, M.-N., Brunel, C., and Jeanteur, P.,** Peptide mapping of *in vivo* and *in vitro* phosphorylated sites of proteins from HeLa hnRPN, *FEBS Lett.,* 90, 54, 1978.
224. **Lam, K. S. and Kasper, C. B.,** Selective phosphorylation of a nuclear envelope polypeptide by an endogenous protein kinase, *Biochemistry,* 18, 307, 1979.
225. **Steer, R. C., Wilson, M. J., and Ahmed, K.,** Protein phosphokinase activity of rat liver nuclear membrane, *Exp. Cell Res.,* 119, 403, 1979.
226. **Steer, R. C., Goueli, S. A., Wilson, M. J., and Ahmed, K.,** Cobalt-stimulated protein phosphokinase activity of the pore complex-lamina fraction from rat liver nuclear envelope, *Biochem. Biophys. Res. Commun.,* 92, 919, 1980.
227. **Steer, R. C., Wilson, M. J., and Ahmed, K.,** Phosphoprotein phosphatase activity of rat liver nuclear membrane, *Biochem. Biophys. Res. Commun.,* 89, 1082, 1979.
228. **Berezney, R. and Coffey, D. S.,** Identification of a nuclear protein matrix, *Biochem. Biophys. Res. Commun.,* 60, 1410, 1974.
229. **Berezney, R. and Coffey, D. S.,** Nuclear protein matrix: association with newly synthesized DNA, *Science,* 189, 291, 1975.
230. **Kleinsmith, L. J., Allfrey, V. G., and Mirsky, A. E.,** Phosphoprotein metabolism in isolated lymphocyte nuclei, *Pro. Natl. Acad. Sci. U.S.A.,* 55, 1182, 1966.
231. **Louie, A. J., Candido, E. P. M., and Dixon, G. H.,** Enzymatic modifications and their possible roles in regulating the binding of basic proteins to DNA and in controlling chromosomal structure, *Cold Spring Harbor Symp. Quant. Biol.,* 38, 803, 1973.
232. **Dixon, G. H., Candido, E. P. M., Honda, B. M., Louie, A. J., MacLeod, A. R., and Sung, M. T.,** The biological roles of post-synthetic modifications of basic nuclear proteins, in *The Structure and Function of Chromatin,* Ciba Found. Symp. 28, Elsevier, New York, 1975, 231.
233. **Ruiz-Carrillo, A., Wangh, L. J., and Allfrey, V. G.,** Processing of newly synthesized histone molecules, *Science,* 190, 117, 1975.
234. **Gurley, L. R., Tobey, R. A., Walters, R. A., Hildebrand, C. E., Hohmann, P. G., D'Anna, J. A., Barham, S. S., and Deaven, L. L.,** Histone phosphorylation and chromatin structure in synchronized mammalian cells, in *Cell Cycle Regulation,* Jeter, J. R., Jr., Cameron, I. L., Padilla, G. M., and Zimmerman, A. M., Eds., Academic Press, New York, 1978, 37.
235. **Isenberg, I.,** Histones, *Annu. Rev. Biochem.,* 48, 159, 1979.
236. **Bustin, M. and Cole, R. D.,** A study of the multiplicity of lysine-rich histones, *J. Biol. Chem.,* 244, 5286, 1969.
237. **Noll, M. and Kornberg, R. D.,** Action of micrococcal nuclease on chromatin and the location of histone H1, *J. Mol. Biol.,* 109, 393, 1977.
238. **Kornberg, R. D.,** Structure of chromatin, *Annu. Rev. Biochem.,* 46, 931, 1977.
239. **Langan, T. A.,** Phosphorylation of liver histone following the administration of glucagon and insulin, *Proc. Natl. Acad. Sci. U.S.A.,* 64, 1276, 1969.
240. **Langan, T. A.,** Cyclic AMP and histone phosphorylation, *Ann. N.Y. Acad. Sci.,* 185, 166, 1971.
241. **Langan, T. A.,** Action of adenosine 3′,5′-monophosphate-dependent histone kinase *in vivo, J. Biol. Chem.,* 244, 5763, 1969.
242. **Adler, A. J., Langan, T. A., and Fasman, G. D.,** Complexes of deoxyribonucleic acid with lysine-rich (fl) histone phosphorylated at two separate sites: circular dichroism studies, *Arch. Biochem. Biophys.,* 153, 769, 1972.
243. **Fasy, T. M., Inoue, A., Johnson, E. M., and Allfrey, V. G.,** Phosphorylation of H1 and H5 histones by cyclic AMP-dependent protein kinase reduces DNA binding, *Biochim. Biophys. Acta,* 564, 322, 1979.
244. **Lamy, F., Lecocq, R., and Dumont, J. E.,** Thyrotropin stimulation of the phosphorylation of serine in the N-terminal region of thyroid H1 histones, *Eur. J. Biochem.,* 73, 529, 1977.

245. **Spaulding, S. W. and Schubart, U. K.,** Time course of thyrotropin-dependent protein phosphorylation in thyroid slices, *Endocrinology,* 103, 2334, 1978.
246. **Morris, S. M., Jr. and Cole, R. D.,** Thyroxine stimulation of tadpole liver histone phosphorylation in vivo, *Dev. Biol.,* 74, 379, 1980.
247. **Yu, M. W., Hori, S., Tolson, N., Huff, K., and Guroff, G.,** Increased phosphorylation of a specific nuclear protein in rat superior cervical ganglia in response to nerve growth factor, *Biochem. Biophys. Res. Commun.,* 81, 941, 1978.
248. **Huff, K. R. and Guroff, G.,** Epidermal growth factor stimulates the phosphorylation of a specific nuclear protein in chick embryo epidermis, *Biochem. Biophys. Res. Commun.,* 85, 464, 1978.
249. **Stevely, W. S. and Stocken, L. A.,** Variations in the phosphate content of histone F1 in normal and irradiated tissues, *Biochem. J.,* 110, 187, 1968.
250. **Ord, M. G. and Stocken, L. A.,** Variations in the phosphate content and thiol/disulphide ratio of histones during the cell cycle. Studies with regenerating rat liver and sea urchins, *Biochem. J.,* 107, 403, 1968.
251. **Gurley, L. R. and Walters, R. A.,** Response of histone turnover and phosphorylation to X-irradiation, *Biochemistry,* 10, 1588, 1971.
252. **Balhorn, R., Rieke, W. O., and Chalkley, R.,** Rapid electrophoretic analysis for histone phosphorylation. A reinvestigation of phosphorylation of lysine-rich histone during liver regeneration, *Biochemistry,* 10, 3952, 1971.
253. **Balhorn, R., Chalkley, R., and Granner, D.,** Lysine-rich histone phosphorylation. A positive correlation with cell replication, *Biochemistry,* 11, 1094, 1972.
254. **Balhorn, R., Balhorn, M., Morris, H. P., and Chalkley, R.,** Comparative high-resolution electrophoresis of tumor histones: variation in phosphorylation as a function of cell replication rate, *Cancer Res.,* 32, 1775, 1972.
255. **Louie, A. J. and Dixon, G. H.,** Kinetics of phosphorylation and dephosphorylation of testis histones and their possible role in determining chromosomal structure, *Nature New Biol.,* 243, 164, 1973.
256. **Bradbury, E. M., Inglis, R. J., Matthews, H. R., and Sarner, N.,** Phosphorylation of very-lysine-rich histone in *Physarum polycephalum.* Correlation with chromosome condensation, *Eur. J. Biochem.,* 33, 131, 1973.
257. **Bradbury, E. M., Inglis, R. J., and Matthews, H. R.,** Control of cell division by very lysine rich histone (F1) phosphorylation, *Nature (London),* 247, 257, 1974.
258. **Bradbury, E. M., Inglis, R. J., Matthews, H. R., and Langan, T. A.,** Molecular basis of control of mitotic cell division in eukaryotes, *Nature (London),* 249, 553, 1974.
259. **Inglis, R. J., Langan, T. A., Matthews, H. R., Hardie, D. G., and Bradbury, E. M.,** Advance of mitosis by histone phosphokinase, *Exp. Cell Res.,* 97, 418, 1976.
260. **Lake, R. S., Goidl, J. A., and Salzman, N. P.,** F-1 histone modification at metaphase in Chinese hamster cells, *Exp. Cell Res.,* 73, 113, 1972.
261. **Lake, R. S. and Salzman, N. P.,** Occurrence and properties of a chromatin-associated F1-histone phosphokinase in mitotic Chinese hamster cells, *Biochemistry,* 11, 4817, 1972.
262. **Lake, R. S.,** Further characterization of the F1-histone phosphokinase of metaphase-arrested animal cells, *J. Cell Biol.,* 58, 317, 1973.
263. **Lake, R. S.,** F-1 histone phosphorylation in metaphase chromosomes of cultured Chinese hamster cells, *Nature New Biol.,* 242, 145, 1973.
264. **Marks, D. B., Paik, W. K., and Borun, T. W.,** The relationship of histone phosphorylation to deoxyribonucleic acid replication and mitosis during the HeLa S-3 cell cycle, *J. Biol. Chem.,* 248, 5660, 1973.
265. **Gurley, L. R., Walters, R. A., and Tobey, R. A.,** The metabolism of histone fractions. VI. Differences in the phosphorylation of histone fractions during the cell cycle, *Arch. Biochem. Biophys.,* 154, 212, 1973.
266. **Gurley, L. R., Walters, R. A., and Tobey, R. A.,** The metabolism of histone fractions. Phosphorylation and synthesis of histones in late G_1-arrest, *Arch. Biochem. Biophys.,* 164, 469, 1974.
267. **Gurley, L. R., Walters, R. A., and Tobey, R. A.,** Cell cycle-specific changes in histone phosphorylation associated with cell proliferation and chromosome condensation, *J. Cell Biol.,* 60, 356, 1974.
268. **Gurley, L. R., Walters, R. A., and Tobey, R. A.,** Sequential phosphorylation of histone subfractions in the Chinese hamster cell cycle, *J. Biol. Chem.,* 250, 3936, 1975.
269. **Hohmann, P., Tobey, R. A., and Gurley, L. R.,** Phosphorylation of distinct regions of F1 histone. Relationship to the cell cycle, *J. Biol. Chem.,* 251, 3685, 1976.
270. **Gurley, L. R., D'Anna, J. A., Barham, S. S., Deaven, L. L., and Tobey, R. A.,** Histone phosphorylation and chromatin structure during mitosis in Chinese hamster cells, *Eur. J. Biochem.,* 84, 1, 1978.
271. **Balhorn, R., Jackson, V., Granner, D., and Chalkley, R.,** Phosphorylation of the lysine-rich histones throughout the cell cycle, *Biochemistry,* 14, 2504, 1975.
272. **Tanphaichitr, N., Moore, K. C., Granner, D. K., and Chalkley, R.,** Relationship between chromosome condensation and metaphase lysine-rich histone phosphorylation, *J. Cell Biol.,* 69, 43, 1976.

273. **Matsumoto, Y., Yasuda, H., Mita, S., Marunouchi, T., and Yamada, M.,** Evidence for the involvement of H1 histone phosphorylation in chromosome condensation, *Nature (London),* 284, 181, 1980.
274. **Hildebrand, C. E. and Tobey, R. A.,** Cell-cycle specific changes in chromatin organization, *Biochem. Biophys. Res. Commun.,* 63, 134, 1975.
275. **Smith, D. L., Chen, C.-C., Bruegger, B. B., Holtz, S. L., Halpern, R. M., and Smith, R. A.,** Characterization of protein kinases forming acid-labile histone phosphates in Walker-256 carcinosarcoma cell nuclei, *Biochemistry,* 18, 3780, 1974.
276. **Chen, C.-C., Smith, D. L., Bruegger, B. B., Halpern, R. M., and Smith, R. A.,** Occurrence and distribution of acid-labile histone phosphates in regenerating rat liver, *Biochemistry,* 18, 3785, 1974.
277. **Chen, C.-C., Bruegger, B. B., Kern, C. W., Lin, Y. C., Halpern, R. M., and Smith, R. A.,** Phosphorylation of nuclear proteins in rat regenerating liver, *Biochemistry,* 16, 4852, 1977.
278. **Sung, M. T. and Dixon, G. H.,** Modification of histones during spermiogenesis in trout: a molecular mechanism for altering histone binding to DNA, *Proc. Natl. Acad. Sci. U.S.A.,* 67, 1616, 1970.
279. **Louie, A. J. and Dixon, G. H.,** Trout testis cells. II. Synthesis and phosphorylation of histones and protamines in different cell types, *J. Biol. Chem.,* 247, 5498, 1972.
280. **Louie, A. J., Sung, M. T., and Dixon, G. H.,** Modification of histones during spermatogenesis in trout. II. Levels of phosphohistone species and kinetics of phosphorylation of histone IIb_1, *J. Biol. Chem.,* 248, 3335, 1973.
281. **Sung, M. T., Dixon, G. H., and Smithies, O.,** Phosphorylation and synthesis of histones in regenerating rat liver, *J. Biol. Chem.,* 246, 1358, 1971.
282. **Marushinge, K., Ling, V., and Dixon, G. H.,** Phosphorylation of Chromosomal basic proteins in maturing trout testis, *J. Biol. Chem.,* 244, 5953, 1969.
283. **Ruiz-Carrillo, A., Wangh, L. J., and Allfrey, V. G.,** Selective synthesis and modification of nuclear proteins during maturation of avian erythroid cells, *Arch. Biochem. Biophys.,* 174, 273, 1976.
284. **Sung, M. T., Harford, J., Bundman, M., and Vidalakas, G.,** Metabolism of histones in avian erythroid cells, *Biochemistry,* 16, 279, 1977.
285. **Garrard, W. T., Kidd, G. H., and Bonner, J.,** Histone phosphorylation during liver regeneration, *Biochem. Biophys. Res. Commun.,* 70, 1219, 1976.
286. **Ord, M. G., Stocken, L. A., and Thrower, S.,** Histone phosphorylations and their disparate roles in interphase, *Sub-Cell. Biochem.,* 4, 147, 1975.
287. **Neumann, J. R., Housman, D., and Ingram, V. M.,** Nuclear protein synthesis and phosphorylation in Friend erythroleukemia cells stimulated with DMSO, *Exp. Cell Res.,* 111, 277, 1978.
288. **Gurley, L. R., Walters, R. A., Barham, S. S., and Deaven, L. L.,** Heterochromatin and histone phosphorylation, *Exp. Cell Res.,* 111, 373, 1978.
289. **Shlyapnikov, S. V., Arutyunyan, A. A., Kurochkin, S. N., Memelova, L. V., Nesterova, M. V., Sashchenko, L. P., and Severin, E. S.,** Investigation of the sites phosphorylated in lysine-rich histones by protein kinase from pig brain, *FEBS Lett.,* 53, 316, 1975.
290. **Gurley, L. R. and Walters R., A.,** Evidence from triton X-100 polyacrylamide gel electrophoresis that histone f2a2, not F2b, is phosphorylated in Chinese hamster cells, *Biochem. Biophys. Res. Commun.,* 55, 697, 1973.
291. **Farago, A., Romnhanyi, T., Antoni, F., Takats, A., and Fabian, F.,** The phosphorylated site of calf thymus F2b histone by the cyclic AMP-dependent protein kinase, *Nature (London),* 254, 88, 1975.
292. **Hashimoto, E., Takeda, M., and Nishizuka, Y., Hamana, K., and Iwai, K.,** Phosphorylated sites of calf thymus histone H2b by adenosine 3′,5′-monophosphate-dependent protein kinase from silkworm, *Biochem. Biophys. Res. Commun.,* 66, 547, 1975.
293. **Marzluff, W. F., Jr. and McCarty, K. S.,** Structural studies of calf thymus F3 histone. II. Occurrence of phosphoserine and ε-N-acetyllysine in thermolysin peptides, *Biochemistry,* 11, 2677, 1972.
294. **Szopa, J., Jacob, G., and Arfmann, H.-A.,** Influence of histone phosphorylation upon histone-histone interactions studied, *in vitro, Biochemistry,* 19, 987, 1980.
295. **Louie, A. J. and Dixon, G. H.,** Synthesis, acetylation, and phosphorylation of histone IV and its binding to DNA during spermatogenesis in trout, *Proc. Natl. Acad. Sci. U.S.A.,* 69, 1975, 1972.
296. **Ruiz-Carrillo, A., Wangh, L. J., Littau, V. C., and Allfrey, V. G.,** Changes in histone acetyl content and in nuclear non-histone protein composition of avian erythroid cells at different stages of maturation, *J. Biol. Chem.,* 249, 7358, 1974.
297. **Sung, M. T.,** Phosphorylation and dephosphorylation of histone V (H5): controlled condensation of avian erythrocyte chromatin. Wagner, T. E., Hartford, J. B., Serra, M., Vandergrift, V., and Sung, M. T., Appendix: phosphorylation and dephosphorylation of histone H5. II. Circular dichroic studies, *Biochemistry,* 16, 286, 1977.
298. **Sung, M. T. and Freedlender, E. F.,** Sites of in vivo phosphorylation of histone H5, *Biochemistry,* 17, 1884, 1978.

299. **Louie, A. J. and Dixon, G. H.,** Kinetics of enzymatic modifications of the protamines and a proposal for their binding to chromatin, *J. Biol. Chem.*, 247, 7962, 1972.
300. **Sanders, M. M. and Dixon, G. H.,** The biosynthesis of protamine in trout testis. V. Sites of phosphorylation, *J. Biol. Chem.*, 247, 851, 1972.
301. **Louie, A. J. and Dixon, G. H.,** Enzymatic modifications of the protamines. II. Separation and characterization of phosphorylated species of protamines from trout testis, *Can. J. Biochem.*, 52, 536, 1974.
302. **Pumo, D. E. and Kleinsmith, L. J.,** Methods for the assessment of nonhistone phosphorylation (acid-stable, alkali-labile linkages), *Methods of Cell Biol.*, 19, 119, 1978.
303. **Smith, D. L., Bruegger, B. B., Halpern, R. M., and Smith, R. A.,** New histone kinases in nuclei of rat tissues, *Nature (London)*, 246, 103, 1973.
304. **Smith, R. A., Halpern, R. M., Bruegger, B. B., Dunlap, A. K., and Fricke, O.,** Chromosomal protein phosphorylation on basic amino acids, *Methods Cell Biol.*, 19, 153, 1978.
305. **Johnson, E. M., Hadden, J. W., Inoue, A., and Allfrey, V. G.,** DNA binding by cyclic adenosine 3′,5′-monophosphate-dependent protein kinase from calf thymus nuclei, *Biochemistry*, 14, 3873, 1975.
306. **Umansky, S. R., Zotova, R. N., and Kovalev, Y. I.,** Comparison of some properties of chromatin nonhistone proteins and nuclear sap proteins, *Eur. J. Biochem.*, 65, 503, 1976.
307. **Kranias, E. G. and Jungmann, R. A.,** Purification and properties of a cyclic AMP-independent protein kinase from calf thymus nuclei, *Biochim. Biophys. Acta*, 517, 447, 1978.
308. **Shoemaker, C. B. and Chalkley, R.,** An H3 histone-specific kinase isolated from bovine thymus chromatin, *J. Biol. Chem.*, 253, 5802, 1978.
309. **Takeda, M., Yamamura, H., and Ohga, Y.** Phosphoprotein kinases associated with rat liver chromatin, *Biochem. Biophys. Res. Commun.*, 42, 103, 1971.
310. **Kamiyama, M. and Dastugue B.,** Rat liver non-histone proteins: correlation between protein kinase activity and activation of RNA synthesis, *Biochem. Biophys. Res. Commun.*, 44, 29, 1971.
311. **Kamiyama, M., Dastugue, B., and Kruh, J.,** Action of phosphoproteins and protein kinase from rat liver chromatin on RNA synthesis, *Biochem. Biophys. Res. Commun.*, 44, 1345, 1971.
312. **Kamiyama, M., Dastugue, B., Defer, N., and Kruh, J.,** Liver chromatin non-histone proteins partial fractionation and mechanism of action on RNA synthesis, *Biochim. Biophys. Acta*, 277, 576, 1972.
313. **Desjardins, P. R., Lue, P. F., Liew, C. C., and Gornall, A. G.,** Purification and properties of rat liver nuclear protein kinases, *Can. J. Biochem.*, 50, 1249, 1972.
314. **Ruddon, R. W. and Anderson, S. L.,** Presence of multiple protein kinase activities in rat liver nuclei, *Biochem. Biophys. Res. Commun.*, 46, 1499, 1972.
315. **Kish, V. M., and Kleinsmith, L. J.,** Nuclear protein kinases. Evidence for their heterogeneity, tissue specificity, substrate specificities, and differential response to cyclic adenosine 3′:5′-monophosphate, *J. Biol. Chem.*, 249, 750, 1974.
316. **Dastugue, B., Tichonicky, L., and Kruh, J.,** Multiple forms of protein kinase in liver cell. II. Nuclear kinases and cytosol phosvitin kinase, *Biochimie*, 56, 491, 1974.
317. **Takeda, M., Matsumura, S., and Nakaya, Y.,** Nuclear phosphoprotein kinases from rat liver, *J. Biochem.*, 75, 743, 1974.
318. **Desjardins, P. R., Liew, C. C., and Gornall, A. G.** Rat liver nuclear protein kinases, *Can. J. Biochem.*, 53, 354, 1975.
319. **Rikans, L. E. and Ruddon, R. W.,** Partial purification and properties of a chromatin-associated phosphoprotein kinase from rat liver nuclei, *Biochim. Biophys. Acta*, 422, 73, 1976.
320. **Stahl, H. and Knippers, R.,** Chromatin-associated protein kinases specific for acidic proteins, *Biochim. Biophys. Acta*, 614, 71, 1980.
321. **Keller, R. K., Socher, S. H., Krall, J. F., Chandra, T., and O'Malley, B. W.,** Fractionation of chick oviduct chromatin: IV. Association of protein kinase with transcriptionally active chromatin, *Biochem. Biophys. Res. Commun.*, 66, 453, 1975.
322. **Spielvogel, A. M., Mednieks, M. I., Eppenberger, U., and Jungmann, R. A.,** Evidence for the identity of nuclear and cytoplasmic adenosine-3′:5′-monophosphate-dependent protein kinase from porcine ovaries and nuclear translocation of the cytoplasmic enzyme, *Eur. J. Biochem.*, 73, 199, 1977.
323. **Gibson, K., Tichonicky, L., and Kruh, J.,** Rat myocardial protein kinases. II. Nuclei, *Biochimie*, 56, 1417, 1974.
324. **Limas, C. J. and Chan-Stier, C.,** Adenosine 3′,5′-monophosphate-dependent protein kinases of myocardial non-histone nucler proteins, *Biochim. Biophys. Acta*, 477, 404, 1977.
325. **Verhaegen, M. and Sand, G.,** Characterization of protein phosphokinase activities in horse thyroid nuclei, *Biochim. Biophys. Acta*, 429, 163, 1976.
326. **Lee, N. M. and Loh, H. H.,** Phosphorylation and methylation of chromatin proteins from mouse brain nuclei, *J. Neurochem.*, 29, 547, 1977.
327. **Desjardins, P. R., Mendelson, I. M., and Anderson, K. M.,** Relationship between RNA polymerase and protein kinase activities in rat mammary gland nuclei, *Can. J. Biochem.*, 53, 591, 1975.

328. **Majumder, G. C.,** Purification and properties of a nuclear protein kinase from rat mammary gland, *Biochem. J.*, 165, 469, 1977.
329. **Hardie, D. G., Matthews, H. R., and Bradbury, E. M.,** Cell-cycle dependence of two nuclear histone kinase enzyme activities, *Eur. J. Biochem.*, 66, 37, 1976.
330. **Schlepper, J. and Knippers, R.,** Nuclear protein kinases from murine cells, *Eur. J. Biochem.*, 60, 209, 1975.
331. **Costa, M., Fuller, D. J. M., Russell, D. H., and Gerner, E. W.,** Cytoplasmic and nuclear protein kinases during the cell cycle, *Biochim. Biophys. Acta,* 479, 416, 1977.
332. **Thomson, J. A., Mon, M. J., Stein, J. L., Duval, K. A., Kleinsmith, L. J., and Stein, G. S.,** Partial fractionation and characterization of nuclear protein kinases in HeLa S_3 cells, *Cell Differ.*, 8, 305, 1979.
333. **Simpson, R. T.,** Protein kinase in HeLa nucleosomes: a reevaluation of the interactions of histones with the ends of core particle DNA, *Nucleic Acids Res.*, 5, 1109, 1978.
334. **Linnala-Kankkunen, A. and Mäenpää, P. H.,** A cyclic GMP-dependent histone kinase bound to liver nucleoli, *Biochim. Biophys. Acta*, 587, 324, 1979.
335. **Faragó, A., Antoni, F., and Fábián, F.,** Histone kinases and cyclic AMP-binding capacity of nuclei of human tonsillar lymphocytes, *Biochem. Biophys. Acta,* 370, 459, 1974.
336. **Neumann, J. R., O'Meara, A. R., and Herrmann, R. L.,** Cyclic AMP-dependent histone-specific nucleoplasmic protein kinase from rat liver, *Biochem. J.*, 171, 123, 1978.
337. **Langan, T.,** Isolation of histone kinases, *Methods Cell Biol.*, 19, 143, 1978,
338. **Sikorska, M., MacManus, J. P., Walker, P. R., and Whitfield, J. F.,** The protein kinases of rat liver nuclei, *Biochem. Biophys. Res. Commun.*, 93, 1196, 1980.
339. **Mednieks, M. I., Jungmann, R. A., and Schweppe, J. S.,** Ontogeny and multiplicity of cyclic AMP-dependent protein kinase in thymic lymphoid cells, *Arch. Biochem. Biophys.*, 197, 185, 1979.
340. **Jungmann, R. A. and Kranias, E. G.,** Nuclear phosphoprotein kinases and the regulation of gene transcription, *Int. J. Biochem.*, 8, 819, 1977.
341. **Walsh, D. A., Ashby, C. D., Gonzalez, C., Calkins, D., Fischer, E. H., and Krebs, E. G.,** Purification and characterization of a protein inhibitor of adenosine 3',5'-monophosphate-dependent protein kinases, *J. Biol. Chem.*, 246, 1977, 1971.
342. **Mitchelson, K., Chambers, T., Bradbury, E. M., and Matthews, H. R.,** Activation of histone kinase in G2 phase of the cell cycle in *Physarum polycephalum, FEBS Lett.*, 92, 339, 1978.
343. **Masaracchia, R. A., Kemp, B. E., and Walsh, D. A.,** Histone-4-phosphotransferase activities in proliferating lymphocytes. Partial purification and characterization of an enzyme specific for ser-47, *J. Biol. Chem.*, 252, 7109, 1977.
344. **Thornburg, W. and Lindell, T. J.,** Purification of rat liver nuclear protein kinase NII, *J. Biol. Chem.*, 252, 6660, 1977.
345. **Thornburg, W., O'Malley, A. F., and Lindell, T. J.,** Purification of rat liver nuclear protein kinase NI, *J. Biol. Chem.*, 253, 4638, 1978.
346. **Thornburg, W., Kieras, R., and Lindell, T. J.,** Selective binding of rat liver nuclear enzymes to histones, phosvitin and casein coupled to sepharose, *Prep. Biochem.*, 8, 133, 1978.
347. **Thornburg, W., Gamo, S., O'Malley, A. F., and Lindell, T. J.,** Properties of rat liver nuclear protein kinases, *Biochim. Biophys. Acta,* 571, 35, 1979.
348. **Jungmann, R. A., Hiestand, P. C., and Schweppe, J. S.,** Mechanism of action of gonadotropin. IV. Cyclic adenosine monophosphate-dependent translocation of ovarian cytoplasmic cyclic adenosine monophosphate-binding protein and protein kinase to nuclear acceptor sites, *Endocrinology,* 94, 168, 1974.
349. **Jungmann, R. A., Lee, S.-G., DeAngelo, A. B.,** Translocation of cytoplasmic protein kinase and cyclic adenosine monophosphate-binding protein to intracellular acceptor sites, *Adv. Cyclic Nucleotide Res.*, 5, 281, 1975.
350. **Jungmann, R. A. and Russell, D. H.,** Cyclic AMP, cyclic AMP-dependent protein kinase, and regulation of gene expression, *Life Sci.*, 20, 1787, 1977.
351. **Palmer, W. K., Castagna, M., and Walsh, D. A.,** Nuclear protein kinase activity in glucagon-stimulated perfused rat livers, *Biochem J.*, 143, 469, 1974.
352. **Castagna, M., Palmer, W. K., and Walsh, D. A.,** Nuclear protein-kinase activity in perfused rat liver stimulated with dibutyryl-adenosine cyclic 3':5'-monophosphate, *Eur. J. Biochem.*, 55, 193, 1975.
353. **Kurosawa, A., Guidotti, A., and Costa, E.,** Induction of tyrosine 3-monooxy-genase in adrenal medulla: role of protein kinase activation and translocation, *Science,* 193, 691, 1976.
354. **Chuang, D.-M. and Costa, E.,** Trans-synaptic regulation of ribonucleic acid biosynthesis in rat adrenal medulla, *Mol. Pharmacol.*, 12, 514, 1976.
355. **Costa, E., Kurosawa, A., and Guidotti, A.,** Activation and nuclear translocation of protein kinase during transsynaptic induction of tyrosine 3-monooxygenase, *Proc. Natl. Acad. Sci. U.S.A.*, 73, 1058, 1976.

356. **Kurosawa, A., Guidotti, A., and Costa, E.,** Induction of tyrosine 3-monooxygenase elicited by carbamylcholine in intact and denervated adrenal medulla: role of protein kinase activation and translocation, *Mol. Pharmacol.,* 12, 420, 1976.
357. **Kurosawa, A., Guidotti, A., and Costa, E.,** Nuclear translocation of cyclic AMP-dependent protein kinase subunits during the transsynaptic activation of gene expression in rat adrenal medulla, *Mol. Pharmacol.,* 15, 115, 1979.
358. **Hollenbeck, R. A., Chuang, D. M., and Costa, E.,** Translocation of cytosol protein kinase into nuclei and the induction of tyrosein hydroxylase in NBD-2 neuroblastoma cells, *Brain Res.,* 171, 481, 1979.
359. **Cho-Chung, Y. S. and Clair, T.,** Altered cyclic AMP-binding and db cyclic AMP-unresponsiveness, *in vivo, Nature (London),* 265, 452, 1977.
360. **Cho-Chung, Y. S., Clair, T., and Porper, R.,** Cyclic AMP-binding proteins and protein kinase during regression of Walker 256 mammary carcinoma, *J. Biol. Chem.,* 252, 6342, 1977.
361. **Cho-Chung, Y. S., Clair, T., and Huffman, P.,** Loss of nuclear cyclic AMP binding in cyclic AMP-unresponsive Walker 256 mammary carcinoma, *J. Biol. Chem.,* 252, 6349, 1977.
362. **Stern, H., Allfrey, V., Mirsky, A. E., and Saetren, H.,** Some enzymes of isolated nuclei, *J. Gen. Physiol.,* 35, 559, 1952.
363. **Das R. and Kanungo, M. S.,** Effects of polyamines on *in vitro* phosphorylation and acetylation of histones of the cerebral cortex of rats of various ages, *Biochem. Biophys. Res. Commun.,* 90, 708, 1979.
364. **Farron-Furstenthal, F. and Lightholder, J. R.,** Effects of polyamines and histones on the phosphorylation of non-histone proteins in isolated rat liver nuclei, *Biochem. Biophys. Res. Commun.,* 83, 94, 1978.
365. **Imai, H., Shimoyama, M., Yamamoto, S., Tanigawa, Y., and Ueda, I.,** Effect of polyamines on phosphorylation of non-histone chromatin proteins from hog liver, *Biochem. Biophys. Res. Commun.,* 66, 856, 1975.
366. **Ahmed, K., Wilson, M. J., Goueli, S. A., and Williams-Ashman, H. G.,** Effects of polyamines on prostatic chromatin- and non-histone-protein-associated protein kinase reactions, *Biochem. J.,* 176, 739, 1978.
367. **Pegg, A. E., Lockwood, D. H., and Williams-Ashman, H. G.,** Concentrations of putrescine and polyamines and their enzymic synthesis during androgen-induced prostatic growth, *Biochem. J.,* 117, 17, 1970.
368. **Liang, T., Mezzetti, G., Chen, C., and Liao, S.,** Selective polyamine-binding proteins. Spermine binding by an androgen-sensitive phosphoprotein, *Biochim. Biophys. Acta,* 542, 430, 1978.
369. **Fuller, D. J. M., Byus, C. V., and Russell, D. H.,** Specific regulation by steroid hormones of the amount of type I cyclic AMP-dependent protein kinase holoenzyme, *Proc. Natl. Acad. Sci. U.S.A.,* 75, 223, 1978.
370. **Talmadge, K. W., Bechtel, E., and Eppenberger, U.,** Protein kinases and proteins binding adenosine 3′:5′-monophosphate in subcellular fractions of calf ovaries, *Eur. J. Biochem.,* 78, 419, 1977.
371. **Rikans, L. and Ruddon, R. W.,** Role of 3′,5′-cyclic AMP in the control of nuclear protein kinase activity, *Biochem. Biophys. Res. Commun.,* 54, 387, 1973.
372. **Lee, P. C. and Jungmann, R. A.,** Ontogeny of cyclic AMP-dependent protein phosphokinase during hepatic development of the rat, *Biochim. Biophys. Acta,* 399, 265, 1975.
373. **Guzik, G. and Anderson, K. M.,** Cyclic nucleotides and other factors affecting the apparent activity of RNA polymerases solubilized from rat mammary gland nuclei, *Int. J. Biochem.,* 7, 607, 1976.
374. **Cho-Chung, Y. S., Bodwin, J. S., and Clair, T.** Cyclic AMP-binding proteins: inverse relationship with estrogen-receptors in hormone-dependent mammary tumor regression, *Eur. J. Biochem.,* 86, 51, 1978.
375. **Shafie, S. M., Cho-Chung, Y. S., and Gullino, P. M.,** Cyclic adenosine 3′:5′-monophosphate and protein kinase activity in insulin-dependent and -independent mammary tumors, *Cancer Res.,* 39, 2501, 1979.
376. **Kwast-Welfeld, J. and Kaniuga, Z.,** Cyclic AMP-binding proteins in chick liver, *Int. J. Biochem.,* 10, 403, 1979.
377. **Prashad, N., Rosenberg, R. N., Wischmeyer, B., Ulrich, C., and Sparkman, D.,** Induction of adenosine 3′,5′-monophosphate binding proteins by $N^6,O^{2'}$-dibutyryladenosine 3′,5′-monophosphate in mouse neuroblastoma cells. Analysis by two-dimensional gel electrophoresis, *Biochemistry,* 18, 2717, 1979.
378. **Richards, J. S. and Rolfes, A. I.,** Hormonal regulation of cyclic AMP binding to specific receptor proteins in rat ovarian follicles. Characterization by photoaffinity labeling, *J. Biol. Chem.,* 255, 5481, 1980.
379. **Friedman, D. and Chambers, D. A.,** Cyclic nucleotide-binding proteins detected by photoaffinity labeling in nucleus and cytoplasm of bovine liver, *Proc. Natl. Acad. Sci. U.S.A.,* 75, 5286, 1978.
380. **Higashino, H. and Takeda, M.,** Changes in adenosine 3′,5′-monophosphate liver and protein kinase activity by glucagon in rat liver nuclei, *J. Biochem.,* 75, 189, 1974.
381. **Liao, S., Lin, A. H., and Tymoczko, J. L.,** Adenyl cyclase of cell nuclei isolated from rat ventral prostate, *Biochim. Biophys. Acta,* 230, 535, 1971.
382. **Soifer, D. and Hechter, O.,** Adenyl cyclase activity in rat liver nuclei, *Biochim. Biophys. Acta,* 230, 539, 1971.

383. **Atmar, V. J., Westland, J. A., Garcia, G., and Kuehn, G. D.,** Adenylate cyclases in *Physarum polycephalum:* inhibition of a nuclear enzyme by polyamines, *Biochem. Biophys. Res. Commun.*, 68, 561, 1976.
384. **Wedner, H. J. and Parker, C. W.,** Adenylate cyclase activity in lymphocyte subcellular fractions. Characterization of a nuclear adenylate cyclase, *Biochem. J.*, 162, 483, 1977.
385. **Jost, J.-P., and Sahib. M. K.,** Role of cyclic adenosine 3′,5′-monophosphate in the induction of hepatic enzymes. II. Effect of N^6,$O^{2'}$-dibutyryl cyclic adenosine 3′,5′-monophosphate on the kinetics of ribonucleic acid synthesis in purified rat liver nuclei, *J. Biol. Chem.*, 246, 1623, 1971.
386. **Varrone, S., Ambesi-Impiombato, F. S., and Macchia, V.,** Stimulation by cyclic 3′,5′-adenosine monophosphate of RNA synthesis in a mammalian cell-free system, *FEBS Lett.*, 21, 99, 1972.
387. **Mendelson, I. S. and Anderson, K. M.,** Rat mammary gland nuclear RNA polymerases in late pregnancy and lactation, *Biochim. Biophys. Acta,* 299, 576, 1973.
388. **Anderson, K. M., Mendelson, I. S., and Guzik, G.,** Solubilized DNA-dependent nuclear RNA polymerases from the mammary glands of late-pregnant rats, *BIochim. Biophys. Acta,* 383, 56, 1975.
389. **Nesher, R., Robinson, W. F., Gibb, L., Bishop, S. P., and Kruger, F. A.,** RNA polymerases in isolated rat heart nuclei and their response to cyclic nucleotides, *J. Mol. Cell. Cardiol.*, 9, 579, 1977.
390. **Schumm, D. E. and Webb, T. E.,** Effect of adenosine 3′:5′-monophosphate and guanosine 3′:5′-monophosphate on RNA release from isolated nuclei, *J. Biol. Chem.*, 253, 8513, 1978.
391. **Pastan, I. and Perlman, R.,** Cyclic adenosine monophosphate in bacteria, *Science,* 169, 339, 1970.
392. **Miyamoto, E., Kuo, J. F., and Greengard, P.,** Cyclic nucleotide-dependent protein kinases. III. Purification and properties of adenosine 3′,5′-monophosphate-dependent protein kinase from bovine brain, *J. Biol. Chem.*, 244, 6395, 1969.
393. **Kuo, J. F., Krueger, B. K., Sanes, J. R., and Greengard, P.,** Cyclic nucleotide-dependent protein kinases. V. Preparation and properties of adenosine 3′,5′-monophosphate-dependent protein kinase from various bovine tissues, *Biochim. Biophys. Acta,* 212, 79, 1970.
394. **Johnson, E. M., Maeno, H., and Greengard, P.,** Phosphorylation of endogenous protein of rat brain by cyclic adenosine 3′,5′-monophosphate-dependent protein kinase, *J. Biol. Chem.*, 246, 7731, 1971.
395. **Kuehn, G. D.,** An adenosine 3′,5′-monophosphate-inhibited protein kinase from *Physarum polycephalum, J. Biol. Chem.*, 246, 6366, 1971.
396. **Kuehn, G. D.,** Cell cycle variation in cyclic adenosine 3′,5′-monophosphate-dependent inhibition of a protein kinase from *Physarum polycephalum, Biochem. Biophys. Res. Commun.*, 49, 414, 1972.
397. **Rubin, C. S., Erlichman, J., and Rosen, O. M.,** Molecular forms and subunit composition of a cyclic adenosine 3′,5′-monophosphate-dependent protein kinase purified from bovine heart muscle, *J. Biol. Chem.*, 247, 36, 1972.
398. **Khandelwal, R. L., Spearman, T. N., and Hamilton, I. R.,** Protein kinase activity in cariogenic and non-cariogenic oral streptococci: activation and inhibition by cyclic AMP, *FEBS Lett.*, 31, 246, 1973.
399. **Huang, L. C.,** Cyclic AMP regulation of protein kinase(s) in rabbit skeletal muscle: nucleoside triphosphate and divalent cation specificity, *Biochim. Biophys. Acta,* 358, 281, 1974.
400. **DeLorenzo, R. J., Walton, K. G., Curran, P. F., and Greengard, P.,** Regulation of phosphorylation of a specific protein in toad-bladder membrane by antidiuretic hormone and cyclic AMP, and its possible relationship to membrane permeability changes, *Proc. Natl. Acad. Sci. U.S.A.*, 70, 880, 1973.
401. **DeLorenzo, R. J. and Greengard, P.,** Activation by adenosine 3′:5′-monophosphate of a membrane-bound phosphoprotein phosphatase from toad bladder, *Proc. Natl. Acad. Sci. U.S.A.*, 70, 1831, 1973.
402. **Maeno, H., Reyes, P. L., Ueda, T., Rudolph, S. A., and Greengard, P.,** Autophosphorylation of adenosine 3′,5′-monophosphate-dependent protein kinase from bovine brain, *Arch. Biochem. Biophys.*, 164, 551, 1974.
403. **Maeno, H., Ueda, T., and Greengard, P.,** Adenosine 3′:5′-monophosphate-dependent protein phosphatase activity in synaptic membrane fractions, *J. Cyclic Nucleotide Res.*, 1, 37, 1975.
404. **Malkinson, A. M., Krueger, B. K., Rudolph, S. A., Casnellie, J. E., Haley, B. E., and Greengard, P.,** Widespread occurrence of a specific protein in vertebrate tissues and regulation by cyclic AMP of its endogenous phosphorylation and dephosphorylation, *Metabolism,* 24, 331, 1975.
405. **Earp, H. S., Smith, P., Ong, S.-H. H., and Steiner, A. L.,** Regulation of hepatic nuclear guanylate cyclase, *Proc. Natl. Acad. Sci. U.S.A.*, 74, 946, 1977.
406. **Ong, S.-H., Whitley, T. H., Stowe, N. W., and Steiner, A. L.,** Immunohistochemical localization of 3′:5′-cyclic AMP and 3′:5′-cyclic GMP in rat liver, intestine, and testis, *Proc. Natl. Acad. Sci., U.S.A.*, 72, 2022, 1975.
407. **Steiner, A. L., Whitley, T. H., Ong, S.-H. H., and Stowe, N. W.,** Cyclic AMP and cyclic GMP: studies utilizing immunohistochemical techniques for the localization of the nucleotides in tissue, *Metabolism,* 24, 419, 1975.

408. **Spruill, W. A., Hurwitz, D. R., Lucchesi, J. C., and Steiner, A. L.,** Association of cyclic GMP with gene expression of polytene chromosomes of *Drosophila melanogaster, Proc. Natl. Acad. Sci. U.S.A.,* 75, 1480, 1978.
409. **Johnson, L. D. and Hadden, J. W.,** Cyclic GMP and lymphocyte proliferation: effects on DNA-dependent RNA polymerase I and II activities, *Biochem. Biophys. Res. Commun.,* 66, 1498, 1975.
410. **Johnson, L. D. and Hadden, J. W.,** Modification of human DNA-dependent RNA polymerase activity by cyclic GMP, *Nucleic Acids Res.,* 4, 4007, 1977.
411. **Keely, S. L., Jr., Corbin, J. D., and Park, C. R.,** On the question of translocation of heart cAMP-dependent protein kinase, *Proc. Natl. Acad. Sci. U.S.A.,* 72, 1501, 1975.
412. **Erlichman, J., Rosenfeld, R., and Rosen, O. M.,** Phosphorylation of a cyclic adenosine 3′:5′-monophosphate-dependent protein kinase from bovine cardiac muscle, *J. Biol. Chem.,* 249, 5000, 1974.
413. **Rosen, O. M. and Erlichman, J.,** Reversible autophosphorylation of a cyclic 3′:5′-AMP-dependent protein kinase from bovine cardiac muscle, *J. Biol. Chem.,* 250, 7788, 1975.
414. **Hofmann, F., Beavo, J. A., Bechtel, P. J., and Krebs, E. G.,** Comparison of adenosine 3′:5′-monophosphate-dependent protein kinases from rabbit skeletal and bovine heart muscle, *J. Biol. Chem.,* 250, 7795, 1975.
415. **Rangel-Aldao, R. and Rosen, O. M.,** Dissociation and reassociation of the phosphorylated and nonphosphorylated forms of adenosine 3′:5′-monophosphate-dependent protein kinase from bovine cardiac muscle, *J. Biol. Chem.,* 251, 3375, 1976.
416. **Rangel-Aldao, R. and Rosen, O. M.,** Mechanism of self-phosphorylation of adenosine 3′:5′-monophosphate-dependent protein kinase from bovine cardiac muscle, *J. Biol. Chem.,* 251, 7526, 1976.
417. **Walter, U., Uno, I., Liu, A. Y.-C., and Greengard, P.,** Study of auto-phosphorylation of isoenzymes of cyclic AMP-dependent protein kinases, *J. Biol. Chem.,* 252, 6588, 1977.
418. **Rangel-Aldao, R. and Rosen, O. M.,** Effect of cAMP and ATP on the reassociation of phosphorylated and nonphosphorylated subunits of the cAMP-dependent protein kinase from bovine cardiac muscle, *J. Biol. Chem.,* 252, 7140, 1977.
419. **Todhunter, J. A. and Purich, D. L.,** Autophosphorylation of cardiac 3′,5′-cyclic AMP-stimulated protein kinase. Kinetic evidence for the regulatory subunit directly acting at the active site in the R_2C_2 complex, *Biochim. Biophys. Acta,* 485, 87, 1977.
420. **Chiu, Y. S. and Tao, M.,** Autophosphorylation of rabbit skeletal muscle cyclic AMP-dependent protein kinase I catalytic subunit, *J. Biol. Chem.,* 253, 7145, 1978.
421. **Ashby, C. D. and Walsh, D. A.,** Characterization of the interaction of a protein inhibitor with adenosine 3′,5′-monophosphate-dependent protein kinases. II. Mechanism of action with the holoenzyme, *J. Biol. Chem.,* 248, 1255, 1973.
422. **Szmigielski, A., Guidotti, A., and Costa, E.,** Endogenous protein kinase inhibitors. Purification, characterization, and distribution in different tissues, *J. Biol. Chem.,* 252, 3848, 1977.
423. **Donnelly, T. E., Jr., Kuo, J. F., Reyes, P. L., Liu, Y.-P., and Greengard, P.,** Protein kinase modulator from lobster tail muscle. I. Stimulatory and inhibitory effects of the modulator on the phosphorylation of substrate proteins by guanosine 3′,5′-monophosphate-dependent and adenosine 3′,5′-monophosphate-dependent protein kinases, *J. Biol. Chem.,* 248, 190, 1973.
424. **Kuo, W.-N. and Kuo, J. F.,** Isolation of stimulatory modulator of guanosine 3′:5′-monophosphate-dependent protein kinase from mammalian heart devoid of inhibitory modulator of adenosine 3′:5′-monophosphate-dependent protein kinase, *J. Biol. Chem.,* 251, 4283, 1976.
425. **Kuo, W.-N., Shoji, M., and Kuo, J. F.,** Stimulatory modulator of guanosine 3′:5′-monophosphate-dependent protein kinase from mammalian tissues, *Biochim. Biophys. Acta,* 437, 142, 1976.
426. **Shoji, M., Brackett, N. L., Tse, J., Shapira, R., and Kuo, J. F.,** Molecular properties and mode of action of homogeneous preparation of stimulatory modulator of cyclic GMP-dependent protein kinase from the heart, *J. Biol. Chem.,* 253, 3427, 1978.
427. **Hashimoto, E., Kuroda, Y., Ku, Y., and Nishizuka, Y.,** Stimulation by polydeoxyribonucleotide of histone phosphorylation by guanosine 3′:5′-monophosphate-dependent protein kinase, *Biochem. Biophys. Res. Commun.,* 87, 200, 1979.
428. **Farron-Furstenthal, F.,** An inhibitor protein of nuclear protein kinases, *Nature (London),* 280, 415, 1979.
429. **Farron-Furstenthal, F.,** Two naturally occurring inhibitors of nuclear protein kinase, *J. Biol. Chem.,* 255, 4589, 1980.
430. **Dabauvalle, M. C., Meggio, F., Creuzet, C., and Loeb, J. E.,** Phosphorylation of most non-histone proteins of chromatin of rat liver by an ATP-GTP-type protein kinase, *FEBS Lett.,* 107, 193, 1979.
431. **Kaplowitz, P. B., Platz, R. D., and Kleinsmith, L. J.,** Nuclear phosphoproteins. III. Increase in phosphorylation during histone-phosphoprotein interaction, *Biochim. Biophys. Acta,* 229, 739, 1971.
432. **Johnson, E. M., Vidali, G., Littau, V. C., and Allfrey, V. G.,** Modulation by exogenous histones of phosphorylation of non-histone nuclear proteins in isolated rat liver nuclei, *J. Biol. Chem.,* 248, 7595, 1973.

433. **Kuroda, Y., Hashimoto, E., and Nishizuka, Y.,** Stimulatory effect of histones on phosphorylation of nuclear phosphoproteins, *J. Biochem.*, 82, 1167, 1977.
434. **McCleary, A. R., Nooden, L. D., and Kleinsmith, L. J.,** Histone affinity chromatography as a tool for fractionating nonhistone chromatin proteins and studying histone-nonhistone protein interactions, *J. Biol. Chem.*, 253, 5199, 1978.
435. **Ohtsuki, K., Shiraishi, H., Yamada, E., Nakamura, M., and Ishida, N.,** A non-histone chromatin protein that is a specific phosphate acceptor of nuclear cAMP-independent protein kinase from mouse spleen cells, *J. Biol. Chem.*, 255, 2391, 1980.
436. **Collett, M. S. and Erikson, R. L.,** Protein kinase activity associated with the avian sarcoma virus *src* gene product, *Proc. Natl. Acad. Sci. U.S.A.*, 75, 2021, 1978.
437. **Levinson, A. D., Oppermann, H., Levintow, L., Varmus, H. E., and Bishop, J. M.,** Evidence that the transforming gene of avian sarcoma virus encodes a protein kinase associated with a phosphoprotein, *Cell,* 15, 561, 1978.
438. **Erikson, R. L., Collett, M. S., Erikson, E., and Purchio, A. F.,** Evidence that the avian sarcoma virus transforming gene product is a cyclic AMP-independent protein kinase, *Proc. Natl. Acad. Sci. U.S.A.*, 76, 6260, 1979.
439. **Maness, P. F., Engeser, H., Greenberg, M. E., O'Farrell, M., Gall, W. E., and Edelman, G. M.,** Characterization of the protein kinase activity of avian sarcoma virus *src* gene product, *Proc. Natl. Acad. Sci. U.S.A.*, 76, 5028, 1979.
440. **Hunter, T. and Sefton, B. M.,** Transforming gene product of Rous sarcoma virus phosphorylates tyrosine, *Proc. Natl. Acad. Sci. U.S.A.*, 77, 1311, 1980.
441. **Collett, M. S., Brugge, J. S., and Erikson, R. L.,** Characterization of a normal avian cell protein related to the avian sarcoma virus transforming gene product, *Cell,* 15, 1363, 1978.
442. **Oppermann, H., Levinson, A., Varmus, H. E., Levintow, L., and Bishop, J. M.,** Uninfected vertebrate cells contain a protein that is closely related to the product of the avian sarcoma virus transforming gene (*src*), *Proc. Natl. Acad. Sci. U.S.A.*, 76, 1804, 1979.
443. **Karess, R. E., Hayward, W. S., and Hanafusa, H.,** Cellular information in the genome of recovered avian sarcoma virus directs the synthesis of transforming protein, *Proc. Natl. Acad. Sci. U.S.A.*, 76, 3154, 1979.
444. **Collett, M. S., Erikson, E., Purchio, A. F., Brugge, J. S., and Erikson, R. L.,** A normal cell protein similar in structure and function to the avian sarcoma virus transforming gene product, *Proc. Natl. Acad. Sci. U.S.A.*, 76, 3159, 1979.
445. **Rohrschneider, L. R., Eisenman, R. N., and Leitch, C. R.,** Identification of a Rous sarcoma virus transformation-related protein in normal avian and mammalian cells, *Proc. Natl. Acad. Sci. U.S.A.*, 76, 4479, 1979.
446. **Hunter, T.,** Proteins phosphorylated by the RSV transforming function, *Cell,* 22, 647, 1980.
447. **Baumann, E. A. and Hand, R.,** Protein kinase activity associated with the D2 hybrid protein related to simian virus 40 T antigen: some characteristics of the reaction products, *Proc. Natl. Acad. Sci. U.S.A.*, 76, 3688, 1979.
448. **Hirsch, J. D. and Martelo, O. J.,** Inhibition of rabbit reticulocyte protein kinases, by hemin, *Biochem. Biophys. Res. Commun.*, 71, 926, 1976.
449. **Kramer, G., Cimadevilla, J. M., and Hardesty, B.,** Specificity of the protein kinase activity associated with the hemin-controlled repressor of rabbit reticulocyte, *Proc. Natl. Acad. Sci. U.S.A.*, 73, 3078, 1976.
450. **Levin, D. H., Ranu, R. S., Ernst, V., and London, I. M.,** Regulation of protein synthesis in reticulocyte lysates: phosphorylation of methionyl-$tRNA_f$ binding factor by protein kinase activity of translational inhibitor isolated from heme-deficient lysates, *Proc. Natl. Acad. Sci. U.S.A.*, 73, 3112, 1976.
451. **Hirsch, J. and Martelo, O. J.,** Inhibition of nuclear protein kinases by adenosine analogues, *Life Sci.*, 19, 85, 1976.
452. **Glazer, R. I. and Kuo, J. F.,** Inhibitory effects of cordycepin on cyclic nucleotide-dependent and cyclic nucleotide-independent protein kinases, *Biochem. Pharmacol.*, 26, 1287, 1977.
453. **Legraverend, M. and Glazer, R. I.,** Inhibition of the phosphorylation of non-histone chromosomal proteins of rat liver by cordycepin and cordycepin triphosphate, *Cancer Res.*, 38, 1142, 1978.
454. **Legraverend, M. and Glazer, R. I.,** Inhibition of the phosphorylation of nonhistone chromosomal proteins by cordycepin and Xylosyladenine in L1210 cells, *in vitro, Mol. Pharmacol.*, 14, 1130, 1978.
455. **Kleinsmith, L. J. and Allfrey, V. G.,** Nuclear phosphoproteins. II. Metabolism of exogenous phosphoprotein by intact nuclei, *Biochim. Biophys. Acta,* 175, 136, 1969.
456. **Meisler, M. H. and Langan, T. A.,** Characterization of a phosphatase specific for phosphorylated histones and protamine, *J. Biol. Chem.*, 244, 4961, 1969.
457. **Griffin, M. J., Cox, R. P., and Grujic, N.,** A chemical method for the isolation of HeLa cell nuclei and the nuclear localization of HeLa cell alkaline phosphatase, *J. Cell Biol.*, 33, 200, 1967.

458. **Cutler, L. S., Chaudhry, A. P., and Montes, M.,** Alkaline phosphatase activity associated with the nuclear pore in normal and neoplastic salivary gland tissue, *J. Histochem. Cytochem.*, 22, 1113, 1974.
459. **Wilson, M. J. and Ahmed, K.,** Presence and androgen control of an alkaline phosphatase in the nucleus of rat ventral prostate, *Biochim. Biophys. Acta,* 429, 439, 1976.
460. **Soriano, R. Z. and Love, R.,** Electron microscopic demonstration of acid phosphatase in nucleoli and nucleoplasm, *Exp. Cell Res.*, 65, 467, 1971.
461. **Tan, E. L., Levine, M. A., Yang, C. S., and Li, H.-C.,** Effects of ATP and metal ions on cytosol and nuclear phosphohistone phosphatases of rat liver, *Int. J. Biochem,*. 7, 21, 1976.
462. **Chou, M. W., Tan, E. L., and Yang, C. S.,** Comparative studies on nuclear and cytosol phosphohistone phosphatases of calf liver, *Int. J. Biochem.*, 9, 27, 1978.
463. **Chou, M. W., Tan, E. L., and Yang, C. S.,** Nuclear phosphoprotein phosphatase from calf liver, *Biochim. Biophys. Acta,* 566, 49, 1979.
464. **Wilson, M. J. and Ahmed K.,** Enzymic properties and effects of androgen on nuclear histone phosphatase activity of rat ventral prostate, *Exp. Cell Res.*, 117, 71, 1978.
465. **Olson, M. O. J. and Guetzow, K.,** Phosphoprotein phosphatase activity of Novikoff hepatoma nucleoli, *Biochem. Biophys. Res. Commun.*, 70, 717, 1976.
466. **Olson, M. O. J. and Guetzow, K. A.,** Nucleolar phosphoprotein phosphatase from Novikoff hepatoma and rat liver: characterization and partial purification, *Biochim. Biophys. Acta,* 526, 174, 1978.
467. **Wilson, M. J., Ahmed, K., and Fischbach, T. J.,** Acidic-phosphoprotein phosphatase activity of rat ventral prostate nuclei. Apparent lack of effect of androgens, *Biochim. Biophys. Acta,* 542, 12, 1978.
468. **Kleinsmith, L. J.,** Assay of nonhistone phosphatase activity, *Methods Cell Biol.*, 19, 161, 1978.
469. **Chou, C.-K., Alfano, J., and Rosen, O. M.,** Purification of phosphoprotein phosphatase from bovine cardiac muscle that catalyzes dephosphorylation of cyclic AMP-binding protein component of protein kinase, *J. Biol. Chem,*. 252, 2855, 1977.

Chapter 5

N-METHYLATION AND *N*-ACETYLATION OF NONHISTONE CHROMOSOMAL PROTEINS

Paul Byvoet

TABLE OF CONTENTS

I. *N*-METHYLATION OF NHC PROTEINS

Methylation of proteins and particularly that of histones, has been the subject of much research, and has been reviewed by Hnilica in *The Structure and Biological Function of Histones*; the first book in this series. Protein methylation in general, including that of histones, has been reviewed comprehensively by Paik and Kim.[1,2] This review will therefore, be confined to *N*-methylation of NHC proteins.

Among the various postsynthetic modifications that have been reported to occur in nuclear proteins, *N*-methylation occupies a special position since, in contrast to acetylation and phosphorylation, it is — at least in histones — not subject to turnover.[3,4] Since it has been established that histones themselves turn over at the same rate as the DNA with which they are complexed,[5,6] their *N*-methylation represents a structural change which is permanent for the life of the cell. In addition, histone methylation occurs exclusively during S-phase[7,8] on close to 100% of "methylatable" histone molecules.[9] Consequently, the resultant daughter cells will come equipped with a full complement of methylated histones H3 and H4. Following the approach used earlier in studies on histone turnover, the decay rates of specific activities of prelabeled NHC proteins and DNA have been compared in a number of studies and found in general to be remarkably similar.[10,11] (Some NHC proteins may turnover once at every cell division.[12-14]) Thus, *N*-methylation of these proteins may also represent a stable modification. As far as the function of *N*-methylation is concerned, it has been shown that increasing methyl substitution at the lysine-ε-amino function increases its affinity for the anionic DNA-phosphates (Figure 1).[15,16] *N*-methylation may therefore function to increase DNA-protein binding at certain critical binding sites.

A. Measurement of Protein *N*-Methylation with the Aid of Radioactively Labeled Precursors

For the assessment of in vitro methylation of histones or NHC proteins, most authors have used methyl-labeled *S*-adenosylmethionine as the precursor. This has the advantage that any detectable uptake of radioactivity into proteins must be due to *N*-methylation. For in vivo experiments, however, this precursor is less effective than methionine, since due to the strongly positive charge of the sulfonium group, it does not tranverse cell membranes as readily.[17,18] Furthermore, it is unstable at nonacid pH and therefore only useful for short pulses when used, for example, to label cultured cells. The only alternate methyl donor for in vivo studies is methyl-labeled methionine. However, this precursor is only partially converted to *S*-adenosylmethionine, the remainder being utilized for incorporation into proteins. This means that in order to measure incorporated radiomethyl groups in histones, it is necessary to separate methionine from the methylated basic amino acids. Ion-exchange chromatographic techniques are available for such separations and have been described.[19] Wherever possible experimental conditions may lend themselves to the application of inhibitors of protein synthesis, such as cycloheximide, which will prevent incorporation of methionine into the peptide backbone.

Others have tried to circumvent this problem by in vivo administration of a mixture of methyl-labeled methionine and methionine labeled elsewhere in the molecule with a different isotope, e.g., ^{35}S, C^{14}. The ratio of the content of the two isotopes present in a protein or protein fraction is then compared with that of an unmethylated protein, such as albumin.[20-23] Although such ratios are not affected by methionine pool sizes, they may conceivably be influenced by changes in activities of competing metabolic pathways of *S*-adenosylmethionine. Differences in these ratios, observed for instance in partially hepatectomized and normal control rats, may therefore reflect changes either in activities of protein *N*-methylation or in utilization of *S*-adenosylmethionine, or both. For instance, during cell proliferation there is a major shift in emphasis towards the pathway leading from *S*-aden-

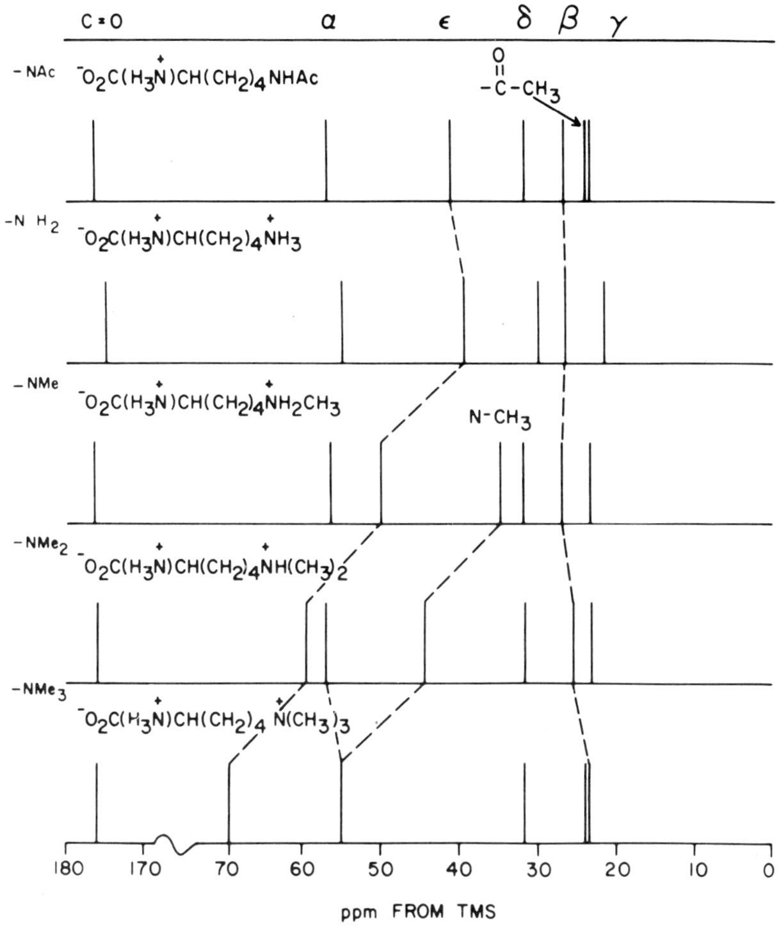

FIGURE 1. Graphic representation of natural abundance Fourier transform Carbon-13 NMR spectra of lysine, ε-*N*-acetyl- and ε-*N*-methylated lysines in water (50 mg/mℓ) at 40° and pH 5. Spectra were obtained with a Brucker® HFX-90 spectrometer at 22.63 MHz with proton broadband decoupling. Hexafluorobenzene was used as external standard, chemical shifts were re-referenced to TMS (tetramethylsilane). As can be seen, methyl substitution results in marked and regular changes in the chemical shifts of CH_2-N and CH_4-N signals only, suggesting that the progressive methyl substitution of lysine residues leads to a progressive decrease in the charge density on the CH_2-N and CH_3-N carbon atoms. This provides good evidence for a regularly increasing positive charge in the region of the ε-amino function with the number of methyl substitutions.[15,16] (From Baxter, C. S. and Byvoet, P., *Biochem. Biophys. Res. Commun.*, 64, 514, 1975. With permission.)

osylmethionine to polyamine synthesis.[24] It is more than likely that such changes may also influence pool sizes of *S*-adenosylmethionine.

B. *N*-Methylation of NHC Proteins

In early studies, uptake of radiolabeled alkyl groups into NHC protein preparations was observed using methyl-labeled methionine or *S*-adenosylmethionine as well as ethyl-labeled ethionine or *S*-adenosylethionine.[20,25-27]

The preparations of NHC proteins in these studies were mainly based on solubility in isotonic saline or insolubility in acid, in contrast to the histone fraction, which is readily solubilized from chromatin with dilute acid, but not with isotonic salt solutions. These experiments indicate very positively that among the NHC proteins there are indeed some that are subject to *N*-methylation, although the type of methylation was not further specified. In addition, radioethylation of NHC proteins was observed when ethyl-labeled ethionine or *S*-adenosylethionine were used as alkyl donor.[20,26]

As already indicated, studies in our laboratory have shown that in addition to *S*-adenosylmethionine, methyltransferases are also able to use *S*-adenosylethionine as substrate in varying degrees.[28,29]

C. *N*-Methylation of NHC Proteins and Proliferation

More recently, increased methylation of NHC proteins and histones was observed in transplantable hepatomas, and in rat liver cells in which proliferation was induced by subtotal hepatectomy[23] or by treatment with triiodothyronine.[30] Protein *N*-methylation in these studies was estimated by determination of the difference in incorporation of label from simultaneously administered (Me-^{3}H)-methionine and ^{14}C-methionine into proteins, using the ratio of the two isotopes present in albumin as a base line. These studies also indicated that proliferation resulted in an increase in the *N*-methylation ratio of NHC proteins as compared to that in normal liver. In contrast to these results, Kim et al.[31] found no differences between *N*-methylation of NHC proteins in liver and in Novikoff hepatoma. All the above studies did, however, show an increased *N*-methylation of histones, which had previously been noted to correlate with mitotic activity.[27,32,34] This would appear plausible, since histone lysine *N*-methylation occurs only during S-phase[7,8] and is irreversible.[3,4]

A correlation was also found between the growth rate of minimal deviation hepatomas and the activity of arginine and lysine *N*-methyltransferase,[35] as well as an inverse correlation between these activities and their age in kidney,[33] brain,[21] liver, and testis[36] of newborn rats. Duerre and Lee[37] administered labeled lysine and methyl-labeled methionine to rats and compared the decay of specific activities of brain histones with respect to both isotopes. They concluded that there was no significant difference, confirming previous findings by Byvoet et al.[3] In addition, their data indicate that methylation of brain histones decreases with increasing age.

This was confirmed by Kanungo and Thakur,[38] who incubated sections of rat brain cortex in the presence of methyl-labeled methionine and cycloheximide. They also reported that uptake of radiomethyl into NHC proteins from immature rat brain was higher than that from adults. Furthermore, their data indicated that administration of estradiol prior to sacrifice only increased uptake of radiomethyl into NHC proteins from adult rat brain but had no effect on radiomethylation of these proteins in immature or aged rat brain.

Finally, using an entirely different approach, Goodman and Benjamin,[39] observed a correlation between DNA synthesis and NHC protein methylation employing radioautography of chromosomal preparations of Diptera salivary glands *(Sciara coprophila)*. Radiomethylation of NHC proteins was achieved by incubating the glands in the presence of (methyl-^{3}H)-methionine and an inhibitor of protein synthesis. The tissue was subsequently sectioned and the sections treated with nuclease, 0.2 *N* HCl and organic solvents to remove nucleic acids, histones, and lipids, respectively. DNA synthesis was followed by exposure of the glands to thymidine-^{3}H, and by radioautography. Using similar techniques, Schmidt and Goodman[40] observed increased methylation of NHC proteins in salivary glands of cortisone-fed larvae as compared to controls.

D. Fractionation of *N*-Methylated NHC Proteins

In an attempt to fractionate radiomethylated NHC proteins, Liew and Suria[41] administered methyl-labeled *S*-adenosylmethionine intraperitoneally to rats in order to obtain exclusive incorporation of labeled *N*-methyl groups into proteins. Interestingly, it was found that radiomethylation of kidney histones and NHC proteins occurred at rates that were one to two orders of magnitude greater than the corresponding protein fractions in liver and heart. Glomerular filtration and subsequent reabsorption by tubular cells of administered *S*-adenosylmethionine may possibly contribute to this phenomenon. Subsequent fractionation of the kidney NHC proteins by polyacrylamide gel electrofocusing revealed the presence of some 20 radioactively labeled bands, of which 2 (with isoelectric points of 6.2 and 6.5) were especially conspicuous. These data indicate that in kidney and probably in other tissues as well there are a number of actively methylated NHC proteins.

E. Specific Types of Protein Methylation

From work on a variety of proteins, including histones, it is now known that there are four major types of postsynthetic methylation:[1,2] (1) ϵ-*N*-methylation of internal lysine residues, (2) guanidino-*N*-methylation of internal arginine residues, (3) *N*-methylation of histidine, and (4) *O*-methylation of carboxyl groups.

Of these four possible modifications only the first two have so far been reported to occur in NHC proteins. The enzymes involved in these various methylations are all of the methyltransferase type and use *S*-adenosylmethionine as the methyl donor.[1,2] The lysine *N*-methyltransferase[42-44] and the arginine-*N*-methyltransferase[45] have been isolated from cell nuclei or have been found to be present in the highest concentration in the cell nucleus. The histidine-*N*-methyltransferase has not been isolated so far, and carboxymethyltransferase activity has only been found in the cytosol.[46]

1. ϵ-N-Methylation of Internal Lysine Residues

Enzymes responsible for *N*-methylation of lysine residues in histones H3 and H4 as well as in other proteins have been isolated and purified from calf thymus,[42] chicken embryo nuclei,[43] and calf brain chromatin.[44] These enzymes have been shown to modify lysine to ϵ-*N*-mono-, di-, or trimethyllysine. The sequence recognized in histones appears to be: X-Arg-Lys-X (X = amino acid).[47,48] Some of the properties of these enzyme preparations are summarized in Table 1.

Radiomethylation of lysine residues in NHC proteins was observed by Kim et al.[31] after incubation of rat liver or Novikoff hepatoma nuclei in vitro with *S*-adenosylmethionine (Me-^{3}H), and by Byvoet et al.[3] after incubating Chinese hamster ovary cells in the presence of methyl-labeled methionine. In both studies incorporation of radiomethyl into methylated lysine derivatives in NHC proteins was found to be much smaller than that into methylated arginines, whereas the reverse was encountered in histones. These findings are in agreement with the results of amino acid analysis, which detects only methylated lysine residues in histones[47,48] and only methylated arginine residues in NHC proteins.[49]

In comparing methylation of NHC proteins in normal liver and Novikoff hepatoma Kim et al.[31] reported a striking difference in the content of methylated amino acids between normal liver and Novikoff hepatomas. In particular they found a relative decrease in methylated arginine and increase in methylated lysine; and a conspicuous relative elevation of monomethyllysine in Novikoff hepatoma.

2. Guanidino-N-Methylation of Internal Arginine Residues

The presently still-unfolding account of arginine guanidino-*N*-methylation in NHC proteins appears to be more interesting than that of ϵ-*N*-methylation of lysine. Arginine-*N*-methyltransferases have been isolated from calf brain, ascites cells, and rat thymus nuclei (Table 2). As can be seen, in all cases a K_m of 2.6×10^{-6} was found for *S*-adenosylmethionine, with a K_I for inhibition by *S*-adenosylhomocysteine of 1.4 to 2.6×10^{-6}. Miyake[50] has described the isolation from rat brain of a histone-methylating enzyme, and a protein arginine-*N*-methyltransferase which methylates arginine residue 107 in myelin basic protein.

Several investigators have carried out in vitro reactions in which the enzyme was incubated

Table 1
SOME CHARACTERISTICS OF DIFFERENT PREPARATIONS OF *S*-ADENOSYLMETHIONINE: PROTEIN-LYSINE *N*-METHYLTRANSFERASE

Source	Calf thymus nuclei	Rat liver nuclei	Chicken embryo nuclei	Calf brain chromatin
K_m *S*-adenosylmethionine	3×10^{-6}	3×10^{-6}	3.1×10^{-6}	ND
K_I *S*-adenosyl homocysteine	ND	1.2×10^{-5}	ND	ND
Histones most efficient substrate	Yes	Yes	Yes	Yes
Methylation of other proteins	Yes	ND	Yes	Yes
Optimal pH	8.5	ND	8.5	8.5[a] 7.7[b]
Ref.	42	29	43	44

Note: ND = not determined.

[a] With H3 histone as substrate.
[b] With H4 histone as substrate.

Table 2
CHARACTERISTICS OF DIFFERENT PREPARATIONS OF *S*-ADENOSYLMETHIONINE: PROTEIN-ARGININE *N*-METHYLTRANSFERASE

Source	Calf brain cytosol	Krebs ascites cells	Rat thymus nuclei	Rat liver cytosol
K_m *S*-adenosylmethionine	2.5×10^{-6}	2.5×10^{-6}	2.5×10^{-6}	4.4×10^{-6}
K_I *S*-adenosyl homocysteine	2.6×10^{-6}	1.4×10^{-6}	—	1.2×10^{-6}
In vitro-derived Me-arginines	N^G-MMA N^G,N^G-DMA $N^{G'},N^G$-DMA	N^G-MMA N^G,N^G-DMA	N^G-MMA[a] DMA	DMA
Ref.	51, 52	53	45	28

Note: N^G-MMA = monomethyl-N^G-arginine, N^G,N^G-DMA = asymmetrical dimethyl-N^G-arginine, $N^{G'},N^G$-DMA = symmetrical dimethyl-N^G-arginine, and DMA = dimethylarginine.

[a] N^G-MMA detection in the hydrolysate of enzyme incubated with methyl-labeled *S*-adenosylmethionine but without protein methylacceptor was assumed to be due to endogenous *N*-methylation of the enzyme.

with various proteins in the presence of *S*-adenosylmethionine. The following methylated arginine derivatives were observed: N^G-monomethylarginine (N^G-MMA), symmetrical dimethylarginine ($N^{G'},N^G$-DMA), and asymmetrical dimethylarginine (N^G,N^G-DMA) (Figure 2). The various data summarized in Table 2 indicate that the generated products varied according to the particular enzyme preparation.

In 1969 Friedman et al.[26] reported that upon administration of methyl-labeled methionine to rats, radiomethylated lysine and arginine derivatives were observed in the hydrolysates of saline-soluble proteins from liver cell nuclei, which were not further identified. Interestingly, if ethyl-labeled ethionine was administered this same protein fraction yielded only

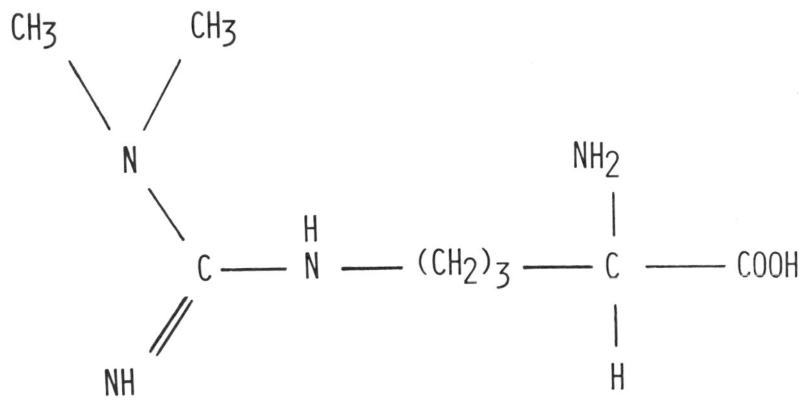

FIGURE 2.

one radioactive peak, which was assumed to be an ethylated arginine derivative. Although the various alkylated derivatives were not further identified, these data would suggest that the arginine methyltransferase was the only methyltransferase capable of modifying NHC proteins with ethionine as alkyl donor.

Presence of dimethyl arginine in 40S HnRNA associated protein — The exclusive presence among nuclear proteins of asymmetrical N^G,N^G-dimethylarginine (in hydrolysates of a protein associated with rapidly labeled heterogeneous HnRNA in 40S ribonucleoproteins) was simultaneously reported in 1977 by Christensen et al.[54] working with *Physarum polycephalum,* and Boffa et al.[49] using rat liver. The diversity of these species suggest that this particular modification must have been present throughout a good part of evolution (Table 3). In both studies the arginine derivative was identified as the asymmetrical N^G,N^G-dimethylarginine; the *Physarum* protein containing 3.1 mol/%, that from rat liver 1.07 mol/% as well as a trace of monomethyl arginine. These findings resolved the identity of an unusual basic residue, which had been noted before by other workers in hydrolysates of these proteins, but never identified.[55-57] In addition, symmetrical and asymmetrical dimethylarginine have been found in hydrolysates of ribosomal proteins.[58] The function of this postsynthetic modification remains unknown, although some studies suggest that guanidinium groups in arginine may represent an important binding site for nucleic acids,[59,60] and methylation of this site might increase its affinity for negatively charged groups in nucleic acids.

3. N-Methylation of Histidine

So far no enzyme has been isolated which is responsible for methylation of histidine residues in proteins. The occurrence of radiomethylated histidine has been reported by Gershey et al. in avian erythrocyte histones[61] and by Kim et al.[31] in Novikoff hepatoma histones and normal rat liver NHC proteins following incubation of nuclei with methyl-labeled *S*-adenosylmethionine. It is interesting to note that the latter authors reported the presence of *N*-methylhistidine in rat liver NHC proteins but not in rat liver histones, whereas the reverse was true for Novikoff hepatoma. This is an interesting switch which is certainly worthy of further investigation. It should be kept in mind that in these studies methylhistidine could only be detected by virtue of its radioactivity. Less sensitive methods such as amino acid analysis have been unable to demonstrate the presence of this modified amino acid in nuclear proteins.

Since the contractile proteins, actin and myosin, are prominent NHC proteins,[62,63] and since these proteins as obtained from other sources have been found to contain methylated

Table 3
AMINO ACID COMPOSITION OF NUCLEAR PROTEINS (HnRNP) IN MOL/100 MOL AMINO ACID

Amino acids	Rat liver nuclei: acid-soluble proteins	Rat liver: 40S HnRNA particles	*Physarum polycephalum*: 40S HnRNA particles
Lysine	11.4	7.7	6.6
Histidine	2.07	2.5	1.0
Arginine	11.1	8.0	3.7
Aspartic acid	6.3	10.8	7.8
Threonine	4.4	3.9	3.0
Serine	5.2	5.0	10.0
Glutamic acid	11.3	10.9	10.6
Proline	4.7	6.8	5.3
Glycine	8.7	17.9	29.0
Alanine	11.2	5.6	7.8
Cystine/2	—	—	—
Valine	6.8	7.0	4.0
Methionine	1.8	3.7	—
Isoleucine	4.0	5.4	2.6
Leucine	7.4	4.7	3.3
Tyrosine	1.6	3.3	Trace
Phenylalanine	1.4	3.3	2.3
ε-*N*-monomethyllysine	0.14	ND	—
ε-*N*-dimethyllysine	0.45	ND	—
N^G,N^G-dimethylarginine	ND	1.07	3.1
Ref.	49	49	54

Note: ND = not detected.

histidine as well as methylated lysine and arginine derivatives, [64-69] one might expect that their nuclear counterparts are methylated also. This question has so far not been pursued.

II. *N*-ACETYLATION OF NHC PROTEINS

In recent years a large amount of information on acetylation of histones has become available (reviewed by Hnilica in *The Structure and Biological Function of Histones*). In addition, Ruiz-Carillo et al.[70] and Isenberg[71] have reviewed more recent work on this subject. Briefly, postsynthetic acetylation occurs on ε-amino groups of several internal lysine residues of histones 3 and 4. The acetyltransferases responsible for this process use acetyl-CoA as acetyl donor. The acetyl groups are continuously removed by a histone deacetylase, creating an active turnover of acetyl groups with half-lives ranging from a few minutes to several hours.[72,73] Histones H1, H2A, and H4 are also acetylated N-terminally on the α-amino of serine, but this occurs during synthesis and these groups do not turn over. A number of substances, such as butyric acid and dimethylsulfoxide, have been found to induce hyper-acetylation of histones. The basis for this phenomenon is the inhibitory effect of these substances on the histone deacetylase.[71] A large amount of circumstantial evidence suggests a connection between histone acetylation and RNA synthesis. In addition, hyperacetylation renders chromatin more susceptible to attack by DNAse I, an enzyme which preferentially digests active genes.[71]

A. Acetylation Of NHC Proteins

Acetylation of NHC proteins in general has been studied by Thakur et al.[74] and Kanungo and Thakur.[75] Sections of rat brain cortex were incubated in the presence of sodium acetate (^{14}C) and cycloheximide. The data obtained indicated that uptake of label into subsequently isolated NHC proteins did occur, and that it decreased with the age of the rats, and was stimulated by the administration of estradiol and epinephrine prior to sacrifice. No preferential uptake of the (^{14}C)-acetate was observed into a particular NHC protein, as indicated by sodium dodecylsulfate acrylamide gel electrophoresis.

B. Acetylation Of High Mobility Group (HMG) Proteins

HMG proteins are discussed in Chapter 2. To briefly summarize, they owe their name to a relatively high mobility in polyacrylamide gel electrophoresis. They are extracted from chromatin by 0.35 *M* NaCl and are soluble in 2% but insoluble in 10% trichloroacetic acid. The four major HMG proteins are numbered 1, 2, 14, and 17. HMG 1 and 2 bind to DNA as well as to histones since these molecules are equipped with a positively charged N-terminus and a negatively charged C-terminus. It is estimated that there are approximately 10^5 to 10^6 HMG molecules per nucleus. In 1978 Sterner et al.[76] reported that HMG proteins 1 and 2 from calf thymus and HMG proteins 1 and E from duck erythrocytes were acetylated and would incorporate label after incubation of the tissues in the presence of ^{14}C-labeled sodium acetate. Amino acid analysis of trypsin/pronase digests of these proteins indicated that 2 to 3% of total lysine, or 1 mol per molecule of protein, was present as *N*-acetyllysine. During further studies,[77] Edman degradation revealed two sites of acetylation in HMG 1: lysine residues at position 2 and 11, both close to the amino-terminus. The configuration of the Lys 2 site was Glycine-Lysine-Glycine; that of the Lys 11 site: Methionine-Lysine-Glycine. Since most acetylated lysines in histones are also flanked by glycine on one or both sides, this might suggest that the enzymes responsible for acetylation and deacetylation of these specific sites are directed by this sequence. It seems obvious however, that there must be other factors that regulate their activity. Sterner et al.[77] also found that HMG 1 and 2 are deacetylated by a histone deacetylase, previously purified and characterized by Vidali et al.[78] Although the deacetylation occurs at a slower rate than that of histones, this rate can be increased by removal of the negatively charged carboxy terminal portion of HMG 1 by enzymatic cleavage.[76] These findings indicate that the affinity between enzyme and substrate is an important factor in this reaction. It still remains a question however, if these charges play an important role in the nucleus where these proteins are integrated in the chromatin structure and where binding to other macromolecules may offset these charges to some degree. These studies also indicated that deacetylation of HMG 1 and 2 was inhibited in the presence of 10 m*M* butyrate, as was found with histones. Several lines of evidence point to a connection between HMG proteins 14 and 17 and gene transcription. First, the preference of DNAse I for active globin genes during limited digestion of avian erythrocyte chromatin is lost following exposure of chromatin to 0.35 *M* NaCl, which removes the HMG proteins. The specificity of the enzyme is restored following reconstitution of the chromatin with HMG 14 and 17.[79] Secondly, 1 mol HMG 14 or 17 per 20 mol of HMG-depleted nucleosomes restores sensitivity to DNAse digestion of actively transcribed genes, such as the globin gene or the endogenous chicken virus RAV-o, but not the ovalbumin gene.[80,81] These data indicate that these proteins can recognize and bind preferentially to actively transcribed nucleosomes among an excess of inactive ones.

Certainly intriguing in this context is a report by Reeves and Candido[82] on inhibition of histone deacetylase by HMG 14 and 17. These authors observed the release of soluble inhibitors of histone deacetylase during limited digestion of Friend erythroleukemic cells with DNAse I,[82] which were subsequently identified as HMG 14 and 17.[83] These findings fit well with the previously mentioned correlation between gene transcriptional activity and histone acetylation. In addition, recent reports indicate that if cells are grown in the presence of 5 to 10 m*M* butyrate and the isolated chromatin is exposed to DNAse I, the enzyme will preferentially digest DNA associated with hyperacetylated histone.[84-86] Moreover, transcriptionally active chromatin fractions have been found to be enriched for acetylated histones.[87]

The picture that emerges from these findings is that, in the nucleosome, *N*-acetylation is a generalized phenomenon probably with a structural function involving histones and HMG proteins 1 and 2. The degree of acetylation is modulated by inhibition of the deacetylase by such factors as HMG 14 and 17. The increased *N*-acetylation combined with the binding of HMG protein alters the structural conformation of the nucleosome so as to make it available for transcription and sensitive to the action of the endonuclease, DNAse I, and micrococcal nuclease.

C. *N*-Acetylation of Protein A-24

Finally, acetylation has been found to occur on a nucleosomal conjugate protein A-24,[88] named after its location in two-dimensional electrophoresis.[89] This protein is a T-configuration conjugate of histone 2A and ubiquitin, the carboxy terminal of which is attached to the ϵ-NH_2 of lysine 119 of H2A through a glycyl-glycine linkage. The protein is discussed here because it does not represent a true histone. The studies by Goldknopf et al.[88,89] on A-24 from Novikoff hepatoma indicate however that it is the histone part of the molecule that is acetylated.

III. ADDENDUM

A. The Question of Reversibility of *N*-Methylation of NHC Proteins

In order for *N*-methylation of proteins to be reversible, a system should exist capable of removing *N*-methyl groups. Earlier reports by Paik and Kim[90,91] indeed demonstrated the presence of such an enzyme which is capable of removing ϵ-*N*-methyl groups from ϵ-*N*-methyllysine residues in calf thymus histones.

The substrate used in these experiments was obtained from calf thymus nuclei incubated with S-adenosyl methionine containing radioactive methyl, followed by preparation of the thus radiomethylated histones.

It was found that during the demethylation process, the labeled methyl-groups were liberated in the form of formaldehyde. Comparison of demethylation in various rat tissues showed the activity in kidney to be at least one order of magnitude greater than in other tissues, and to be confined to the mitchondrial fraction. The enzyme, called ϵ-alkyllysinase by the authors, was further purified from rat kidney mitochondrial fraction and shown to require the presence of an electronacceptor and to be stimulated by FAD. Under optimal conditions, the enzyme consumed half a mole of oxygen per mole of substrate during demethylation of ϵ-*N*-mono(^{14}C)methyllysine, producing equimolar amounts of lysine and formaldehyde.

More recently, Gagnon[92] and Chene[93] reported the presence of a proteincarboxyl methylesterase in various rat tissues. Interestingly, the activity of this enzyme in rat kidney, was also found to be at least one order of magnitude greater than that in other tissues, similar to the ϵ-alkyllysinase described by Paik and Kim.

The reversibility of *N*-methylation of NHC proteins was studied in the author's laboratory by comparing the turnover of *N*-methyl groups with that of the peptide backbone (unpublished). Chinese hamster ovary cells were randomly grown in the presence of tritiated amino acids to label the peptide backbone, and methionine(methyl-^{14}C) to label *N*-methyl groups. Cells were collected by centrifugation. After resuspension in unlabeled medium, aliquots were removed at 4 time points which were separated by approximately one-half doubling time. Cells were harvested and histones, NH and HMG chromosomal proteins were isolated according to 3 different extraction procedures. Ratios of halflives for ^{14}C and ^{3}H were calculated from the first order rate decay curves of the specific activities calculated on the basis of total DNA content. Ratios at various stages of preparation were shown to be close to unity. Total basic amino acids containing incorporated *N*-methyl-^{14}C were obtained from hydrolysates of the final purified protein fractions by ion-exchange column chromatography. Average ratios of *N*-methyl-^{14}C over total ^{3}H-amino acids eluted were: 1.025, 1.068, and 1.143 respectively for histones, NH and HMG chromosomal proteins.

These data indicate tthat similar to our previous findings on turnover of histone *N*-methyl groups,[3,4] *N*-methylation of NH and HMG chromosomal proteins is practically irreversible.

In addition, the results of these studies also indicated that arginine is the main amino acid methylated in NH and HMG chromosomal proteins. This finding was recently confirmed for the highly methylated nucleolar acidic phosphoprotein C23 by Lischwe et al.[94]

ACKNOWLEDGMENTS

I am grateful to J. Lynn Hubbard for expert preparation of this manuscript, and to Dr. Ronald K. Keller for his careful editing. Research reported from my laboratory was supported by the Veterans Administration Medical Research Service and grants from the National Cancer Institute, National Institutes of Health.

REFERENCES

1. **Paik, W. K. and Kim, S.,** *Protein Methylation,* (Monograph), John Wiley & Sons, New York, 1980.
2. **Paik, W. K. and Kim, S.,** Protein methylation, *Science,* 174, 114, 1971.
3. **Byvoet, P., Shepherd, G. R., Hardin, J. M., and Noland, B. J.,** The distribution and turnover of labeled methyl groups in histone fractions of cultured mammalian cells, *Arch. Biochem. Biophys.,* 148, 558, 1972.
4. **Byvoet, P.,** In vivo turnover and distribution of radio-N-methyl in arginine-rich histones from rat tissues, *Arch. Biochem. Biophys.,* 152, 887, 1972.
5. **Byvoet, P.,** Metabolic integrity of deoxyribo-nucleohistone, *J. Mol. Biol.* 17, 311, 1966.
6. **Hancock, R.,** Conservation of histones in chromatin during growth and mitosis *in vitro, J. Mol. Biol.,* 40, 457, 1969.
7. **Tidwell, T., Allfrey, V. G., and Mirsky, A. E.,** The methylation of histones during regeneration of the liver, *J. Biol. Chem.,* 243, 707, 1968.
8. **Shepherd, G. R., Hardin, J. M., and Noland, B. J.,** Methylation of lysine residues of histone fractions in synchronized mammalian cells, *Arch. Biochem. Biophys.,* 143, 1, 1971.
9. **Delange, R. J., Fambrough, D. M., Smith, E. L., and Bonner, J.,** Calf and pea histone. IV. III. Complete amino acid sequence of pea seedling histone. IV. Comparison with the homologous calf thymus histone, *J. Biol. Chem.,* 244, 5669, 1969.
10. **Srebreva, L., Russev, G., and Tsanev, R.,** Metabolic stabilization of non-histone chromosomal proteins in Ehrlich ascites tumour cells, *Int. J. Biochem.,* 10, 691, 1979.
11. **Seale, R. L.,** Conservation of non-histone chromatin proteins during growth in Hela cells, *Biochem. Biophys. Res. Commun.,* 63, 140, 1975.
12. **Djondjurov, L. P., Ivanova, E. C., and Tsanev, R. G.,** Metabolic behaviour of non-histone chromosomal proteins in proliferating and in resting fibroblasts, *Eur. J. Biochem.,* 77, 545, 1977.
13. **Russev, G., Anachkova, B., and Tsanev, R.,** Fractionation of rat-liver chromatin nonhistone proteins into two groups with different metabolic rates, *Eur. J. Biochem.,* 58, 253, 1975.
14. **Tsanev, R., Djondjurov, L., and Ivanova, E.,** Cell cycle dependent turnover of non-histone proteins in chromatin, *Exp. Cell Res.,* 84, 137, 1974.
15. **Baxter, C. S. and Byvoet, P.,** CMR studies of protein modification. Progressive decrease in charge density at the ϵ-amino function of lysine with increasing methyl substitution, *Biochem. Biophys. Res. Commun.,* 64, 514, 1975.
16. **Byvoet, P. and Baxter, C. S.,** in *Chromsomal Proteins and Their Role in the Regulation of Gene Expression,* Stein, G. S., Ed., Academic Press, New York, 1975, 286.

17. **Stekol, J. A., Anderson, E. I., and Weiss, S.,** S-adenosylmethionine in the synthesis of choline, creatinine and cysteine in vivo and in vitro, *J. Biol. Chem.*, 233, 425, 1958.
18. **Zappia, V., Galletti, P., Porcelli, M., Ruggiero, G., and Andreana, A.,** Uptake of adenosylmethionine and related sulfur compounds by isolated rat liver, *FEBS Lett.*, 90, 331, 1978.
19. **Byvoet, P., Sayre, D., and Baxter, C. S.,** Methods for the assessment of histone methylation, in *Methods in Cell Biology*, Stein, G. S., Stein, J., and Kleinsmith, L. J., Eds., Academic Press, New York, 1978, chap. 9.
20. **Orenstein, J. M. and Marsh, W. H.,** Incorporation in vivo of methionine and ethionine into, and the methylation and ethylation of, rat liver nuclear proteins, *Biochem. J.*, 109, 697, 1968.
21. **Paik, W. K., Kim, S., and Lee, H. W.,** Protein methylation during the development of brain, *Biochem. Biophys. Res. Commun.*, 46, 933, 1972.
22. **Borun, T. W., Pearson, D., and Paik, W. K.,** Studies of histone methylation during the Hela S-3 cell cycle, *J. Biol. Chem.*, 247, 4288, 1972.
23. **Short, J., Kibert, L., Wedmore, R., Ove, P., and Zemel, R.,** Chromatin protein methylation in proliferating liver and hepatoma cells, *Cytobios*, 25, 115, 1979.
24. **Pegg, A. E. and Hibasami, H.,** The role of s-adenosylmethionine in mammalian polyamine synthesis, in *Transmethylation*, Usdin, E., Borchardt, R. T., and Creveling, C., Eds., Elsevier, New York, 1979, 105.
25. **Paik, W. K. and Kim, S.,** Enzymatic methylation of protein fractions from calf thymus nuclei, *Biochem. Biophys. Res. Commun.*, 29, 14, 1967.
26. **Friedman, M., Shull, K. M., and Farber, E.,** Highly selective in vivo ethylation of rat liver nuclear protein by ethionine, *Biochem. Biophys. Res. Commun.*, 34, 857, 1969.
27. **Lee, H. W. and Paik, W. K,.** Histone methylation during hepatic regeneration in rat, *Biochim. Biophys. Acta*, 277, 107, 1972.
28. **Baxter, C. S. and Byvoet, P.,** Effects of carcinogens and other agents on histone methylation by a histone arginine methyltransferase purified from rat liver cytoplasm, *Cancer Res.*, 34, 1418, 1974.
29. **Baxter, C. S. and Byvoet, P.,** Effects of carcinogens and other agents on histone methylation in rat liver nuclei by endogenous histone lysine methyltransferase, *Cancer Res.*, 34, 1424, 1974.
30. **Short, J. and Kibert, L.,** Enhanced hepatic chromatin protein methylation induced by triiodothyronine treatment of the rat, *Endocrine Res. Commun.*, 7, 113, 1980.
31. **Kim, Y. S., Lou, M. F., Paik, M. K., Kim, S., and Paik, W. K.,** Methylation of nuclear proteins of Novikoff hepatoma, *Carcinogenesis*, 1, 129, 1980.
32. **Turner, G. and Hancock, R. L.,** Histone methylase activity of adult, embryonic and neoplastic liver tissues, *Life Sci.*, 9, 917, 1970.
33. **Paik, W. K., Benditt, M., and Kim, S.,** Protein methylase III of rat kidney during the early development, *FEBS. Lett.*, 35, 236, 1973.
34. **Lee, C. T. and Duerre, J. A.,** Changes in histone methylase activity of rat brain and liver with ageing, *Nature (London)*, 251, 240, 1974.
35. **Paik, W. K., Kim, S., Ezirike, J., and Morris, H. P.,** S-adenosylmethyltransferases in hepatomas, *Cancer Res.*, 35, 1159, 1975.
36. **Paik, W. K., Lee, H. W., and Lawson, D.,** Age dependency of various protein methylases, *Exp. Gerontol.*, 6, 271, 1971.
37. **Duerre, J. A. and Lee, C. T.,** In vivo methylation and turnover of rat brain histones, *J. Neurochem.*, 23, 541, 1974.
38. **Kanungo, M. S. and Thakur, M. K.,** Steroids and aging. Effects of oestradiol on covalent modifications of chromosomal proteins and transcription of chromatin of the brain of rats of various ages, *J. Steroid Biochem.*, 11, 879, 1979.
39. **Goodman, R. M. and Benjamin, W. B.,** Nucleoprotein methylation in salivary gland chromosomes of *Sciara coprophila:* Correlation with DNA synthesis, *Exp. Cell Res.*, 77, 63, 1973.
40. **Schmidt, E. C. and Goodman, R. M.,** Hormone induced alterations in non-histone protein. Methylation and phosphorylation in *Sciara coprophila, Differentiation*, 6, 177, 1976.
41. **Liew, C. C. and Suria, D.,** Assessment of methylation in nonhistone chromosomal protein, in *Methods in Cell Biology*, Stein, G., Stein, J., and Kleinsmith, L. J., Eds., Academic Press, New York, 1978, chap. 8.
42. **Paik, W. K. and Kim, S.,** Solubilization and partial purification of protein methylase III from calf thymus nuclei, *J. Biol. Chem.*, 245, 6010, 1970.
43. **Greenaway, P. J. and Levine, D.,** Identification of a soluble protein methylase in chicken embryo nuclei, *Biochim. Biophys. Acta*, 350, 374, 1974.
44. **Wallwork, J. C., Quick, D. P., and Duerre, J. A.,** Properties of soluble rat brain histone lysine methyltransferase, *J. Biol. Chem.*, 252, 5977, 1977.
45. **Gallwitz, D.,** Histone methylation: partial purification of two histone-specific methyltransferases from rat thymus nuclei preferentially methylating histones F2a, and F3, *Arch. Biochem. Biophys.*, 245, 650, 1971.

46. **Kim, S. and Paik, W. K.,** Purification and properties of protein methylase. II, *J. Biol. Chem.*, 245, 1806, 1970.
47. **Delange, R. J. and Smith, E. L.,** The structure of histones, *Acc. Chem. Res.*, 5, 368, 1972.
48. **Delange, R. J., Fambrough, D. M., Smith, E. L., and Bonner, J.,** Calf and pea histone IV. II. The complete amino acid sequence of calf thymus histone IV, presence of ε-N-acetyllysine, *J. Biol. Chem.*, 244, 319, 1969.
49. **Boffa, L. C., Karn, J., Vidali, G., and Allfrey, V. G.,** Distribution of N^G,N^G-dimethylarginine in nuclear protein fractions, *Biochem. Biophys. Res. Commun.*, 74, 969, 1977.
50. **Miyaki, M.,** Methylases of myelin basic protein and histone in rat brain, *J. Neurolog. Chem.*, 24, 909, 1975.
51. **Paik, W. K. and Kim, S.,** Protein methylase. I. Purification and properties of the enzyme, *J. Biol. Chem.*, 243, 2108, 1968.
52. **Lee, H. W., Kim, S., and Paik, W. K.,** S-adenosylmethionine: protein-arginine methyltransferase. Purification and mechanism of the enzyme, *Biochemistry*, 16, 78, 1977.
53. **Casellas, P. and Jeanteur, Ph.,** Protein methylation in animal cells. I. Purification and properties of S-adenosyl-L-methionine: protein (arginine) N-methyltransferase from Krebs II ascites cells, *Biochim. Biophys. Acta*, 519, 243, 1978.
54. **Christensen, M. E., Beyer, A. L., Walker, B., and Lestourgeon, W. M.,** Identification of N^G, N^G-dimethylarginine in a nuclear protein from the lower eukaryote *Physarum polycephalum* homologous to the major proteins of mammalian 40s ribonucleoprotein particles, *Biochem. Biophys. Res. Commun.*, 74, 621, 1977.
55. **Sarasin, A.,** Particules ribonucleoproteiques 40s des noyaux de foie de rat. Proprietes des proteines de ces particules, *FEBS Lett.*, 4, 3(27), 1969.
56. **Krichevskaya, A. A. and Georgiev, G. P.,** Further studies on the protein moiety in nuclear DNA-like RNA containing complexes, *Biochim. Biophys. Acta*, 194, 619, 1969.
57. **Martin, T., Billings, P., Levey, A., Ozarslan, S., Quinlan, T., Swift, H., and Urbas, L.,** Some properties of RNA-protein complexes from the nucleus of eukaryotic cells, *Cold Spring Harbor Symp. Quant. Biol.*, 38, 921, 1973.
58. **Klagstromm, M. and Furano, A. V.,** Methylated amino acids in the proteins of bacterial and mammalian cells, *Arch. Biochem. Biophys.*, 169, 529, 1975.
59. **Mansy, S., Engstrom, S. K., and Peticolas, W. L.,** Laser Raman identification of an interaction site on DNA for arginine containing histones in chromatin, *Biochem. Biophys. Res. Commun.*, 68, 1242, 1976.
60. **Seeman, N. C., Rosenberg, J. M., and Rich, A.,** Sequence specific recognition of double helical nucleic acids by proteins, *Proc. Natl. Acad. Sci. U.S.A.*, 73, 804, 1976.
61. **Gershey, E. L., Haslett, G. W., Vidali, G., and Allfrey, V. G.,** Chemical studies of histone methylation. Evidence for the occurrence of 3-methyl-histidine in avian erythrocyte histone fractions, *J. Biol. Chem.*, 244, 4871, 1969.
62. **Lestourgeon, W. M., Forer, A., Yang, Y. Z., Bertram, J. S., and Rusch, H. P.,** Contractile proteins. Major components of nuclear and chromosomal non-histone proteins, *Biochim. Biophys. Acta*, 379, 529, 1975.
63. **Douvas, A. S., Harrington, C. A., and Bonner, J.,** Major nonhistone proteins of rat liver chromatin: preliminary identification of myosin, actin, tubulin and tropomyosin, *Proc. Natl. Acad. Sci. U.S.A.*, 10, 3902, 1975.
64. **Johnson, P., Harris, C. I., and Perry, S. V.,** 3-Methylhistidine in actin and other muscle proteins, *Biochem. J.*, 105, 361, 1967.
65. **Asatoor, A. M. and Armstrong, M. D.,** 3-Methylhistidine, a component of actin, *Biochem. Biophys. Res. Commun.*, 26, 168, 1967.
66. **Kuehl, W. M. and Adelstein, R. S.,** The absence of 3-methylhistidine in red, cardiac and fetal myosins, *Biochem. Biophys. Res. Commun.*, 39, 956, 1970.
67. **Reporter, M. and Corbin, J. L.,** N^G,N^G-dimethylarginine in myosin during development, *Biochem. Biophys. Res. Commun.*, 43, 644, 1971.
68. **Huszar, G.,** Amino acid sequence around the two ε-N-trimethyllysine residues in rabbit skeletal muscle myosin, *J. Biol. Chem.*, 247, 4057, 1972.
69. **Elzinga, M., Collins, J. H., Kuehl, W. M., and Adelstein, R. S.,** Complete amino acid sequence of actin of rabbit skeletal muscle, *Proc. Natl. Acad. Sci. U.S.A.*, 70, 2687, 1973.
70. **Ruiz-Carillo, A., Wangh, L. J., and Allfrey, V. G.,** Processing of newly synthesized histone molecules, *Science*, 190, 117, 1975.
71. **Isenberg, I.** Histones, *Annu. Rev. Biochem.* 48, 159, 1979.
72. **Jackson, V., Shires, A., Chalkley, R., and Granner, D. K.,** Studies on highly metabolically active acetylation and phosphorylation of histones, *J. Biol. Chem.*, 250 4856, 1975.

73. **Covault, J. and Chalkley, R.,** The identification of distinct populations of acetylated histone, *J. Biol. Chem.*, 255, 9110, 1980.
74. **Thakur, M. K., Das, R., and Kanungo, M. S.,** Modulation of acetylation of chromosomal proteins of the brain of rats of various ages by epinephrine and estradiol, *Biochem. Biophys. Res. Commun.*, 81, 828, 1978.
75. **Kanungo, M. S. and Thakur, M. K.,** Phosphorylation and acetylation of non-histone chromosomal proteins of the brain of rats of various ages and their modulation by calcium and estradiol, *Biochem. Biophys. Res. Commun.*, 86, 14, 1979.
76. **Sterner, R., Vidali, G., Heinrikson, R. L., and Allfrey, V. G.,** Postsynthetic modification of high mobility group proteins. Evidence that high mobility group proteins are acetylated, *J. Biol. Chem.*, 253, 7601, 1978.
77. **Sterner, R., Vidali, G., and Allfrey, V. G.,** Studies of acetylation and deacetylation of high mobility group proteins. Identification of the sites of acetylation in HMG-1, *J. Biol. Chem.*, 254, 11577, 1979.
78. **Vidali, G., Boffa, L. C., and Allfrey, V. G.,** Properties of an acidic histone binding protein fraction from cell nuclei. Selective precipitation and deacetylation of histones F2A1 and F3, *J. Biol. Chem.*, 247, 7365, 1972.
79. **Weisbrod, S. and Weintraub, H.,** Isolation of a subclass of nuclear proteins responsible for conferring a DNAse I-sensitive structure on globin chromatin, *Proc. Natl. Acad. Sci. U.S.A.*, 76, 630, 1979.
80. **Weisbrod, S., Groudine, M., and Weintraub, H.,** Interaction of HMG 14 and 17 with actively transcribed genes, *Cell,* 19, 289, 1980.
81. **Sandeen, G., Wood, W. I., and Felsenfeld, G.,** The interaction of high mobility proteins HMG 14 and 17 with nucleosomes, *Nucleic Acid Res.*, 8, 3757, 1980.
82. **Reeves, R. and Candido, E. P. M.,** Effect of DNase I on histone acetyl-transferase and deacetylase enzyme activities in Friend cell nuclei: evidence for a putative inhibitor of the deacetylase enzymes in transcriptionally active chromatin, *Biochem. Biophys. Res. Commun.*, 89, 571, 1979.
83. **Reeves, R. and Candido, E. P. M.,** Partial inhibition of histone deacetylase in active chromatin by HMG 14 and HMG 17, *Nucleic Acid Res.*, 8, 1947, 1980.
84. **Vidali, G., Boffa, L. C., Bradbury, E. M., and Allfrey, V. G.,** Butyrate suppression of histone deacetylation leads to accumulation of multiacetylated forms of histones H3 and H4 and increased DNAse I sensitivity of the associated DNA sequences, *Proc. Natl. Acad. Sci., U.S.A.*, 75, 2239, 1978.
85. **Sealy, L. and Chalkley, R.,** DNA associated with hyperacetylated histone is preferentially digested by DNAse I, *Nucleic Acid Res.*, 5, 1863, 1978.
86. **Simpson, R. T.,** Structure of chromatin containing extensively acetylated H3 and H4, *Cell,* 13, 691, 1978.
87. **Levy-Wilson, B., Watson, D. C., and Dixon, G. H.,** Multiacetylated forms of H4 are found in a putative transcriptionally competent chromatin fraction from trout testis, *Nucleic Acid Res.*, 6, 259, 1979.
88. **Goldknopf, I. L., French, M. F., Musso, R., and Busch, H.,** Presence of protein A24 in rat liver nucleosomes, *Proc. Natl. Acad. Sci. U.S.A.*, 74, 5492, 1977.
89. **Goldknopf, I. L., Rosenbaum, F., Sterner, R., Vidali, G., Allfrey, V. G., and Busch, H.,** Phosphorylation and acetylation of chromatin conjugate protein A24, *Biochem. Biophys. Res. Commun.*, 90, 269, 1979.
90. **Paik, W. K. and Kim, S.,** Enzymatic demethylation of calf thymus histones, *Biochem. Biophys. Res. Commun.*, 51, 781, 1973.
91. **Paik, W. K. and Kim, S.,** Alkyllysinase. New assay method, purification and biological significance, *Arch. Biochem. Biophys.*, 165, 369, 1974.
92. **Gagnon, C.,** Presence of a protein methylesterase in mammalian tissues, *Biochem. Biophys. Res. Commun.*, 88, 847, 1979.
93. **Chene, L., Bourget, L., Vinay, P., and Gagnon, C.** Preferential localization of a protein methylating-demethylating system in proximal tubules of rat kidney, *Arch. Biochem. Biophys.*, 213, 299, 1982.
94. **Lischwe, M. A., Roberts, K. D., Yeoman, L. C., and Busch, H.,** Nucleolar specific acidic phosphoprotein C23 is highly methylated, *J. Biol. Chem.*, 257, 14600, 1982.

Chapter 6

POLY (ADP-RIBOSE) POLYMERASE — A NONHISTONE NUCLEAR PROTEIN

Najma Malik and Mark E. Smulson

TABLE OF CONTENTS

I. INTRODUCTION

An understanding of the interactions of nonhistone proteins in the specialized domains within chromatin is of primary importance in assigning functional roles for these macromolecules in the regulation of eukaryotic genes. Limited progress has been accomplished in this field because of the sparse availability of purified and well-characterized nonhistone proteins for structure/function studies. In contrast to the abundant and essentially homogeneous histones, the nonhistone proteins are diverse and heterogeneous and they differ markedly from tissue to tissue. These characteristics have prevented the extensive purification of many individual nonhistone proteins (with the exception of the HMG proteins) for use in elucidating their interaction with domains of chromatin. However, more recently, using the heat shock system of *Drosophila,* Elgin et al.[1] have provided considerable data on the association within chromatin and chromosomes of specialized classes of nonhistone proteins related to the heat shock system of *Drosophila.* These studies were made possible by the availability of specific antisera to certain nonhistone proteins. Current models for chromatin structure rarely take into consideration the interaction of nonhistone protein with DNA; however, before a final physical or topographic determination of higher-ordered folding of the genome is possible, more information must be obtained on the binding of such proteins within the polynucleosomal structural domains of chromatin.

Recently, considerable information from a number of laboratories has been provided on the interaction within chromatin of one specific nonhistone protein, the enzyme poly (ADP-ribose) polymerase. These studies have been facilitated due to this protein being bound to oligonucleosomes and having a defined locus of enzymatic activity, which can be assayed both conveniently and sensitively with its radioactive substrate (NAD).

The enzyme catalyzes the successive transfer of the ADP-ribose moiety of nicotinamide adenine-dinucleotide (NAD) to generate a covalently bound homopolymer of poly ADP-ribose on nuclear protein acceptors. The enzyme has an absolute requirement for DNA, histone, and the divalent cation, Mg^{2+} and cleaves the NAD at the nicotinamide-ribose bond (bond energy $\sim$ 34 kJ/mol).

Poly ADP-ribose is degraded in whole nuclei by two classes of enzymes: phosphodiesterases and glycohydrolases; the former cleave the pyrophosphate bridge and the latter cleave the $1'' \rightarrow 2'$ glycosidic linkages between adjacent ribose moieties.[2,3] Thus, these enzymes, in conjunction with polymerase, provide a system for polymerization and depolymerization of poly (ADP-ribose).

The poly ADP-ribosylation reaction was first demonstrated in rat liver nuclei.[4,6] Recently several review articles have considered the historical developments in this area.[7-13]

Since the discovery of poly (ADP-ribose) polymerase and ADP-ribosylation, many roles have been assigned to this reaction, yet the precise biological function for poly (ADP-ribose) is still not known. Functional roles for both include the maintenance of chromatin superstructure, regulation of DNA synthesis and DNA repair, cell multiplication, transformation, and differentiation.

II. ROLE OF POLY(ADP-RIBOSE) POLYMERASE IN DNA REPLICATION/ DNA REPAIR

A. DNA Replication

The implication of poly (ADP-ribose) polymerase in DNA replication arises from observations where its enzymatic activity was found to be higher in proliferating cells than in quiescent cells. Such observations have included studies with mitogen-stimulated lympho cytes, SV40 transformed cells, hormone stimulated oviducts, hepatomas, and regenerating rat liver cells.[14,19] In several studies the increase in polymerase activity was paralleled by an increase in DNA content, however, in others the increase far exceeded the rate of DNA synthesis. In studies by Lehmann et al.[15] the use of inhibitors of DNA synthesis suggests that during semiconservative replication the synthesis of poly (ADP-ribose) and DNA are two independent events occurring during S-phase.

B. DNA Repair

Evidence indicating the involvement of poly ADP-ribosylation in DNA repair has been seen in observations where alkylating agents and other DNA-damaging treatments result in the depletion of intracellular NAD pools and the increase in poly (ADP-ribose) synthesis.[20] Gill[21] reported that poly (ADP-ribose) polymerase activity was considerably higher in radiosensitive tissue containing fragmented DNA. Miller et al.[22] demonstrated that the antitumor drug Bleomycin which causes DNA fragmentation elevated poly (ADP-ribose) polymerase activity. Further it has been shown by Smulson et al.[23] and Whish et al.[24] that in isolated nuclei from HeLa cells and *Physarium polycephalum,* respectively, higher polymerase activity was observed following treatment with streptozotocin, the 2-deoxy-D-glucose derivative of alkylating agent, *N*′-methyl-*N*′-nitrosourea (MNU).

Considerable contributions have been made by our group in attributing the decrease in cellular NAD levels to the activation of poly (ADP-ribose) polymerase and the subsequent modification of specific nuclear proteins.[25,27] Smulson et al.[25] showed that when HeLa cells were treated with MNU: (1) the intracellular levels of NAD were lowered, (2) the poly (ADP-ribose) polymerase activity was stimulated, and (3) the fragmentation of DNA occurred. During repair periods, repair of damaged DNA was shown to be dependent on DNA polymerase activity and enhanced by poly ADP-ribosylation.[25] Treatment of the cells with nicotinamide, an inhibitor of poly (ADP-ribose) polymerase activity, resulted in an inhibition of DNA repair and an increase in DNA fragmentation. These observations suggest that poly (ADP-ribose) polymerase plays a role in the repair of cellular DNA damaged by alkylating agents.

Berger et al.[28] using normal human lymphoblasts demonstrated an increase in poly (ADP-ribose) synthesis following treatment with alkylating agents and UV irradiation. However, in DNA repair deficient fibroblasts with the genetic defect of *Xeroderma pigmentosum* (XP), there was little or no such increase in poly (ADP-ribose) synthesis in response to UV irradiation.[29] However, permeabilization of XP cells followed by treatment with DNase I elevated poly (ADP-ribose) levels.[30]

Benjamin and Gill[31] showed that fragmented double-stranded DNA with flush ends containing 3′-OH residues was the most effective in stimulating poly ADP-ribosylation. Shall et al.,[32] using L1210 cells treated with dimethyl sulfate (DMS), demonstrated that by the use of 3-methylbenzamide, an inhibitor of poly (ADP-ribose) polymerase, the rejoining of single stranded breaks could be inhibited.

Recently, McCurry and Jacobson[33] have reported that in normal human fibroblasts following UV irradiation or treatment with *N*-methyl-*N*′-nitrosoguanidine (MNNG), the increase in cellular poly (ADP-ribose) levels was concomitant with a decrease in cellular NAD levels. However, in fibroblasts homozygous for the XP genotype, no such drop in NAD levels was observed. A partial reduction of cellular NAD levels was noted in the heterozygotes; all the excision repair-competent cells showed reduced NAD pools on treatment with MNNG. These varied approaches all suggest that poly ADP-ribosylation of nuclear proteins participates in DNA excision repair in eukaryotic cells.

III. POLY (ADP-RIBOSE) ACCEPTOR PROTEINS

Considerable progress has been made in cataloging the acceptor proteins for poly ADP-ribosylation and in characterizing the linkage of (ADP-ribose) to these acceptors. Both in vivo as well as in vitro studies have indicated that histones as well as nonhistone chromosomal proteins are ADP-ribosylated.

A. Histones

Histone H1 and core histones are major acceptors of poly (ADP-ribose) among this class of nuclear proteins. The initial observation that the histones are ADP-ribosylated arose from the work of Nishizuka et al.,[34] where it was observed that histone H1, H2a, H2b, and H3 are ADP-ribosylated. Subsequently, numerous workers have confirmed and extended these initial observations in HeLa cells, rat liver cells, and trout testis nuclei.[35,43]

The (ADP-ribose) linkage to proteins is labile to alkali and neutral hydroxylamine. This leads to the suggestion that (ADP-ribose) is covalently attached to acceptor proteins through an ester linkage between the C′1 of the ribose and carboxyl group of aspartic or glutamic acid in histones.[10,11] Dixon et al.,[44] Burzio et al.,[42] and Riquelme et al.[41] initially reported on the peptide residues involved in the modification in histones H2b and H1. They isolated an ADP-ribosylated peptide generated upon tryptic digestion of histone H2b, with a sequence similar to the five terminal residues of the NH_2-terminus of histone H2b. Furthermore, it was suggested that glutamic acid 2 residue of histone H2b was ADP-ribosylated. Recently Ogata et al.,[45] using [ribose (NMN)-^{14}C] NAD as the (ADP-ribose) donor labeled the terminal ribose group of modified histone H2b. The results of this analysis conclusively demonstrated that the (ADP-ribose)-histone H2b bond is between glutamic acid residue 2 and reducing terminal ribose.

Histone H1 has been shown to be a principal acceptor of ADP-ribose in rat liver nuclei.[41,46] The most important aspect of the modification of H1 has been the demonstration that poly ADP-ribosylation leads to a cross-linked dimer of this histone (see below). Moreover, experiments have indicated that the ADP-ribose was covalently bound to both the –COOH and NH2 terminal regions of H1. Dixon et al.[46] reported a similar observation for the modification of H1 in trout testis.

Riquelme et al.[41] reported that on chymotryptic digestion of histone H1 it could be cleaved into –COOH and $-NH_2$ terminal halves. The two fragments can be isolated by separation on Bio-Gel® P-10. Figure 1 shows the separation of [^{14}C] ADP-ribosylated H1. In the top panel is the separation of calf thymus H1 after ADP-ribosylation with 0.1 m*M* [^{14}C] NAD and in the bottom panel is the separation of ADP-ribosylated H1 using 1 m*M* [^{14}C] NAD. The two peaks represent the –COOH and $-NH_2$ termini chymotryptic fragments. The result of Figure 1 indicates that on treatment with low NAD concentration the amount of radioactive (ADP-ribose) attached to the NH_2-terminal half was about three to five times more than that of the COOH-terminal half. However, on treatment with high NAD concentration both halves of H1 were equally ADP-ribosylated.

Recently Ogata et al.,[47] using ADP-[^{14}C] ribosylated histone H1, demonstrated that glutamic acid residues 2 and 14 and the –COOH terminal lysine residue 213 are modified in rat liver histone H1. It is significant that the sites of ADP-ribosylation in histone H1 are localized in the extended polar regions of the molecule, and so may contribute in the interactions of higher ordered polynucleosomes.[48]

Moreover, studies on chromatin structure have revealed that histone H1 is associated with the internucleosomal DNA, and plays an important role in the organization of higher ordered or solenoid structures of polynucleosomes.[49,50] It was suggested by these works that when the $-NH_2$ terminal region of a given H1 molecule is in close proximity to the –COOH terminal of another H1 molecule, which is located in the next helical turn of the solenoid, that such H1-H1 interactions may stabilize the higher order structure of the polynucleosomes. It was, therefore, of considerable significance when Stone et al.[51] demonstrated the existence of a dimer of H1 stabilized by a polymer of poly (ADP-ribose), which was 15 residues in

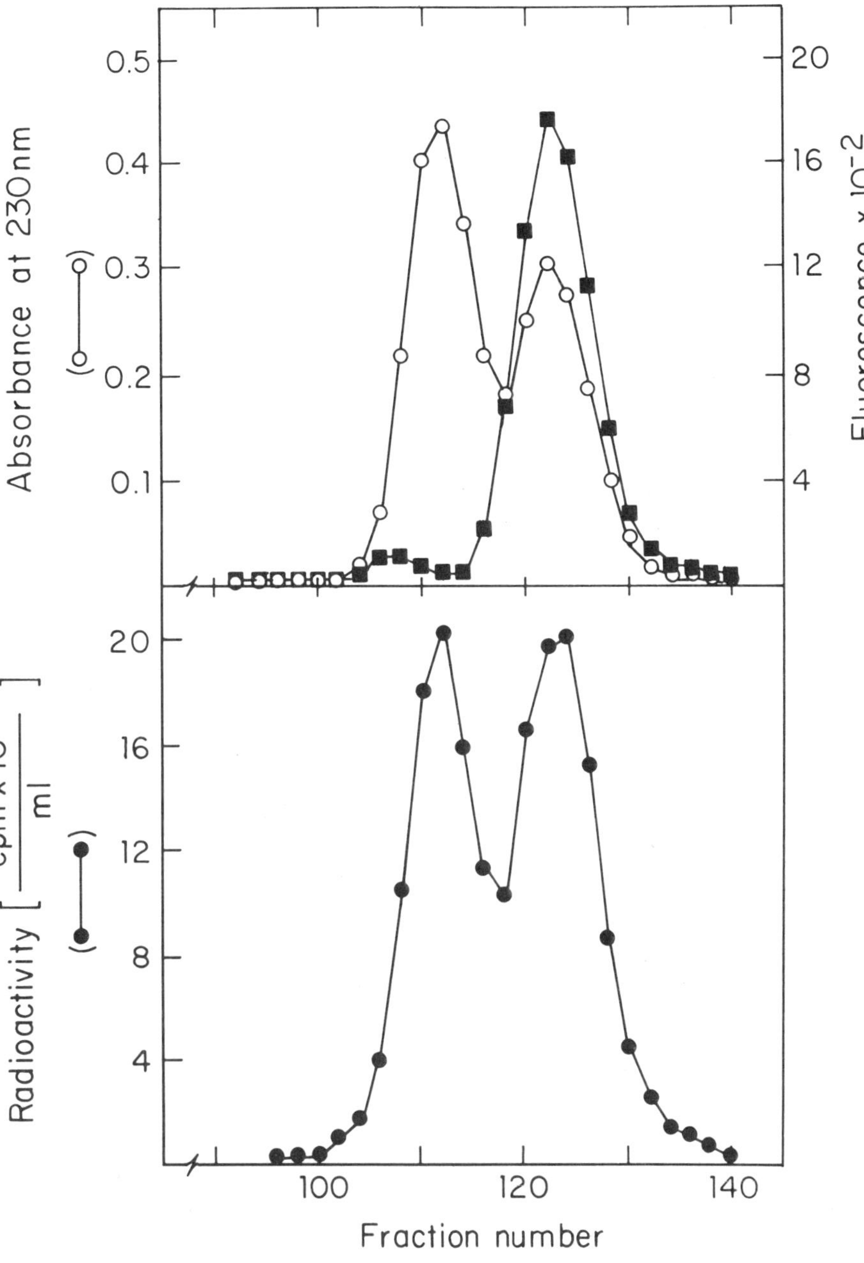

FIGURE 1. Separation of the NH_2-terminal and the COOH-terminal chymotryptic fragments of [^{14}C] ADP-ribosylated H1. The fractions containing the mixture of both fragments were separated on Sephadex® G-50 column. Fractions containing peptides were collected. The upper panel shows the distribution of peptides, ○—○, absorbance at 230 nm, and ■—■, the fluorescence of tryosine. The bottom panel shows ●—●, the distribution of radioactivity. The NH_2-terminal fragments correspond to fractions 119 to 132, and the COOH-terminal fragment is contained in fractions 103 to 118. (From Riquelme, P. T., Burzio, L., and Koide, S. S., *J. Biol. Chem.*, 254, 3018, 1979. With permission.)

length. These observations indicate that poly (ADP-ribose) may play a role in the condensation of chromatin. The mechanism of synthesis of H1 dimer has been studied in detail in isolated nucleosomal particles.[62] Figure 2 shows the result when chromatin fragments 8- to

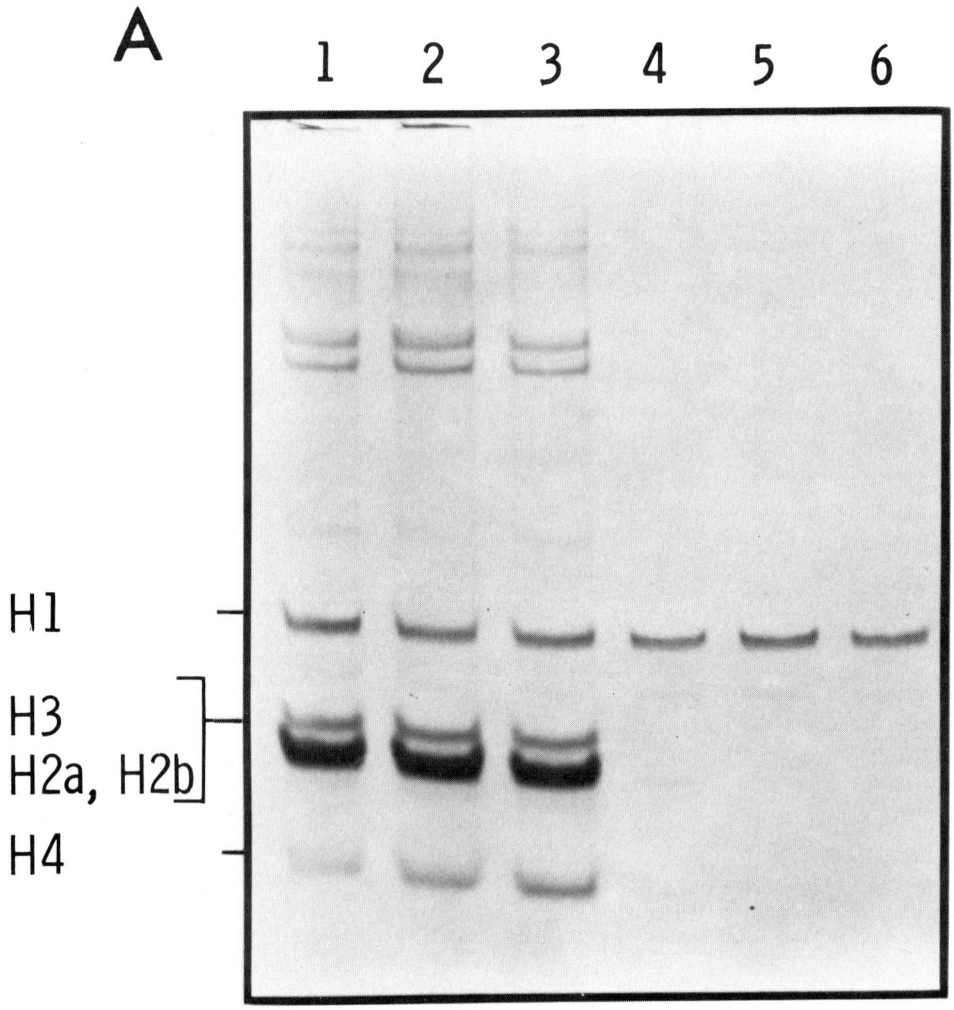

FIGURE 2A

FIGURE 2. Analysis of acid-soluble proteins and histone H1 during a chase with 100 μ*M* NAD. Chromatin fragments of 8 to 12-nucleosome repeat length (2.0 A_{260} units) were incubated under standard ADP-ribosylation conditions. The reaction was initiated by the addition of [^{32}P]-NAD (25 nm) and after 30 sec two equal aliquots were removed (Lanes 1 and 4). The rest of the sample was chased with a final concentration of 100 μ*M* for 1 min (Lanes 2 and 5) and 3.5 min (Lanes 3 and 6). In one series of samples (Lanes 1 to 3), total histones were extracted, while in the other series (Lanes 4 to 6) the samples were extracted for histone H1. The extracted histone samples were electrophoresed on 15% acid-urea gels. (A) Stained gel; (B) autoradiograph. (From Butt, T. R. and Smulson, M., *Biochemistry,* 19, 5234, 1980. With permission.)

10-nucleosomes in length were treated with low concentration of [^{32}P] NAD for 30 sec followed by increasing the concentration of NAD to 100 μ*M* using nonradioactive NAD for 1 and 3.5 min. The alteration in the level of H1 modification was more evident when histone H1 was selectively extracted (Lanes 4 to 6). On treatment with low concentration of NAD the majority of the modification represented mono (ADP-ribosylation) of H1. However, on chase with 100 μ*M* NAD, the ADP-ribose chains were elongated, as shown by 12 to 15 intermediate bands (Lane 5) to the final formation of H1-complex (Lane 6).[62] Finally, it

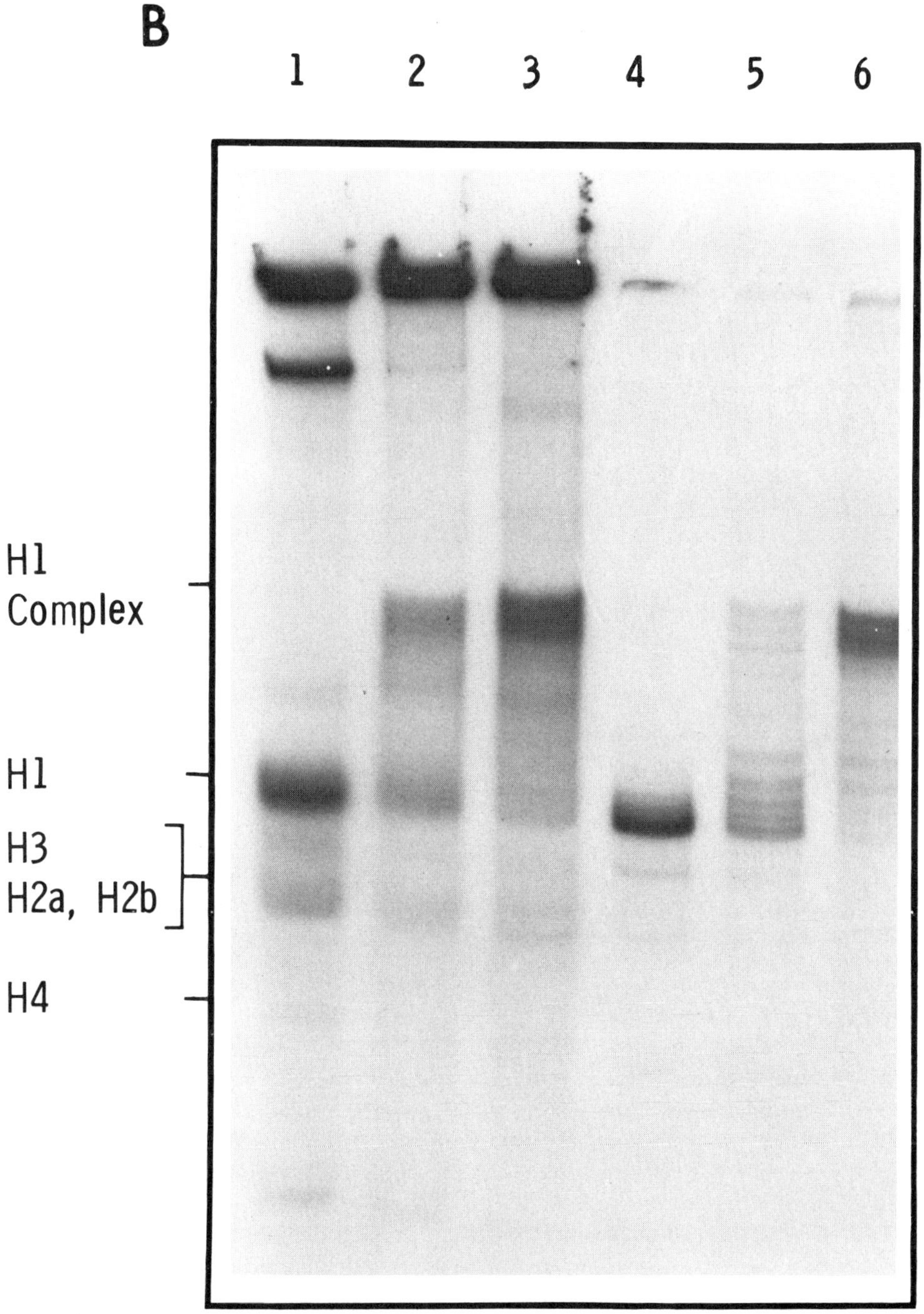

FIGURE 2B.

was shown that the H1-complex migrates in SDS-polyacrylamide gels at a position corresponding to a molecular weight twice that of H1 while the intermediates migrate close to the unmodified H1 species.[67]

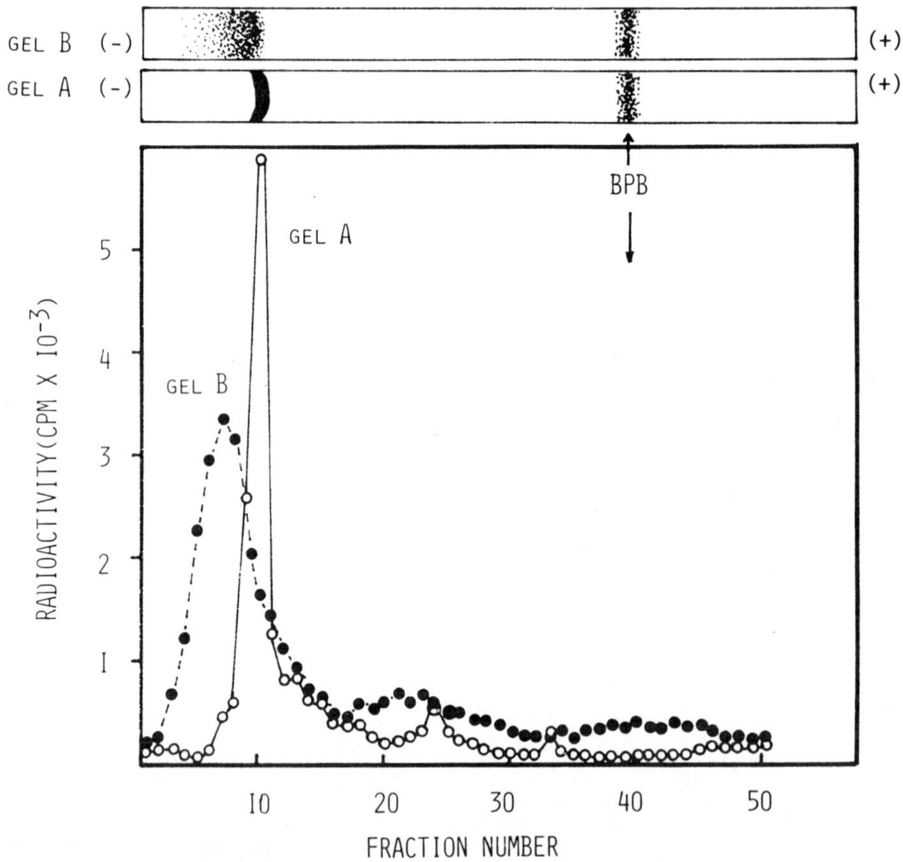

FIGURE 3. SDS-polyacrylamide gel electrophoresis of [^{3}H]-ADP-ribosylated poly (ADP-ribose) polymerase. Gel A represents the ADP-ribosylation of polymerase using 1 μM NAD for 20 sec. Gel B represents further 1 min incubation after addition of 0.5 μmol cold NAD to the sample in gel A. The acid-insoluble samples were collected and electrophoresed. The gels were cut into 1.5 mm thick slices and counted in a scintillation counter. (From Yoshihara, K., Hashida, T., Tanaka, Y., Yoshihara, H., and Ohgushi, H., *Biochem. Biophys. Res. Commun.*, 78, 1281, 1977. With permission.)

B. Nonhistone Proteins (NHP)

Considerable evidence has been accumulated showing that poly ADP-ribose is distributed in both histones and nonhistone proteins. Yoshihara et al.[52] initially reported tat a Ca^{2+}-, Mg^{2+}-dependent DNA endonuclease, isolated from rat liver, was inhibited by mono-ADP-ribosylation of the enzyme protein. Partial restoration of the enzymatic activity was achieved by mild alkaline treatment to remove covalently bound (ADP-ribose).

A significant observation in the area of nonhistone protein ADP-ribosylation was made by Yoshihara et al.[53] They demonstrated that purified poly (ADP-ribose) polymerase isolated from calf thymus nuclei catalyzes its modification under in vitro conditions.[53] Figure 3 shows the SDS-polyacrylamide gel electrophoresis of ADP-ribosylated purified poly (ADP-ribose) polymerase. The SDS gel was subsequently sliced into 1.5 mm thick slices and the radioactivity associated with each slice was measured. The acid insoluble activity was found to be at the same position on the gel as the enzyme protein. They further showed that when the [^{32}P] labeled early reaction product was chased by cold NAD and subsequently electrophoresed, the peak of the radioactivity shifted toward origin and became broader with a simultaneous shift and a broadening of the enzyme band representing the elongation of the

Table 1
PURIFICATION OF HeLa CELL POLY (ADP-RIBOSE) POLYMERASE[a]

Purification step	Protein (mg)	Enzyme act. (units)	Sp act. (units/mg of protein)	X-fold purified	% Recovered
Nuclei	177.00	43.6	0.25	1.0	110
0.3 *M* KCl extract	88.00	49.5	0.56	2.3	114
DNA-agarose	1.02	34.5	33.82	137.5	79
Hydroxylapatite	0.34	19.1	55.85	227.0	44
Sephadex® G-200	0.09	10.5	111.70	454.0	24

[a] The enzyme was purified from 2.5 × 10^9 HeLa cells (787 mg of total protein).

From Jump, D. B. and Smulson, M., *Biochemistry*, 19, 1024, 1980. With permission.

bound ADP-ribose polymer. This observation was also seen in HeLa cells by Jump et al.[37] who showed that 40 to 80% of the total chromatin poly ADP-ribosylation was found to be associated with an internucleosomal protein with molecular weight of 112,000. Subsequently this major acceptor was purified 454-fold from HeLa nuclei and shown to copurify with poly (ADP-ribose) polymerase.[54] Table 1 gives the specific activities at various steps of purification. Figure 4 shows the protein composition and acceptors for poly (ADP-ribose) at each step of the poly (ADP-ribose) polymerase purification from HeLa cells. Nuclear proteins at various steps of purification were ADP-ribosylated using [^{32}P] NAD followed by ethanol precipitation and electrophoresis on SDS-polyacrylamide gels. Table 2 shows that in HeLa nuclei 43% of the total ADP-ribosylation was associated with the polymerase, 40% with the histones, and the remaining incorporated with the other nuclear NHPs.[56] Other NHPs found to be ADP-ribosylated in the nuclei include the high mobility group proteins (HMG); H6 (or HMG 17) in trout testis.[43]

The availability of a very sensitive radioactive assay for poly (ADP-ribose) polymerase and the success in the purification of this enzyme, has aided the elucidation of its association with chromatin and its possible role in the modulation of chromatin structure and function. This has been achieved by (1) the reconstitution of the polymerase with chromatin, (2) studies on the poly (ADP-ribose) polymerase-dependent nucleosomal aggregation, and (3) the development of specific antibodies to this nonhistone protein.

IV. RECONSTITUTION OF POLY (ADP-RIBOSE) POLYMERASE WITH CHROMATIN

The in vitro analysis of poly (ADP-ribose) polymerase activity associated with different oligonucleosome chain lengths has indicated a unique relationship which is demonstrated in Figure 5.[55] It was found that the specific activity of the enzyme increased with nucleosomal size, reaching a maximum on particles of approximately 8N periodicity. Subsequently, Jump et al.[37] analyzed in vitro the acceptors of the poly (ADP-ribose) in different sized oligonucleosomes. Figure 6A demonstrates the straining pattern of the nucleosomal proteins separated on SDS polyacrylamide gels from such an experiment, while Figure 6B shows the autoradiographic representation of the modified proteins in the various classes of oligonucleosomes. Quantitative analysis of the distribution of ADP-ribose incorporation between protein C (polymerase) and histones with regard to the different sized nucleosomal fragments indicated a significant relationship (Figure 7). With the smaller fragments 60 to 85% of the (ADP-ribose) incorporation was noted to be bound to poly (ADP-ribose) poly-

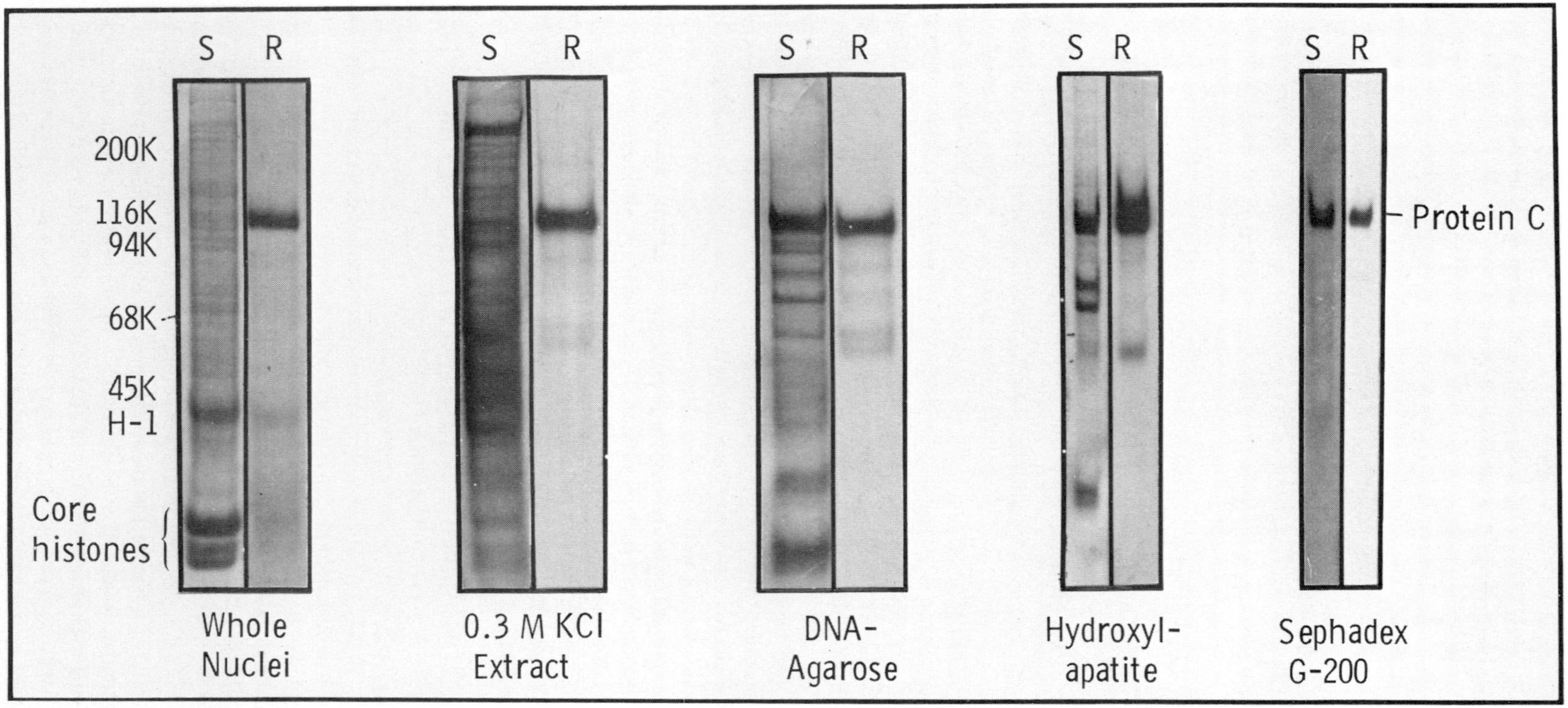

FIGURE 4. Analysis of the protein composition and acceptors for poly (ADP-ribose) at various steps of poly (ADP-ribose) polymerase purification. Protein samples from each step of enzyme purification were taken and [^{32}P]-ADP-ribosylated followed by ethanol precipitation and electrophoretic analysis on SDS-polyacrylamide gels. Proteins were stained (S) with Coomassie® blue and dried for autoradiography (R). (From Jump, D. B. and Smulson, M., *Biochemistry*, 19, 1024, 1980. With permission.)

Table 2
DISTRIBUTION OF IN VITRO SYNTHESIZED POLY (ADP-RIBOSE) AMONG VARIOUS HeLa NUCLEAR PROTEINS[a]

	Protein-bound ADP-ribose	
Protein fraction	**pmol of ADP-ribose**[a]	**% of total**
Poly (ADP-ribose) polymerase	0.57	43.5
Other nonhistone proteins	0.21	16.0
Histone H1	0.15	11.5
Core histones	0.38	29.0

[a] HeLa nuclei (5 × 10^7/mℓ) were incubated with [^{32}P] NAD (1 μM, 1 Ci/mmol) for 10 min at 25°C. The reaction was terminated by the addition of thymidine to 50 μM, and micrococcal nuclease was added to 200 units/mℓ; nuclei were digested for 5 min at 37°C, and proteins were precipitated in ethanol. Proteins were separated by NaDodSO$_4$-polyacrylamide gel electrophoresis. Protein-bound [^{32}P] ADP-ribose was quantitated by excising labeled proteins from the gel, followed by liquid scintillation counting.

From Jump, D. B., Butt, T. R., and Smulson, M., *Biochemistry*, 19, 1031, 1980. With permission.

merase. However, as the nucleosome chain length increased from 1N to 12N, the modification of histones increased from 8 to 25%, the latter being the level detected in nuclei.[56] These data emphasize the importance of the intramolecular interactions of oligonucleosomes in determining the extent of ADP-ribosylation of histones.

The reconstitution of defined components isolated from a multicomponent system aids in better appreciation of their interaction within such a complex of molecules. This approach has been utilized to better understand the association of the polymerase with chromatin. Jump et al.[56] were able to deplete a preparation of octanucleosomes of poly (ADP-ribose) polymerase by mild salt treatment; resulting in no endogenous enzymatic activity being found in association with the such depleted chromatin (Figure 8). However, when the stripped octanucleosomes were reconstituted (via gradual dialysis) with a precise amount of purified enzyme, the chromatin was capable of extensive poly ADP-ribosylation. The enzyme was shown to be reassociated directly with the chromatin particles and sedimented coincident with the reconstituted oligonucleosomes. Furthermore, through staph nuclease digestion studies, it was found that the enzyme was reconstituted with internucleosomal DNA.[56-59] The most significant aspect of this study was the observation that the purified enzyme was incapable of histone modification, yet upon reconstitution with chromatin it regained this activity (Figure 9).[56] These types of reconstitution experiments and those described below have begun to define the interactions of one specific nonhistone protein with domains of chromatin. The advantage of the use of poly (ADP-ribose) polymerase experimentally has been the ability to detect its presence by its enzymatic reaction. Hopefully, the approaches developed in these studies will act as models for future studies on other nonhistone proteins. For example, poly (ADP-ribose) polymerase from HeLa cell nuclei has been located in the linker DNA region between adjacent nucleosomes.[57-59] The enzymatic activity has been shown not to coincide with the position of nucleosomal core particles on sucrose gradients

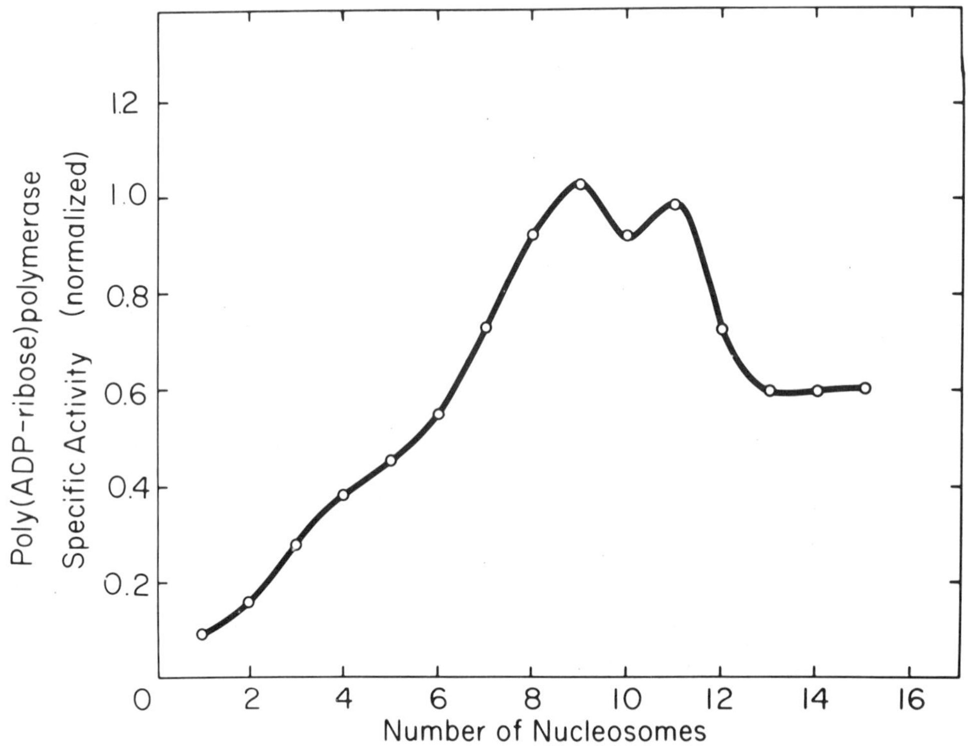

FIGURE 5. The relationship between poly (ADP-ribose) polymerase specific activity and the number of nucleosomes on a chromatin fragment. (Taken from Butt, T. R., Brothers, J. F., Giri, C. P., and Smulson, M. E., *Nucleic Acid Res.*, 5, 2775, 1978. With permission.)

but at a position where mononucleosomes possessing linker region DNA sediment.[59] Jump et al.[37] investigated the micrococcal nuclease digestion of S-phase nuclei and demonstrated that the nuclease selectively digested chromatin at or near the DNA replicating fork.[60,61] Poly (ADP-ribose) polymerase activity was found to be associated with nucleosomes preferentially digested by micrococcal nuclease. These results suggest that the polymerase is associated with the extended forms of chromatin undergoing DNA replication or DNA repair.[37]

A. Poly (ADP-Ribose) Induced Aggregation of Chromatin

An NAD concentration-dependent aggregation of chromatin particles has recently been demonstrated that may prove useful in describing the effect of ADP-ribosylation on chromatin structure.[62,63] Octanucleosomes were incubated with increasing concentrations of the substrate NAD (25 n*M* to 1 m*M*) under conditions optimal for poly (ADP-ribose) polymerase activity. Analysis of the products of such an experiment on polyacrylamide-agarose gels is illustrated in Figure 10. The data indicate that these concentrations of NAD did not affect the overall mobility of the bulk of the nucleosomes (as observed by ethidium bromide staining pattern). However, autoradiography of the incorporated ADP-ribosylated chromatin revealed a progressive increase in the size of the modified products. The aggregation was found to be progressive with successive increases in NAD concentration and occurred in discrete steps with respect to time at a fixed NAD concentration.[62] The results suggested that this aggregation could be due to an enzyme catalyzed cross-linking of nuclear proteins within chromatin particles.

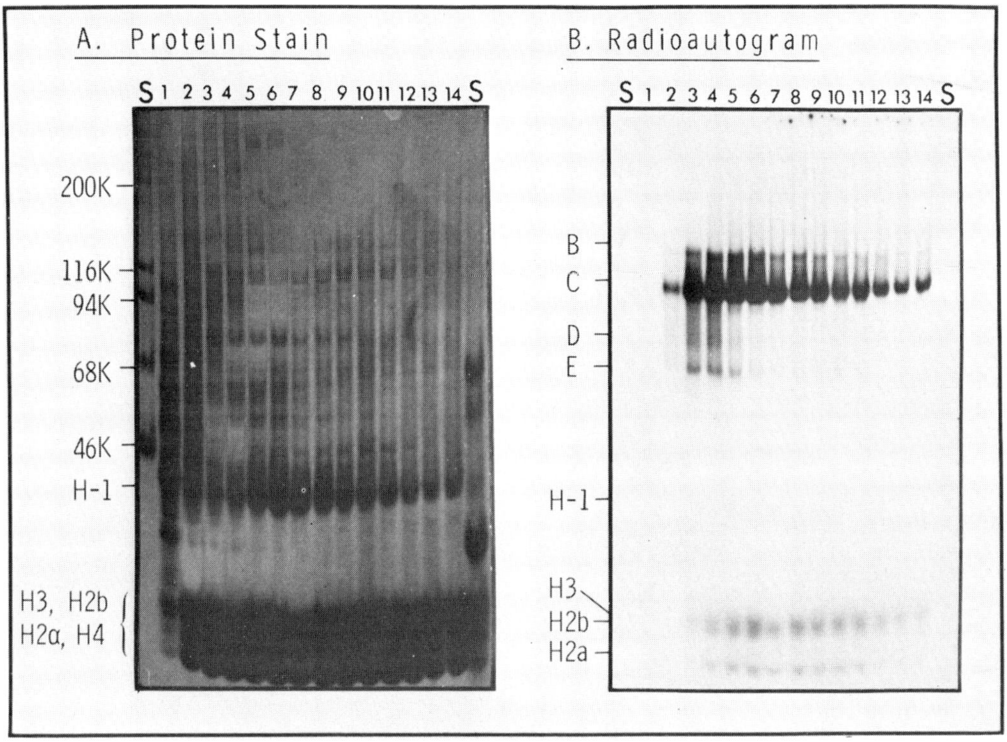

FIGURE 6. Relationship between oligonucleosome size and degree of ADP-ribosylation of various nuclear protein acceptors. Oligonucleosomes of various repeat units separated on sucrose gradients were ADP-ribosylated using [^{32}P]-NAD. Subsequently the nuclear proteins were ethanol precipitated and run on SDS-polyacrylamide gels. (A) Represents the protein staining pattern using Coomassie® blue and (B) autoradiographic analysis. Lane S represents the protein standards run on the gel. Lanes 1 through 14 represent oligonucleosomes obtained from different sucrose gradient fractions. (From Jump, D. B., Butt, T. R., and Smulson, M., *Biochemistry*, 19, 1031, 1980. With permission.)

However, these data could not be used alone to prove that the aggregated material contained chromatin per se, since there was no ethidium bromide staining in the region of the gel where the ADP-ribosylated aggregates migrated. However, when nucleosomes were incubated in a similar manner and were run on sucrose gradients, an increase in their sedimentation velocity was observed.[62] Moreover, upon incubation with increasing concentrations of NAD, the poly (ADP-ribose) polymerase activity was found to co-sediment with the aggregated material. These data suggested that besides the nuclear protein acceptors, the enzyme itself and DNA were also present in the aggregated material.[63]

The presence of poly (ADP-ribose) polymerase in such aggregates was also confirmed by assay *in situ*.[62] This enzyme assay procedure is performed directly on gels of electrophoretically separated units of chromatin; the general method might have application for other nonhistone proteins. In this experiment (Figure 11), the polynucleosomal aggregation was accomplished by the incubation of the nucleosomes with increasing concentrations of nonradioactive NAD, followed by electrophoresis on polyacrylamide gels. *In situ,* the poly (ADP-ribose) polymerase assay was then performed directly on the gel using [^{32}P] NAD. Autoradiography of the gel confirmed that the enzymatic activity was indeed associated with the chromatin aggregates. In Lane 1 where the nucleosomes were incubated with a low NAD concentration, the stainable bands representing bulk of the chromatin coincided with the labeled chromatin. However, in Lanes 2 and 3 where the nucleosomes were incubated with

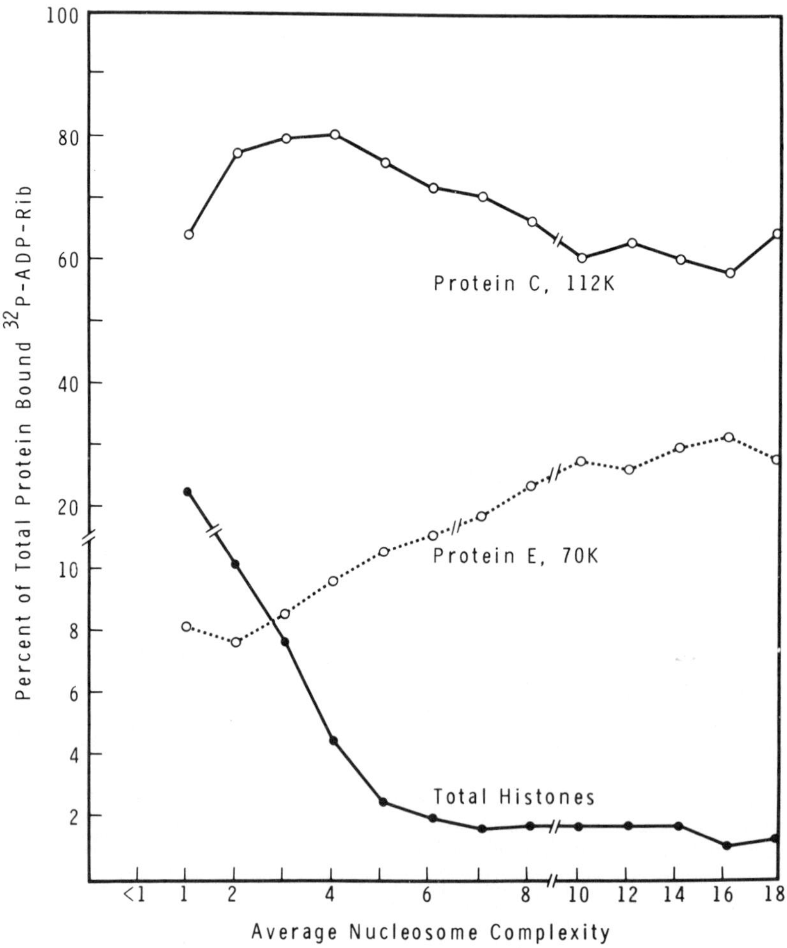

FIGURE 7. Percent protein-bound [^{32}P]-poly (ADP-ribose) associated with poly (ADP-ribose) polymerase and total histones in chromatin fragments of varying complexity. Total [^{32}P] ADP-ribose covalently bound to protein was recovered by excising each labeled region from SDS-polyacrylamide gel containing acceptor proteins. The gel slices were counted on liquid scintillation counter. The fraction of the total protein-bound [^{32}P] ADP-ribose bound to various acceptors was determined and plotted. The average nucleosome size was determined by analysis on DNA gel. (From Jump, D. B., Butt, T. R., and Smulson, M., *Biochemistry*, 19, 1031, 1980. With permission.)

high NAD concentrations, the incorporated label was found in the aggregate, thus confirming that poly (ADP-ribose) polymerase, DNA, and histones were present in the the chromatin aggregate.

The chain lengths of poly (ADP-ribose) were also determined in both unaggregated and highly aggregated samples of oligonucleosomes by the extraction of polymer and their subsequent electrophoresis on 20% polyacrylamide gels (Figure 12). The data revealed a broad range of poly (ADP-ribose) chain lengths ranging from mono (ADP-ribose) to chains of approximately 60 to 90 residues in length.[62] It is of considerable interest that the length of predominant species of chains in each sample was dependent on NAD concentration and proportional to the extent of aggregation of the nucleosomes. The poly (ADP-ribose) chains formed by the nucleosomes incubated with 100 μM NAD were of such lengths to preclude entry into thc polyacrylamide gel. Thus, it is possible that within the aggregates poly (ADP-

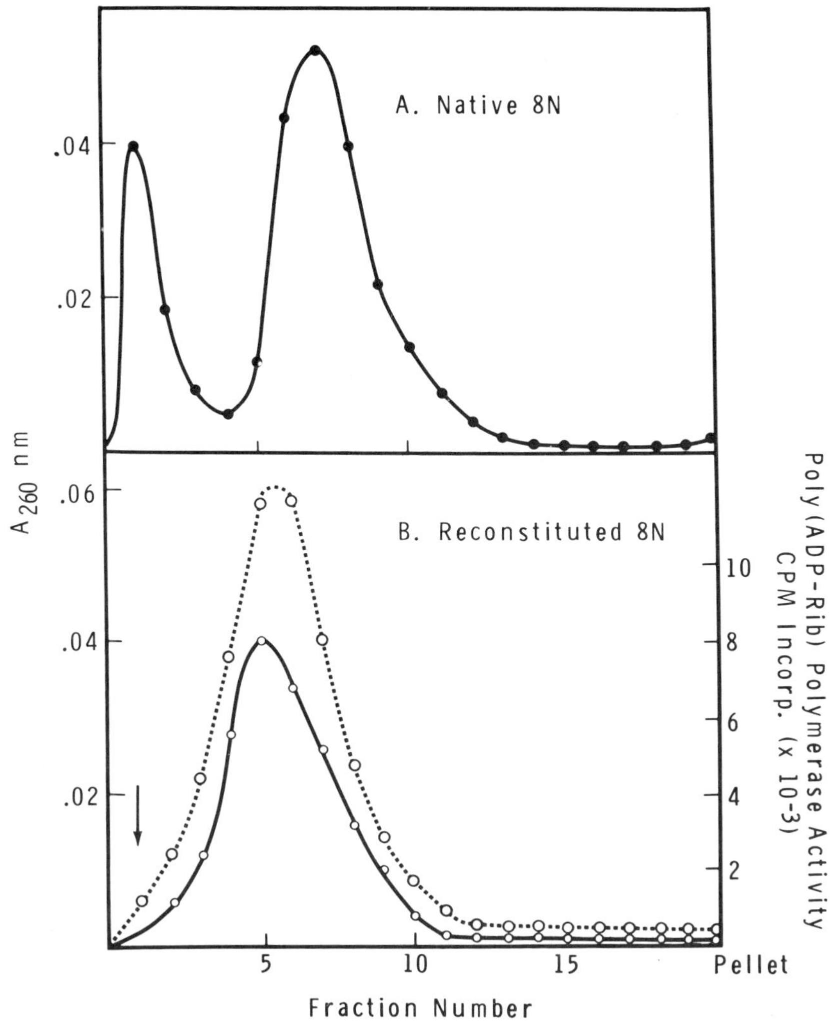

FIGURE 8. Velocity sedimentation of native and reconstituted HeLa chromatin. Native (A) and reconstituted (B) HeLa octanucleosomes were sedimented in 10 to 30% sucrose gradients for 6 hr at 38K in SW40 rotor. Gradients were fractionated and analyzed for absorbance at 260 nm (●—●, ○—○) and poly (ADP-ribose) polymerase activity (○----○). (From Jump, D. B., Butt, T. R., and Smulson, M., *Biochemistry*, 19, 1031, 1980. With permission.)

ribose) chains cross-linked regions of nucleosomes thereby causing the condensation of chromatin. Moreover, Butt et al.[63] demonstrated that nuclei incubated with 1 m*M* NAD were more resistant to micrococcal nuclease digestion than untreated nuclei. These results indicate that poly ADP-ribosylation may play a role in eliciting structural changes at or in different domains of chromatin.

The presence of histone H1 in the aggregated chromatin has been established, albeit indirectly, by Butt et al.[63] who showed that in nucleosomes at lower concentrations of NAD (where aggregation is not observed), H1 dimer synthesis is not seen. However, under conditions of chromatin aggregation considerable H1 dimer formation is observed. This significant point is illustrated in the data of Figure 2.

Data thus far available suggest that the cross-linking of H1 (and perhaps other nuclear proteins, including the enzyme itself) may be the major factor contributing to the condensation

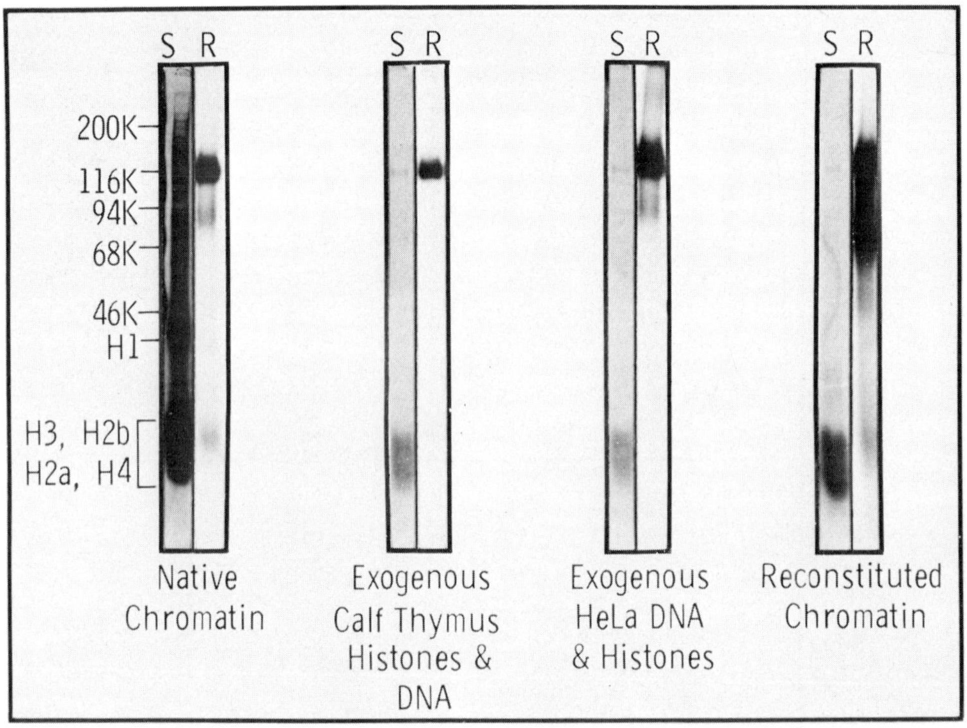

FIGURE 9. Acceptor analysis of purified poly (ADP-ribose) polymerase and native and reconstituted oligonucleosomal chromatin. Native (0.10 unit of A_{260} nm and reconstituted (0.05 A_{260} nm) chromatin was incubated in the standard acceptor assay (minus calf thymus histones and DNA) for 10 min at 25°C. Samples were precipitated with ethanol and analyzed by SDS-polyacrylamide gel electrophoresis. Purified poly (ADP-ribose) polymerase (0.3 μg/0.2 mℓ) was incubated with either calf thymus histones and DNA (12.5 μg/mℓ each) in the standard acceptor assay for 10 min at 25°C followed by ethanol precipitation of proteins and electrophoretic analysis. (From Jump, D. B., Butt, T. R., and Smulson, M., *Biochemistry*, 19, 1031, 1980. With permission.)

of nucleosomes described above. This hypothesis was confirmed when upon incubation with a high NAD concentration, no aggregation was observed following the selective removal of histone H1 from oligonucleosomes by treatment with Dowex®.[62] However, as shown in Figure 13, when histone H1 was reconstituted with the H1-depleted chromatin and subsequently incubated with 100 μ*M* NAD, the poly (ADP-ribose) promoted aggregation was restored.[64]

B. Development of a Specific Antibody to Poly (ADP-Ribose) Polymerase

In order to probe the location of poly (ADP-ribose) polymerase within nuclei and to further understand its function, antibodies were elicited against purified HeLa poly (ADP-ribose) polymerase.[65] Thus the antibody would serve as a specific probe for the nonhistone protein poly (ADP-ribose) polymerase, and help in understanding its interaction with chromatin structure.

The specificity of the antibody to poly (ADP-ribose) polymerase was established by using the gel transfer technique.[66] Nuclear proteins extracted from nuclei using 0.3 *M* KCl were separated on SDS polyacrylamide gels, and electrophoretically transferred to nitrocellulose (Figure 14). The nitrocellulose strips were then incubated with an appropriate dilution of immune serum and subsequently treated with [^{125}I]-protein A to locate the antigen-antibody complex. Autoradiography of the nitrocellulose strips showed that only one nuclear protein band reacted with the antibody and that the position of migration of the immune-reactive

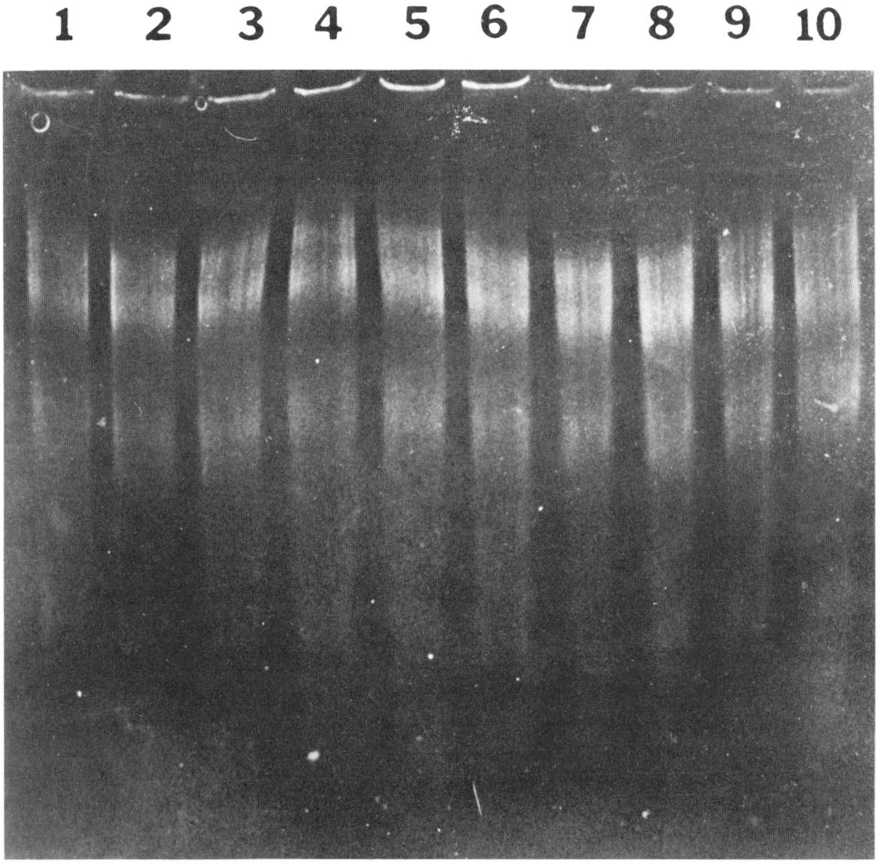

A.

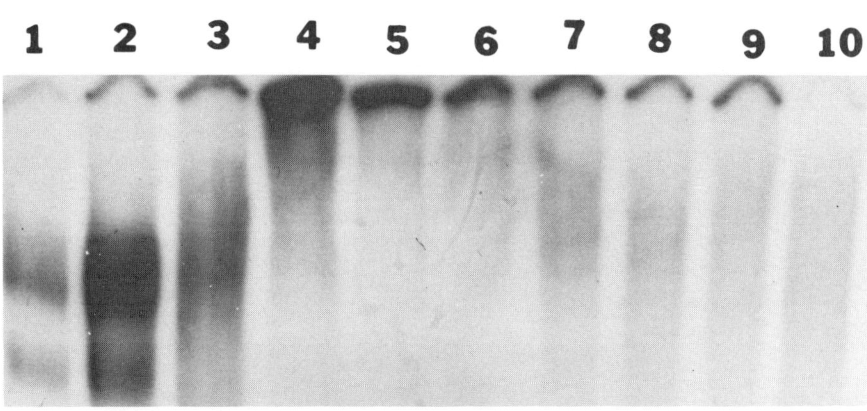

B.

FIGURE 10. NAD concentration-dependent complex formation of polynucleosomes, as analyzed by chromatin gel electrophoresis. (A) Ethidium bromide stain; (B) autoradiograph. (Lane 1) chromatin incubated with 25 μ*M* NAD for 5 min at 5°C. (Lanes 2 to 6) chromatin incubated at 20°C with 25 n*M*, 1 μm, 10 μ*M*, and 1 n*M* NAD, respectively. In Lanes 7 to 10, the samples were incubated with 100 μ*M* NAD for 5 min and terminated with nicotinamide, and 10 μg of proteinase K was added; the samples were incubated for an additional 1, 5, 10, and 20 min at 20°C, respectively. (From Butt, T. R. and Smulson, M., *Biochemistry,* 19, 5234, 1980. With permission.)

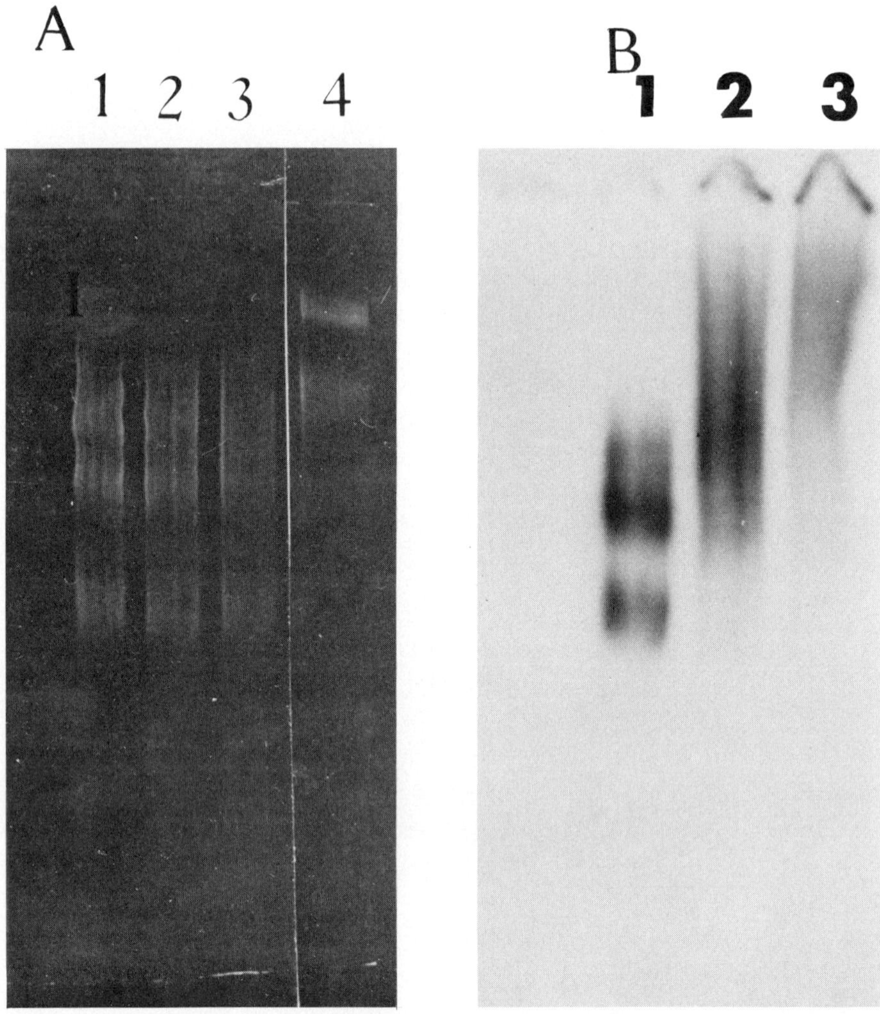

FIGURE 11. Presence of poly (ADP-ribose) polymerase in aggregated forms of chromatin by *in situ* enzyme analysis performed directly on polyacrylamide gels. Oligonucleosomes (0.1 A_{260} unit) were either preincubated in the absence of NAD (Lane 1) or in the presence of 10 μ*M* and 100 μ*M* unlabeled NAD (Lanes 2 and 3). The reaction was terminated with nicotinamide, and the samples were applied to a polyacrylamide gel. The samples were electrophoresed. The slab gel was subsequently incubated with a polymerase assay mixture containing [^{32}P] NAD. The gel was dried and autoradiographed. (From Butt, T. R. and Smulson, M., *Biochemistry,* 19, 5234, 1980. With permission).

protein was similar to that of the [^{32}P] NAD labeled poly (ADP-ribose) polymerase.[65] The immune serum was found to react with isolated poly (ADP-ribose) polymerase extracted from chromatin (0.3 *M* KCl) as well as when it is in association with polynucleosomes. Furthermore, 81% activity was inhibited when 0.3 *M* KCl extract was treated with immune serum whereas when nucleosomes were treated with anti-polymerase serum, 75% of the enzyme activity was inhibited (Table 3).

The results indicated that the antipolymerase serum recognized the denatured form of the enzyme (SDS transfer reactivity) as well as the protein in its native state. Furthermore, inhibition of enzymatic activity on immunoprecipitation suggests that the active site of the polymerase is an antigenic determinant.

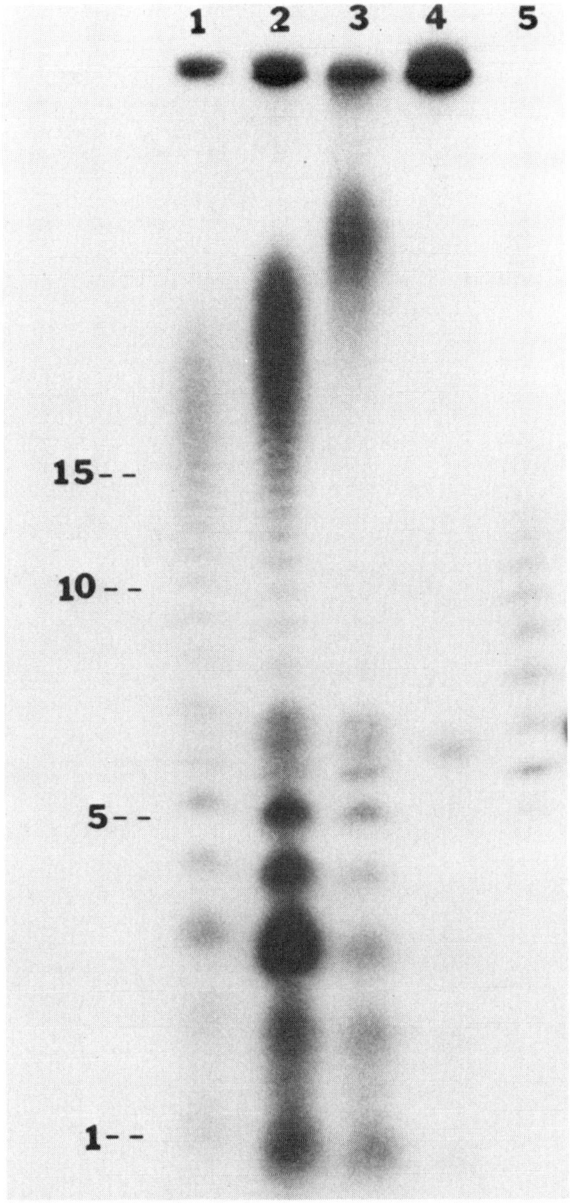

FIGURE 12. Correlation between poly (ADP-ribose) chain length, nucleosome aggregation, and NAD concentration. Lanes 1 to 4 represent polymer from oligonucleosomes incubated with 25 μ*M*, 1 μ*M*, 10 μ*M*, and 100 μ*M* NAD, respectively, and Lane 5, marker polymer with average chain length of 10 N. (From Butt, T. R. and Smulson, M., *Biochemistry*, 19, 5234, 1980. With permission.)

In further studies, the antipolymerase serum was shown to react with the nitrocellulose fingerprint of electrophoretically separated oligonucleosomes. Figure 15A shows the ethidium bromide staining pattern of oligonucleosomes resolved on a polyacrylamide-agarose gel, and Figure 15B represents the autoradiograph of the nitrocellulose fingerprint of Figure 15A treated with antipolymerase serum followed by treatment with [^{125}I] protein A. These

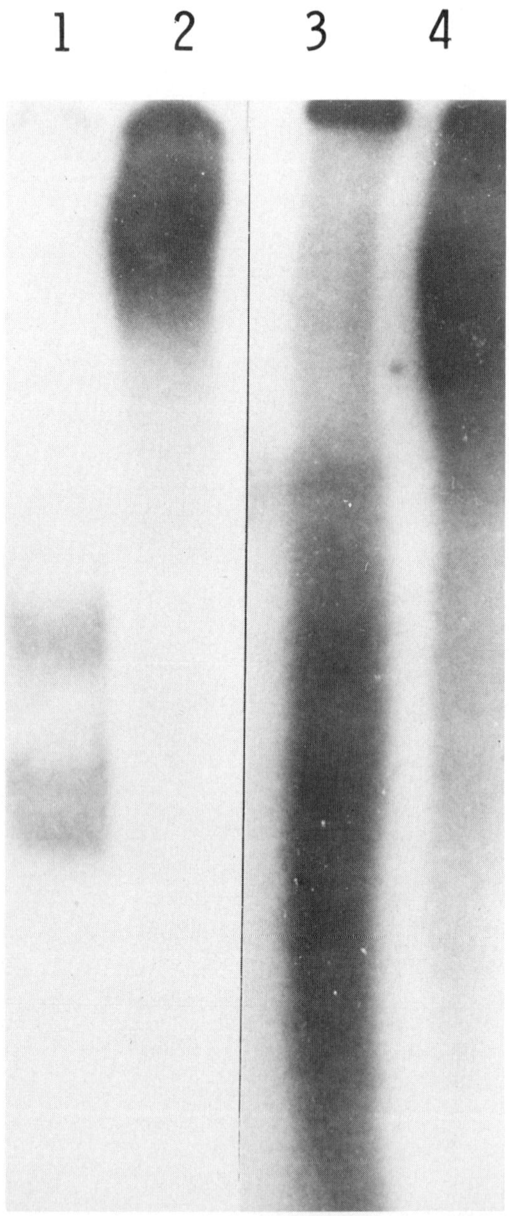

FIGURE 13. Reconstitution of histone H1 restores poly (ADP-ribose) induced nucleosomal aggregation. Oligonucleosomes were treated with low (0.1 μ*M*) and high (500 μ*M*) concentrations of [^{32}P]-NAD and analyzed by native 2.5% polyacrylamide gel electrophoresis. Autographic pattern is shown. A. Lanes 1 and 2: untreated oligonucleosome hypo- and hyper-ADP-ribosylation, respectively; Lanes 3 and 4: H1 and (ADP-ribose) polymerase reconstituted oligonucleosomes, hypo- and hyper-ADP-ribosylated, respectively. (From Wong, M., Malik, N., and Smulson, M., *Eur. J. Biochem.*, 128, 209, 1982. With permission.)

results should be compared with the reactivity of antihistone H3 serum to a similar transfer of oligonucleosomes (Figure 16). The data would suggest that the polymerase is not associated with all nucleosomes within a given nucleosome repeat size. These immunological techniques

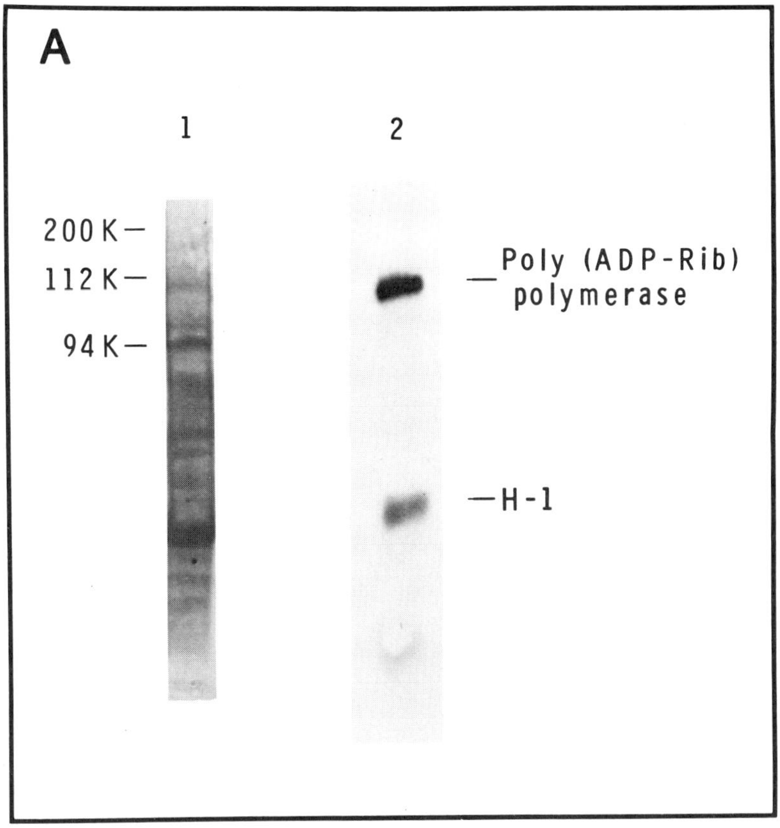

A.

FIGURE 14. Correlation between antipolymerase sera reactivity with protein band on nitrocellulose blot and electrophoretically separated [^{32}P] automodified poly (ADP-ribose) polymerase. A 0.3 *M* KCl extract of HeLa nuclei was subjected to SDS-polyacrylamide gel electrophoresis. After electrophoresis, the proteins separated on the gels were electrophoretically transferred to nitrocellulose sheets. A. Lane 1 represents an amido black stain of typical 0.3 *M* KCl extract nitrocellulose transfers. Lane 2: the 0.3 *M* KCl extract was incubated with [^{32}P] NAD, DNA, and histones. The sample was electrophoretically resolved on gel, transferred to nitrocellulose, and subsequently autoradiographed. B. In a separate transfer to those shown in A, fingerprinted proteins on nitrocellulose were incubated with antipolymerase serum, followed by detection of antigen-antibody complex with [^{125}I] protein A. An autoradiographic representation is shown. Lane 1: reaction with preimmune sera; Lane 2: reaction with antipolymerase serum. (From Malik, N., Bustin, M., and Smulson, M., *Nucl. Acid Res.*, 10, 2939, 1982. With permission).

may aid in a better appreciation of the domains of chromatin bounded by poly (ADP-ribose) polymerase; similar approaches may be applicable for the study of other nonhistone proteins.

V. CONCLUSIONS

A considerable amount of new information has emerged from a variety of groups aimed at the elucidation of the function and interaction with chromatin of the nonhistone protein, poly (ADP-ribose) polymerase. It is evident that the nonhistone proteins, although present

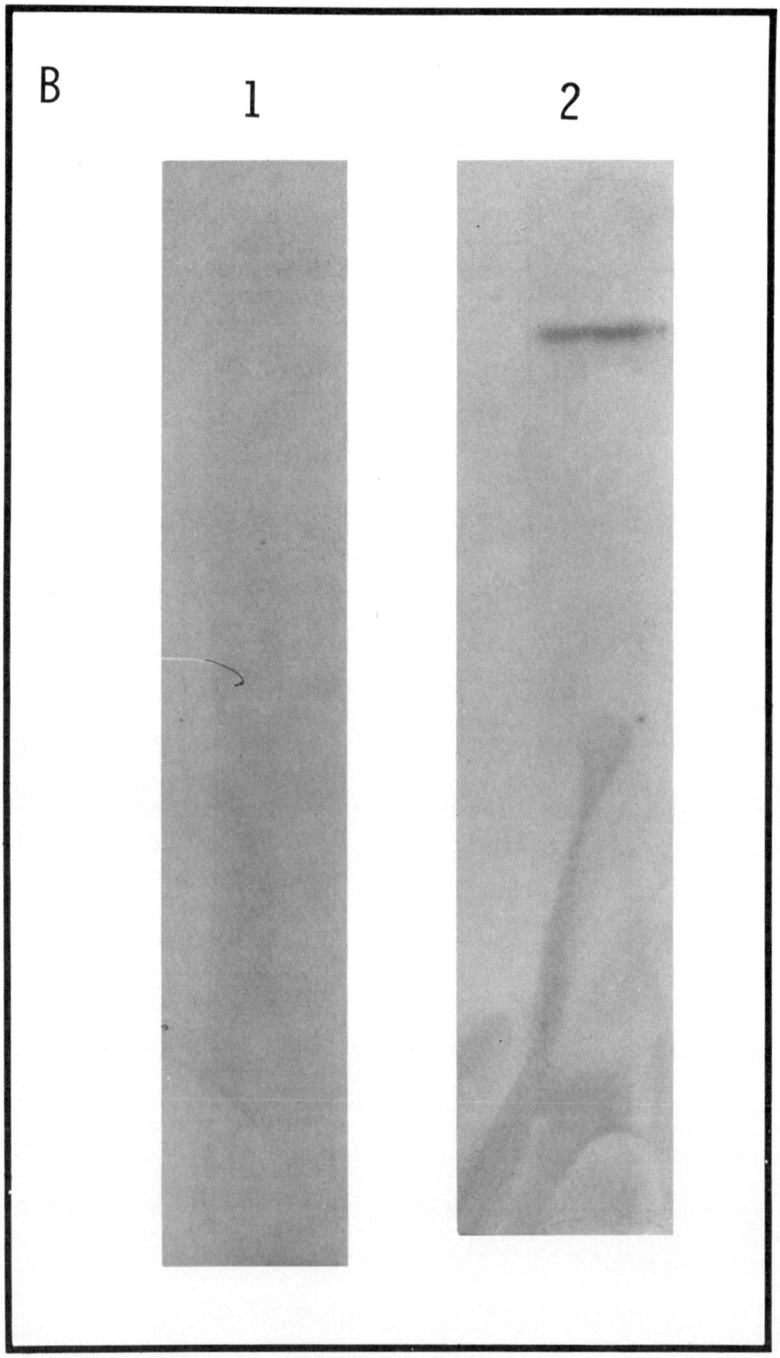

FIGURE 14B.

in minor amounts, play a significant role in the biological activity as well as the structure of specific domains of chromatin. Poly (ADP-ribose) polymerase has the distinction, among nonhistone proteins, of possessing an enzymatic activity which can be conveniently meas-

Table 3
CORRELATION BETWEEN OLIGONUCLEOSOMAL POLY (ADP-RIBOSE) POLYMERASE SPECIFIC ACTIVITY AND ANTIPOLYMERASE SERA REACTIVITY

S-value of oligonucleosomes	Polymerase activity (cpm ^{32}P incorporated/ A_{260}) × 10^3		Antibody reactivity (cpm [^{125}I] protein A bound/A_{260}) × 10^3	
12 (2-2N)	60	(1)	44	(1)
40 (6-8N)	550	(9)	134	(3)

Note: Polymerase activity was determined using 0.02 A_{260} per assay. Antibody reaction was determined by using solid phase radioimmunoassay. Microtiter plates were coated with 0.05, 0.1, and 0.2 A_{260} of oligonucleosomes from sucrose gradient fractions. 1:50 dilution of antipolymerase sera was used. After treatment with [^{125}I] protein A the wells were cut and counted in a gamma counter. Wells treated with chromatin and incubated with the preimmune serum were used to obtain the background. The numbers in parentheses represent fold increase in either polymerase activity or enzyme mass.

From Malik, N., Bustin, M., and Smulson, M., *Nucl. Acid Res.*, 10, 2939, 1982. With permission.

ured. It can be isolated in pure form in reasonable quantities, and antibody has been produced against it. Thus, the various experimental approaches developed for the study of polymerase might act as models for future studies of other nonhistone proteins. By applying the new tools of DNA technology, huge advances in our understanding of the programmed structure of the eukaryotic genome have been accomplished in a relatively short time. To complement this explosion of information it will be equally important in the future to ascertain the precise interaction of chromatin-associated proteins with the genome.

VI. ADDENDUM

Because of unforeseen delays in the publication of these volumes since the original submisssion of the chapter (Fall, 1981), a brief updating of selected advancements in this area, both from our laboratory as well as others, was considered desirable. As a follow-up to those studies described in Figures 1 and 2, immunological and gel blotting techniques have been utilized to demonstrate the natural occurrence of the poly ADP-ribose-H1 dimer complex in actively dividing unlabeled Hela cells.[68] Additionally, a specific immuno-affinity column has been developed utilizing antibody to the modified moiety, poly ADP-ribose, per se to allow the fractionation of those limited regions of chromatin enriched in poly ADP-ribosylated domains (CA. 10% of chromatin) from the bulk of chromatin.[69] The latter fraction is essentially devoid of poly ADP-ribosylated nucleosomes, acceptors, or polymerase activity. The characterization of the oligonucleosomal DNA proximal to the ADP-ribosylated chromatin domains was compared with the bulk chromatin. The immuno-fractionated, poly ADP-ribosylated nucleosomal DNA, contained significant amounts of internal single-strand breaks compared with bulk chromatin.

Poly ADP-ribosylated histone H1 could be conveniently fractionated from bulk H1 by this new immuno-fractionation column method;[70] in vitro phosphorylated species of histone

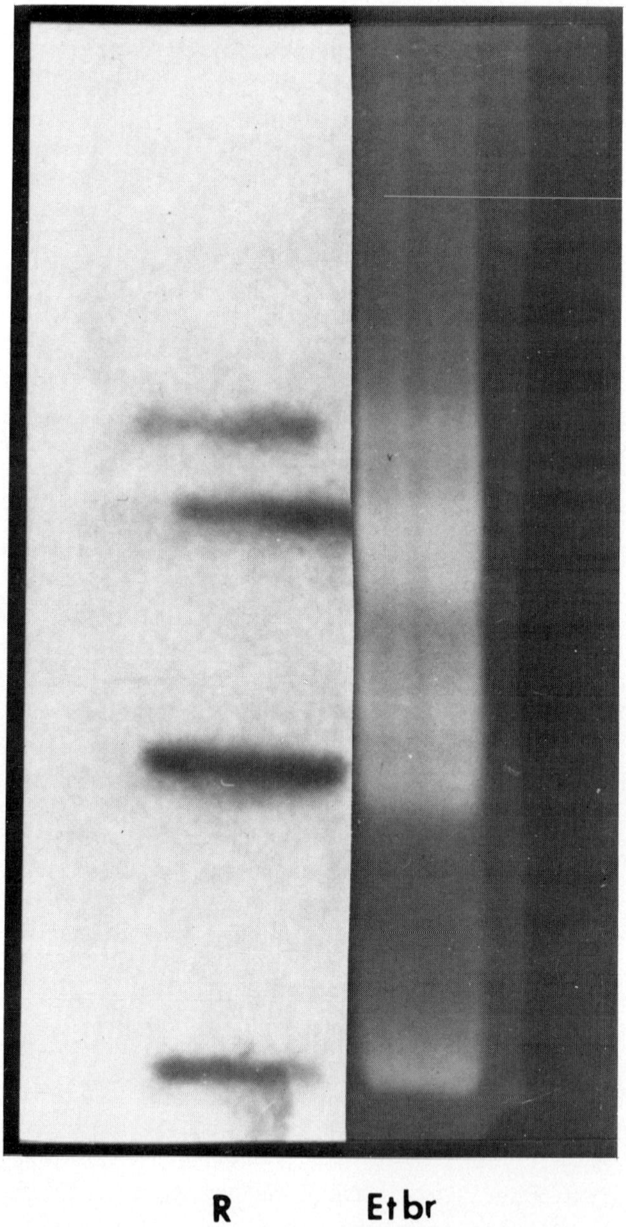

FIGURE 15. The reactivity of antipoly (ADP-ribose) polymerase sera with differing subsets of nucleosome oligomers. Etbr. A micrococcal nuclease digest of HeLa cells nuclei was separated on native chromatin gels and stained with ethidium bromide. R. The gel was transferred by electrophoresis to nitrocellulose. The transfer was subsequently incubated with antipolymerase sera and [^{125}I] protein A and exposed for autoradiography. (From Malik, N., Bustin, M., and Smulson, M., *Nucl. Acid Res.*, 10, 2939, 1982. With permission.)

H1 appeared to be particularly accessible to poly ADP-ribosylation. Nuclei were incubated with [^{32}P]ATP and the phosphorylated H1 extracted. Considerable amounts of phosphorylated H1 showed affinity to the poly ADP-ribose immuno-column; this could be reversed by

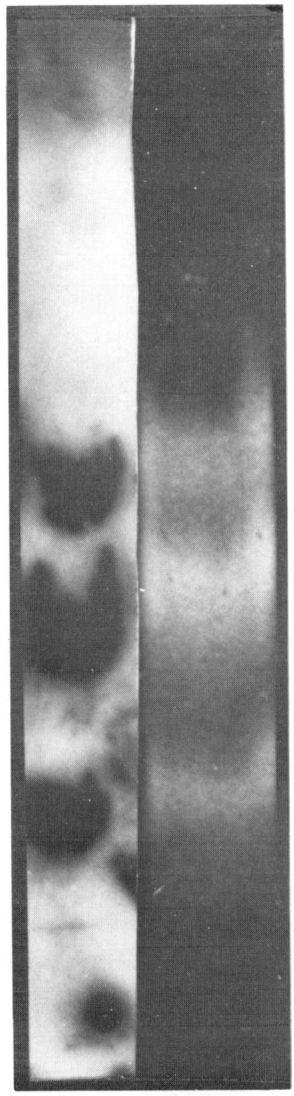

FIGURE 16. Reactivity of antihistone-H3 sera with fingerprints of oligonucleosomes of differing repeat size. Etbr. A micrococcal nuclease digest of HeLa nuclei was separated on native chromatin gel and stained with ethidium bromide. R. Chromatin from the gel was transferred to nitrocellulose. The nitrocellulose strip was subsequently incubated with anti-H3 sera and [^{125}I] protein A followed by autoradiography. (From Malik, N., Bustin, M., and Smulson, M., *Nucl. Acid Res.*, 10, 2939, 1982. With permission.)

treatment of the phosphorylated H1 samples with phosphodiesterase which cleaves the poly ADP-ribose chain.

Finally host cell poly (ADP-ribose) polymerase was shown to be associated with polyoma viral minichromosomes.[71] It was of interest that those forms of polyoma minichromatin containing significant amounts of replicative intermediate forms of chromatin showed the highest affinity to the anti-poly (ADP-Rib) immunoaffinity column.[71]

Two major review volumes have been more recently compiled on the overall area of poly ADP-ribosylation. Ueda and Hayishi[72] have recently published a comprehensive volume which should be of great value to workers interested in this particular area of research. More recently (November, 1982), the 13th Annual Symposium of The Princess Takamatsu Cancer Research Fund was devoted entirely to studies dealing with poly ADP-Ribosylation and its relationship to biological functions such as DNA repair and cellular transformation. A review volume on this meeting has recently been published;[73] a brief summary of several of the highlights of this meeting, particularly relevant to the present chapter are briefly outlined below. Evidence was provided that poly ADP-Rib polymerase activity is required for the ligation reaction of DNA strand rejoining, subsequent to DNA damage in vivo (S. Shall). Monoclonal antibodies, which recognize both a linear as well as a branched form of poly ADP-ribose, were characterized (Kawamitsu, *et al.*).

Branched-chain poly (ADP-ribose) polymers were visualized by electron microscopy occurring via automodification on purified polymerase (P. Mandel). Fifteen sites for poly (ADP-ribose) per molecule of polymerase enzyme, some with greater than 80 units of poly (ADP-ribose), were described by Kawaichi, *et al.* In addition, an enzyme has been characterized which cleaves the linkage of poly (ADP-ribose) to histones (Kawaichi *et al*). It is interesting that this enzyme was shown to be active only with mono-ADP-ribosylated units attached to the histone in contrast to when longer poly (ADP-ribose) chains are attached to acceptors.

With respect to the modification of histone H1 in vivo, the T 1/2 of polymer attached to histone H1 was calculated to be 1.3 min, whereas the mono-ADP-ribosylated adduct to this histone was shown to have a halflife much longer (10 min) (H. Hilz). This same type of stoichiometry was also established during DNA repair (M.K. Jacobson); protein-bound mono-ADP-ribosylated acceptors were generally found to be in excess over polymer during normal (i.e. nonrepair) growth of cells. However, poly-ADP-ribosylated chains are noted to be greatly increased subsequent to DNA strand breaks in cells.

Polynucleotide ligase has been found to possess specific affinity for free poly (ADP-ribose) (K. Ueda). Poly (ADP-ribose) polymerase was found to be differentially sensitive to heat shock in intact cells (W. Kidwell). The synthesis of (ADP-ribose) has been noted to be decreased during differentiation of HL-60 cells by TPA (M. Miwa), whereas increased synthesis of poly (ADP-ribose) appears to be significant in various induction processes, leading to transformation in cultured hepatocytes (H.C. Pitot). Finally, the conference featured a number of studies on the effects of poly (ADP-ribose) polymerase inhibitors (especially benzamide analogues) on DNA repair mechanisms and DNA strand breaks in vivo.

REFERENCES

1. **Elgin, S. R. C., Serunian, L. A., and Silver, L. M.,** Internucleosomal localization of poly (ADP-rib) polymerase, *Cold Spring Harbor Symp. Quant. Biol.*, 839, 1977.
2. **Futai, M., Muzano, D., and Sugimura, T.,** Hydrolysis of the polymer formed from NAD with rat liver phosphodiesterase yielding nucleoside 5′-monophosphate, *Biochem. Biophys. Res. Commun.*, 28, 395, 1967.
3. **Ueda, K., Oka, J., Narumiya, S., Miyakawa, N., and Hayaishi, O.,** Poly ADP-ribose glycohydrolase from rat liver nuclei, a novel enzyme degrading the polymer, *Biochem. Biophys. Res. Commun.*, 46, 516, 1972.

4. **Chambon, P., Weill, J. D., Doly, J., Strosser, M. R., and Mandel, P.,** On the formation of a novel adenylic compound by enzymatic extracts of liver nuclei, *Biochem. Biophys. Res. Commun.*, 25, 638, 1966.
5. **Nishizuka, Y., Ueda, K., Nakazawa, K., and Hayaishi, O.,** Studies on the polymer of adenosine diphosphate ribose. I. Enzymic formation from nicotinamide adenine dinucleotide in mammalian nuclei, *J. Biol. Chem.*, 242, 3164, 1967.
6. **Hasegawa, S., Fugimura, S., Shimizu, Y., and Sugimura, T.,** The polymerization of adenosine 5'-diphosphate-ribose moiety of NAD by nuclear enzyme. II. Properties of the reaction product, *Biochem. Biophys. Acta*, 149, 369, 1967.
7. **Sugimura, T.,** Poly (adenosine diphosphate-ribose), *Prog. Nucleic Acid Res. Mol. Biol.*, 13, 127, 1973.
8. **Honjo, T. and Hayaishi, O.,** Enzymatic ADP-ribosylation of proteins and regulation of cellular activity, *Curr. Top. Cell Regul.*, 7, 87, 1973.
9. **Hayaishi, O.,** *Trends Biochem. Sci.*, 1, 9, 1976.
10. **Hayaishi, O. and Ueda, K.,** Poly (ADP-ribose) and ADP-ribosylation of proteins, *Annu. Rev. Biochem.*, 46, 95, 1977.
11. **Hilz, H. and Stone, P.,** Poly (ADP-ribose) and ADP-ribosylation of proteins, *Rev. Physiol. Biochem. Pharmacol.* 76, 1, 1976.
12. **Smulson, M. E. and Shall, S.,** Poly (ADP-Ribose), *Nature (London)*, 263, 14, 1976.
13. **Smulson, M. E. and Sugimura, T.,** *Novel ADP-Ribosylations of Regulatory Enzymes and Proteins*, Elsevier/North Holland, New York, 1980.
14. **Smulson, M. E.,** Influence of adenosine diphosphoribosylation of histones on the replication of human cells, *Natl. Cancer Inst. Monogr.*, 38, 205, 1973.
15. **Lehmann, A. R., Kirk-Bell, S., Shall, S., and Whish, W. J. D.,** The relationship between cell growth, macromolecular synthesis and poly (ADP-ribose) polymerase in lymphoid cells, *Exp. Cell. Res.*, 83, 63, 1974.
16. **Sugimura, T., Miwa, M., Kanai, Y., Oda, K., Segawa, K. Tanaka, M., and Sakura, H.,** Recent progress in studies on poly ADP-Ribosylation, in *Control Mechanisms in Cancer*, Criss, W. E., Ono, T., and Sabine, J. R., Eds., Raven Press, New York, 1976, 231.
17. **Muller, W. E. G., Totnuka, A., Nusser, I., Obermeier, J., Rohde, H. J., and Zahn, R. K.,** Poly (adenosine diphosphate-ribose) polymerase in quail oviduct, *Nucleic Acid Res.*, 1, 1317, 1974.
18. **Hilz, H., Adamietz, P., Bredoehorst, R., and Leiber, U.,** *Poly ADP Rib — an Int. Symp.*, Proc. No. 26(Abstr.), Harris, M., Ed., Washington, D.C., 1973, 47.
19. **Leiber, U., Kittler, M., and Hilz, H.,** Enzymes of poly (ADP-ribose) metabolism in proliferating and nonproliferating liver tissues, *Hoppe-Seyler's Z. Physiol. Chem.*, 354, 1347, 1973.
20. **Roitt, I. M.,** *Biochem. J.*, 63, 300, 1956.
21. **Gill, D. M.** The soluble polymerizing system, *J. Biochem.*, 77, 1, 1975.
22. **Miller, M. R., Castellot, J. J., and Pardee, A. B.,** A permeable animal cell preparation for studying macromolecular synthesis, DNA synthesis and the role of deoxyribonucleotides in S phase initiation, *Biochemistry*, 17, 1073, 1978.
23. **Smulson, M. E., Stark, P., Gazzoli, M., and Roberts, J.,** Release of template restriction for DNA synthesis by Poly (ADP-Rib) polymerase during the HeLa cell cycle, *Exp. Cell Res.*, 90, 175, 1975.
24. **Whish, W. J. D., Davies, M. I., and Shall, S.,** Stimulation of poly (ADP-ribose) polymerase activity by the antitumor antibiotic streptozotocin, *Biochem. Biophys. Res. Commun.*, 65, 722, 1975.
25. **Smulson, M. E., Schein, P., Mullins, D. W., Jr., and Sudhakar, S.,** A putative role for nicotinamide adenine dinucleotide-promoted nuclear protein modification in the antitumor activity of N-methyl-N-nitrosourea, *Cancer Res.*, 37, 3006, 1977.
26. **Sudhakar, S., Tew, K. D., and Smulson, M. E.,** Effect of 1-methyl-1-nitrosourea on poly (ADP-ribose) polymerase activity at the nucleosomal level, *Cancer Res.*, 39, 1405, 1979.
27. **Sudhakar, S., Tew, K. D., Schein, P. S., Wooley, P. V., and Smulson, M. E.,** Nitrosourea interaction with chromatin and effect on poly (ADP-ribose) polymerase activity, *Cancer Res.*, 39, 1411, 1979.
28. **Berger, N. A., Sikorski, G. W., Petzold, S. J., and Kurohara, K. K.,** Association of poly (adenosine diphosphoribose) synthesis with DNA damage and repair in normal human lymphocytes, *J. Clin. Invest.*, 63, 1164, 1979.
29. **Berger, N. A., Sikorski, G. W., Petzold, S. J., and Kurohara, K. K.,** Defective poly (adenosine diphosphoribose) synthesis in xeroderma pigmentosum, *Biochemistry*, 19, 289, 1980.
30. **Berger, N. A. and Sikorski, G. W.,** Proc. NATO Meet. Chromosome Damage and Repair, Norway, 1980.
31. **Benjamin, R. C. and Gill, D. M.,** Poly (ADP-ribose) synthesis in vitro programmed by damaged DNA, *J. Biol. Chem.*, 255, 10502, 1980.
32. **Durkacz, B. W., Omidiji, O., Gray, D. A., and Shall, S.,** (ADP-ribose) participates in DNA excision repair, *Nature (London)*, 283, 593, 1980.

33. **McCurry, L. S. and Jacobson, M. K.**, Poly (ADP-ribose) synthesis following DNA damage in cells heterozygous or homozygous for the xeroderma pigmentosum genotype, *J. Biol. Chem.*, 256, 551, 1981.
34. **Nishizuka, Y., Ueda, K., Honjo, T., and Hayaishi, O.**, Enzymic adenosine diphosphate ribosylation of histone and poly (adenosine diphosphate ribose) synthesis in rat liver nuclei, *J. Biol. Chem.*, 243, 3765, 1968.
35. **Tanuma, S. I., Enomoto, T., and Yamada, M. A.**, Distribution of poly (ADP-ribose) in histones of HeLa cell nuclei, *Biochem. Biophys. Res. Commun.*, 74, 599, 1977.
36. **Giri, C. P., West, M. H. P., and Smulson, M. E.**, Differential poly ADP-ribosylation of chromosomal proteins in nuclei versus isolated nucleosomes, *Biochemistry*, 17, 1492, 1978.
37. **Jump, D. B., Butt, T. R., and Smulson, M.**, Relationship between poly ADP-ribosylation and different functional forms of chromatin, *Biochemistry*, 18, 983, 1979.
38. **Tanigawa, Y., Kawamura, M., and Shimoyama, M.**, Effect of polyamines on ADP-ribosylation of nuclear proteins from rat liver, *Biochem. Biophys. Res. Commun.*, 76, 406, 1977.
39. **Okayama, H., Ueda, K., and Hayaishi, O.**, Purification of ADP-ribosylated nuclear proteins by covalent chromatography on dihydroxyboryl polyacrylamide beads and their characterization, *Proc. Natl. Acad. Sci. U.S.A.*, 75, 1111, 1978.
40. **Ueda, K., Fukushima, M., Okayama, H., and Hayaishi, O.**, Nicotinamide adenine dinucleotide glycohydrolase from rat liver nuclei, *J. Biol. Chem.*, 250, 7541, 1975.
41. **Riquelme, P. T., Burzio, L., and Koide, S. S.**, ADP ribosylation of rat liver lysine-rich histone in vitro, *J. Biol. Chem.*, 254, 3018, 1979.
42. **Burzio, L. O., Riquelme, P. T., and Koide, S. S.**, ADP ribosylation of rat liver nucleosomal core histones, *J. Biol. Chem.*, 254, 3029, 1979.
43. **Wong, N. C. W., Poirier, G. G., and Dixon, G. H.**, Adenosine diphosphoribosylation of certain basic chromosomal proteins in isolated trout testis nuclei, *Eur. J. Biochem.*, 77, 11, 1977.
44. **Dixon, G. H., Wong, N., and Poirier, G. G.**, Adenosine diphosphoribosylation of basic chromosomal proteins in trout testis nuclei, *Fed. Proc.*, 35, 1623, 1976.
45. **Ogata, N., Kunihiro, N., Ueda, K., and Hayaishi, O.**, ADP-ribosylation of histone H2 B, *J. Biol. Chem.*, 255, 7610, 1980.
46. **Okayama, H. and Hayaishi, O.**, ADP-ribosylation of nuclear protein A24, *Biochem. Biophys. Res. Commun.*, 84, 744, 1978.
47. **Ogata, N., Ueda, K., Kagamiyama, H., and Hayaishi, O.**, ADP-ribosylation of histone H1, *J. Biol. Chem.*, 255, 7616, 1980.
48. **Rattle, H. W. E., Langan, T. A., Danby, S. E., and Bradbury, E. M.**, Studies on the role and mode of operation of the very lysine-rich histones in eukaryote chromatin, *Eur. J. Biochem.*, 81, 499, 1977.
49. **Whitlock, J. P., Jr. and Simpson, R. T.**, Removal of histone H1 exposes a fifty base pair DNA segment between nucleosomes, *Biochemistry*, 15, 3307, 1976.
50. **Worcel, A. and Benyajati, C.**, Higher order coiling of DNA in chromatin, *Cell*, 12, 83, 1977.
51. **Stone, P. R., Lorimer, W. S., and Kidwell, W. R.**, Properties of the complex between histone H1 and poly (ADP-ribose) synthesized in HeLa cell nuclei, *Eur. J. Biochem.*, 81, 9, 1977.
52. **Yoshihara, K., Tanigawa, Y., Burzio, L., and Koide, S. S.**, Evidence for adenosine diphosphate ribosylation of Ca-, Mg-dependent endonuclease, *Proc. Natl. Acad. Sci., U.S.A.*, 72, 289, 1974.
53. **Yoshihara, K., Hashida, T., Tanaka, Y., Yoshihara, H., and Ohgushi, H.**, Enzyme bound early product of purified poly (ADP-ribose) polymerase, *Biochem. Biophys. Res. Commun.*, 78, 1281, 1977.
54. **Jump, D. B. and Smulson, M.**, Purification and characterization of the major non-histone protein acceptor for poly ADP-ribose in HeLa cell nuclei, *Biochemistry*, 19, 1024, 1980.
55. **Butt, T. R., Brothers, J. F., Giri, C. P., and Smulson, M. E.**, A nuclear protein-modifying enzyme is responsive to ordered chromatin structure, *Nucleic Acid Res.*, 5, 2775, 1978.
56. **Jump, D. B., Butt, T. R., and Smulson, M.**, Reconstitution of HeLa cell poly (ADP-ribose) polymerase with purified oligonucleosomal chromatin, *Biochemistry*, 19, 1031, 1980.
57. **Mullins, D. W., Jr., Giri, C. P., and Smulson, M. E.**, Poly (adenosine diphosphate ribose) polymerase: the distribution of a chromosome-associated enzyme within the chromatin substructure, *Biochemistry*, 16, 506, 1977.
58. **Giri, C. P., West, M. H. P., and Smulson, M. E.**, Nuclear protein modification and chromatin substructure. I. Differential poly (adenosine disphosphate) ribosylation of chromosomal proteins in nuclei versus isolated nucleosomes, *Biochemistry*, 17, 3495, 1978.
59. **Giri, C. P., West, M. H. P., Ramirez, M. L., and Smulson, M. E.**, Internucleosomal localization of poly (ADP-rib) polymerase, *Biochemistry*, 17, 3501, 1978.
60. **Seale, R. L.**, Studies on the mode of segregation of histone Nu bodies during replication in HeLa celis, *Cell*, 9, 423, 1976.
61. **Hildebrand, C. E. and Walters, R. A.**, Rapid assembly of newly synthesized DNA into chromatin subunits prior to joining of small DNA replication intermediates, *Biochem. Biophys. Res. Commun.*, 73, 157, 1976.

62. **Butt, T. R. and Smulson, M.,** Relationship between nicotinamide adenine dinucleotide concentration and in vitro synthesis of poly ADP-ribose on purified nucleosomes, *Biochemistry,* 19, 5234, 1980.
63. **Butt, T. R., De Coste, B., Jump, D. B., Nolan, N., and Smulson, M.,** Characterization of a putative poly ADP-ribose-chromatin complex, *Biochemistry,* 19, 5243, 1980.
64. **Wong, M., Malik, N., and Smulson, M.,** The participation of poly ADP-ribosylated histone H1 in oligonucleosomal condensation, *Eur. J. Biochem.,* 128, 209, 1982.
65. **Malik, N., Bustin, M., and Smulson, M.,** Antibody to poly (ADP-rib) polymerase and its use in chromatin analysis, *Nucl. Acid Res.,* 10, 2939,1982.
66. **Towbin, H., Staehelin, T., and Gordon,** Electrophoretic transfer of proteins from polyacrylamide gels to nitrocellulose sheets: procedure and some applications, *Proc. Natl. Acad. Sci. U.S.A.,* 76, 4350, 1979.
67. **Nolan, N. L., Butt, T. R., Wong, M., Lambrianidou, A., and Smulson, M. E.,** Characterization of poly (ADP-rib)-histone H1 complex formation in purified polynucleosomes and chromatin, *Eur. J. Biochem.,* 113, 15, 1980.
68. **Wong, M, Kanai, Y, Miwa, M., Bustin, M., and Smulson, M.,** Immunological evidence for the *in vivo* occurrence of crosslinked complex of poly ADP-ribosylated histone H1, *Proc. Natl. Acad. Sci., U.S.A.,* 80, 205, 1983.
69. **Malik, N., Miwa, M., Sugimura, T., Thraves, P., and Smulson, M.,** Immunoaffinity fractionation of the poly (ADP-ribosylated) domains of chromatin, *Proc. Natl. Acad. Sci. U.S.A.,* 80, May, 1983.
70. **Wong, M., Miwa, M., Sugimura, T., and Smulson, M.,** Relationship Between histone H1 poly (ADP-ribosylation) and histone H1 phosphorylation using anti-poly (ADP-rib) antibody, *Biochemistry,* 22, in press.
71. **Prieto-Soto, A., Gourlie, B., Miwa, M., Pigiet, V., Sugimura, T., Malik, N., and Smulson, M.,** Polyoma virus minichromosomes: Poly ADP-ribosylation of Associated chromatin proteins, *J. Virol.,* 45, 600, 1983.
72. **Hayaishi, O. and Ueda, K.,** *ADP-ribosylation Reactions, Biology and Medicine,* Academic Press, New York, 1982.
73. **Miwa, M., Smulson, M., Shall, S., and Sugimura, T.,** ADP-Ribosylation, DNA repair, and cancer, *Proc. 13th Int. of The Princess Takamatsu Cancer Research Fund, Tokyo,* Japan Scientific Societies Press, Tokyo, 1983.

Chapter 7

DNA-BINDING PROTEINS AND DNA-SYNTHESIZING ENZYMES IN EUKARYOTES

Jen-Fu Chiu

TABLE OF CONTENTS

I. INTRODUCTION

The structural unit of eukaryotic chromosomes is a complex of DNA with an equal weight of histones. These subunits, nucleosomes, are each made up of every 200 or so base pairs of DNA wrapped around a core of 8 histones (two each of H2A, H2B, H3, and H4).[1,2] Histone H1 and the nonhistone chromosoal proteins apparently interact with the outer surface of the DNA-histone bead. The basic chromatin fiber must be folded or organized into a higher-order structure. Replication of eukaryotic chromosomes involves duplication of both DNA and nucleosomal units. The mechanism of chromosomal replication must accurately duplicate each structure so that concurrent gene expression is unaltered.

Chromosomal DNA synthesis in the nucleus of eukaryotic cell proceeds in a regulated transitory pattern.[3-7] The DNA synthesis is followed by semiconservative mechanism. The resultant daughter helixes carry one parental DNA strand and one synthesized *de novo*. Once initiated, DNA polymerases incorporate nucleotides progressively in the 5′ to 3′ direction along each template strand with the aid of many other enzymes and other proteins that interact with the DNA. Many components of the replication complexes in prokaryotes have been extensively characterized and functionally analyzed.[8] The analogous systems in eukaryotes are not yet well developed. However, many enzymes and proteins that are involved in normal and repair replication of DNA in eukaryotic cells continue to be identified, isolated, and characterized. Putative replication complexes comprise at least the following enzymes: unwinding enzyme,[9] DNA polymerase,[10-14] DNA-ligase,[15,16] single-strand DNA-dependent ATPase,[17-20] DNA binding protein,[9,19,21,22] endodeoxyribonuclease,[23-27] exodeoxyribonuclease,[28] RNase H and/or other ribonuclease,[29-31] RNA polymerase,[32] and structural proteins of the chromatin.[1,2,33,34] The DNA polymerases have been most extensively studied because of their clear involvement in DNA synthesis. Only some of these proteins will be discussed in this chapter.

II. DNA POLYMERASES

Since the first description of DNA polymerase activity in vitro in the late 1950s, a tremendous number of investigations have been done to elucidate the properties and mechanism of action of these enzymes. Most of the research centered on bacterial DNA polymerases. It was not until a decade ago that separation and characterization of the various eukaryotic DNA polymerases was achieved and some firm ideas about their structure and function gained. One major reason for this delay in the progress is that low concentrations of these enzymes are present in the cells. Modern techniques of cell fractionation and enzyme purification have helped to overcome these difficulties.

Cells characteristically contain multiple forms of DNA polymerases. In prokaryotic cells, three activities are known: DNA polymerase I, II, and III.[13,35] Eukaryotic cells have also at least three enzymes: DNA polymerase α, β, and γ.[10-14] These DNA polymerases can easily be distinguished from one another by their size, chromatographic behavior, primer template specificities, optimal ion conditions, sensitivity to *N*-ethylmaleimide,[10-14] and immunological comparison.[14a] A new type of polymerase activity called δ was recently described. This enzyme has properties similar to DNA polymerase α, but has associated exonuclease activity.[36] The classification and some properties of eukaryotic DNA polymerases are listed in Table 1.

The location of the DNA polymerases in the cell has raised some questions about their possible roles. When cells are fractionated into nuclear and cytoplasmic fractions in aqueous solutions, most of the DNA polymerase α appears in cytoplasm. However, this may be an artifact since other techniques, which involved nonaqueous breakage of cells or drug-induced enucleation, indicate that 80 to 90% of the total DNA polymerase activity is, in fact, associated with the nucleus.[37] Using immunofluorescence techniques, it has recently been shown that the enzyme is located mainly in the perinuclear region.[38] Mitochondria also contain DNA polymerase activity. Recent evidence has suggested that the mitochondria-associated DNA polymerase may be a form of DNA polymerase Γ.[39-41] The processes of viral infection of induction from the lysogenic or transformed state generally lead either to a change in the cellular DNA polymerase levels or to the appearance of a new virus-induced, and presumably virus-encoded, DNA polymerase.[10] These enzymes may be autonomous or may require functional or structural components from the host.

Intracellular DNA synthesis is regulated by distinct multienzyme systems[13,20,42] controlling initiation, propagation, and termination. Although the overall mechanisms differ greatly, all known DNA polymerases share some common features, as shown in Figure 1; that is

1. The template consists of a primer with a 3′-OH terminus
2. The monomeric substrates are deoxynucleoside 5′-triphosphates
3. All enzymes need divalent cations (Mg^{2+} or Mn^{2+})
4. Newly synthesized DNA chain grows from 5′ → 3′ direction
5. Deoxynucleoside monophosphate is incorporated according to Watson-Crick base-pairing rules
6. All DNA polymerases are metallo-enzymes containing Zn^{++} in their molecules

At least three different types of DNA polymerization can be classified: (1) replicative nuclear DNA synthesis, which is characterized by being semiconservative, symmetrical, bidirectional, and with short RNA-primed intermediates in the lagging strand; (2) repair DNA synthesis, occurring on both strands and restricted to short DNA gaps; (3) mitochondrial and adenovirus DNA synthesis, which is known to be continuous and proceeds in an asymmetrical way. DNA synthesis catalyzed by DNA polymerase α is a typical replicative type and is suggested to carry out replication of nuclear DNA. DNA polymerase α produces a short DNA fragment on ribo-primers in a quasiprocessive way. It may also be able to synthesize a longer DNA chain with the aid of other enzymes. On the contrary, DNA polymerase β works in a distributive way and incorporates nucleotides to the gaps on the activated DNA. Its properties suggest that DNA polymerase β performs a repair type of DNA synthesis. DNA polymerase α synthesizes DNA in vitro in a highly processive fashion. It can replicate mitochondrial and adenovirus DNAs, as well as synthesize the long, single-stranded DNA regions which might represent displaced strands of parental DNA being replicated asymmetrically. The details of the function of DNA polymerase α, β, and γ will be discussed later.

A. DNA Polymerase α

DNA polymerase α is the major fraction of cellular DNA polymerase activity. It is found largely in the cytoplasmic fraction during cell fractionation by aqueous procedures. This observation is in conflict with its possible role in nuclear DNA replication. It is probable that the in vivo intracellular location of the enzyme is in the nucleus. One can explain that this enzyme leaks from nuclei during the process of cell fractionation. Using cytochalasin-induced enucleation, Herrick and associates[43] have demonstrated that greater than 85% of the DNA polymerase activity is associated with the nucleus. Siebert et al.[44] have also obtained similar results using nonaqueous techniques with organic solvents to isolate nuclei. Foster and Gurney have prepared mouse fibroblast nuclei in glycerol and found only 10% of the cellular DNA polymerase activity to be outside the nucleus.[37]

DNA polymerase α has been isolated and purified from a number of animal cells as well as from lower eukaryotes. It seems to be ubiquitous in growing cells and has been found

Table 1
CLASSIFICATION OF EUKARYOTIC DNA POLYMERASES

DNA polymerase	Molecular weight	Sensitive to inhibitors: NEM	Aphidicolin	ddNTPs	Salt effect	Template primer	Type of DNA synthesis
DNA polymerase α	130,000—280,000	+	+	−	Inhibited by 25 m*M* or higher concentration	RNA-primed DNA	Quasiprocessive type
DNA polymerase β	30,000—50,000	−	−	+	Stimulated by 100—200 m*M* NaCl; Inhibited by 50 m*M* PO^{-3}_4	Deoxy-primer on ribo-templates in the absence of phosphate	Distributive type
DNA polymerase γ	150,000—300,000	+	−	+	Stimulated by 100—250 m*M* NaCl and 50 m*M* PO^{-3}_4	Deoxy-primer on ribo-template in the presence of phosphate	Highly processive type
DNA polymerase δ	190,000	+	+	−			

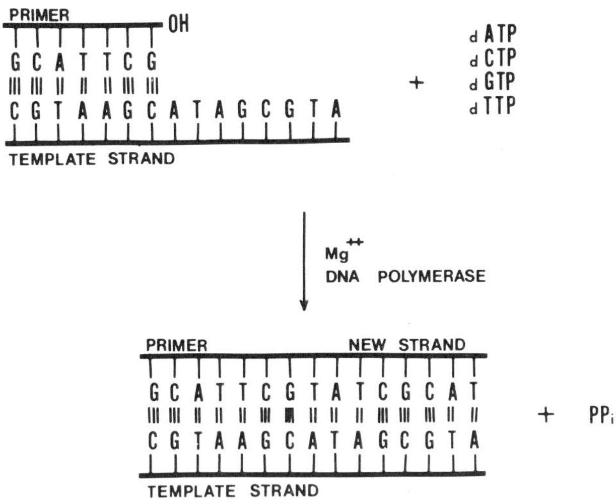

FIGURE 1. Enzymatic reaction of DNA polymerases α, β, and γ.

in calf thymus,[45-48] human,[49,50] murine,[51-56] guinea pig,[57] hamster cells,[58,59] avian tissues,[60-62] sea urchin,[63,64] protozoan,[65] yeast,[66,67] and slime mold.[68]

Molecular weight determinations or determination of sedimentation velocity are widely used to recognize the species of DNA polymerases. DNA polymerase α has a sedimentation coefficient of 6 to 8S, indicative of a high molecular weight enzyme. However, the ionic strength strongly affects sedimentation properties of proteins. At low ionic strength the DNA polymerase α exists in aggregate form with molecular weights in the range of 300,000.[69-71] Under high ionic strength of the solution, this polymerase has a molecular weight range from 160,000 to 180,000.[66,70-72] The DNA polymerase α purified from HeLa cell and calf thymus has a molecular weight of 155,000 to 175,000 daltons and seems to contain polypeptide subunit of 50,000 to 87,000 daltons.[10,72a] This indicates that the active enzyme with a molecular weight around 155,000 to 175,000 might be a dimer. DNA polymerase α in KB-cells and yeast can dissociate into monomer of 80,000 to 90,000 daltons. Similar situations also happen to DNA polymerase α in calf thymus.[46,72b] The active form of the major DNA polymerases of *Drosophila* embryos and of *Tetrahymena* appears to be a monomer of 80,000 to 87,000 daltons.[73,74] There is some discrepancy in the values for the molecular weight obtained from sedimentation analyses and from gel filtration, indicating a considerable asymmetry in the shape of this protein.[75] Craig and Keir[59] have investigated the DNA polymerases of BHK-21C13 cells and tentatively identified four subspecies of the DNA polymerase α ranging in molecular weight from 140,000 to 1,000,000. They have noted the molecular asymmetry of DNA polymerase α and lack of correlation of molecular weights as derived from gel filtration or sedimentation coefficients, as did Holmes et al.[75] previously. DNA polymerase α can be identified as a heterogeneous population, from their molecular weight, their behavior on ionic column chromatography, and other properties. However, it is not yet clear whether these are different enzymes or they are complexes of DNA polymerase α with various other cellular proteins which may or may not be involved in DNA replication. The possibility that the DNA polymerase α may form complexes with necessary proteins such as DNA-unwinding proteins from calf thymus has been raised by Herrick et al.[76] The accessory proteins which form a complex with DNA polymerases have their specific function in DNA synthesis. For instance, two helix-unwinding proteins isolated from calf thymus will stimulate the activity of DNA polymerase α but not that of DNA polymerase β or γ.[76-78] These proteins have properties similar to the gene 32 product of phage T4 for which genetic evidence indicates an involvement in the replication of T4 DNA and which is known specifically to stimulate the activity of T4 DNA polymerase and will not stimulate DNA polymerase α.

DNA polymerase α requires sulfhydryl groups for activity. It is much more susceptible to inhibition by the thiol-active reagents, *p*-mercuribenzoate and *N*-ethylmaleimide than DNA polymerase β. For example, DNA polymerase α is inhibited to the extent of 80 to 100% by *p*-mercuribenzoate at concentrations of 10 to 40 μ*M*.[49,54,79-81] Unlike the situation with the DNA polymerase β, no protection against the inhibitory effect of *p*-mercuribenzoate was afforded by prior incubation of the DNA polymerase α with DNA.[81] *N*-ethylmaleimide is also strongly inhibitory with the DNA polymerase α relative to the DNA polymerase β.[57,78,82-85] Ethidium bromide also exerts a more potent inhibitory effect on the activity of the DNA polymerase α than it does on that of the DNA polymerase β.[49]

It is well known that replicative DNA polymerase activity requires a DNA template in order to incorporate deoxyribonucleotides into polymer chains. However, their ability depends on the availability of 3′-OH termini on the template. Hence, maximum incorporation of deoxynucleotides catalyzed by DNA polymerase α is not on native or denatured DNA primer templates, which contain relatively few 3′-OH termini, but on "activated DNA".[49,78-81,86] Activated DNAs are formed by limited digestion of DNase I on duplex DNA. This procedure can produce a large number of 3′-OH termini. Digestion with micrococcal nuclease to the same extent, which produces 3′-phosphoryl termini, results in the abolition of template activity, indicating the absolute requirement of 3′-OH terminal for DNA polymerase activity.[49,54,69,87] It has been demonstrated that the binding of polymerase α to DNA is absolutely dependent on the presence of single-stranded regions, and that the polymerase has no measurable affinity for duplex DNA. The interaction of the enzyme with 3′-hydroxyl termini can be demonstrated only subsequent to, or in concert with, its binding to single strand, and only base-paired termini can be recognized.[87a] The product of the reaction when DNA polymerase α acts on activated DNA was found to be double-stranded DNA with hairpin structures. DNA polymerase α is rather specific for polydeoxyribonucleotide templates and shows little activity with RNA or synthetic ribohomopolymers (e.g., poly(A)). Fisher et al.[87b] demonstrated that the apparent affinity of DNA polymerase α for a variety of homopolymeric and heteropolymeric template molecules can vary over two to three orders of magnitude. These results support the interpretation that the purified polymerase protein is capable of discriminating polydeoxynucleotide templates on the basis of signals that appear to be most reasonably determined by base composition or sequence.

It is now well established that the synthesis of RNA is a prerequisite for DNA synthesis. The association of RNA with nascent DNA was first demonstrated in prokaryotes both in vivo and in vitro.[88,89] Since then there has been considerable evidence for the direct involvement of RNA in the initiation of DNA replication in *Escherichia coli*,[90] in replication from DNA of bacteriophage φX174,[91] and in replicative form DNA of bacteriophage M13.[92] Evidence linking RNA synthesis with the initiation of DNA synthesis in eukaryotic systems was also demonstrated. Sato et al.[93] and Magnusson et al.[94] have isolated RNA-linked, short, nascent strands of DNA from nuclei of polyoma-infected cells. In vitro evidence for a coupling of replication to transcription has been presented by Keller[95] and Stavrianopoulos et al.[96] The initiation of new chains in vivo may also proceed via the limited transcription of the parental DNA strands by an RNA polymerase producing an oligonucleotide initiator. This directly implicates the DNA polymerase α enzyme activity with DNA replication but not DNA polymerase β, as chain propagation by the DNA polymerase β enzyme using an oligoribonucleotide as an initiator appears to be limited. Possible functions for DNA polymerases α and β will be discussed in more detail later.

Recently, Byrnes and Black[36] described a new type of DNA polymerase activity called δ with properties similar to the α polymerase, but having associated exonuclease activity.

B. DNA Polymerase β

DNA polymerase β occurs primarily in nuclei, although small amounts of this enzyme occasionally have been found in the cytoplasm. It appears that whatever techniques of cell disruption, subcellular fractionation, and enzyme extraction are applied to a cell or tissue system, DNA polymerase β is always found to be present in the cellular nucleus. It is firmly bound to chromatin and may be solubilized by high ionic strength buffer.[82,86] It was isolated and purified to apparent homogeneity from calf thymus,[97] human KB cells,[98] and chick embryos.[99] The enzyme sediments in buffers of high ionic strength at 3 to 4S, but like DNA polymerase α, it also forms a large aggregate under low ionic strength conditions. According to sedimentation value, the purified DNA polymerase β from both calf thymus KB cells or rat ascites cells has a molecular weight of about 45,000 daltons.[97,98,100] Upon electrophoresis on SDS-denaturing polyacrylamide gels the enzyme appears as a single band indicating one component with a molecular weight of 43,000 to 45,000. This value is in agreement with sedimentation data. It was speculated that additional proteins present in the cell are required to allow the DNA polymerases to catalyze DNA synthesis. Chiu and Sung[101] described a protein factor isolated from developing rat brain which stimulates the activity of DNA polymerase β but not that of DNA polymerase α. When an enzyme preparation is treated to remove this protein a partially inactivated DNA polymerase β results. This partially inactivated DNA polymerase β can be stimulated somewhat by either spermidine or the protein factor but is stimulated much more strongly by the two together. The protein fraction by itself had no DNA polymerase activity in the presence or absence of spermidine and neither the protein factor nor spermidine had any stimulatory effect on DNA polymerase α. Recently, Probst et al.[102] have reported on several proteins from Novikoff hepatoma that markedly stimulate the isolated Novikoff hepatoma DNA polymerase β. The mechanism on interaction of accessory proteins with DNA polymerase is still unknown.

Unlike DNA polymerase α, DNA polymerase β does not require sulfhydryl groups for its activity. It is not inhibited by SH-blocking reagents. DNA polymerase β activity is not markedly inhibited by 25 μ*M* of *p*-chloromercuribenzoate while DNA polymerase α is completely inhibited.[49,54,79-81] The higher concentrations (0.3 m*M* or higher) of *p*-chloromecuribenzoate or $HgCl_2$ that will inhibit DNA polymerase β ranges from 20 to 94%.[103,104] Haines et al.[81] observed that the inhibition of the DNA polymerase β activity by *p*-mercuribenzoate is reduced by half if the enzyme is first bound to DNA before exposure to the mercurial. *N*-ethylmaleimide at concentrations of 1.5 to 3 m*M* shows only a little inhibition on DNA polymerase β activity.[82,84,85] DNA polymerase β exhibits a high resistance to a number of chemical reagents. For example, DNA polymerase β was inhibited to a small extent by 4 m*M* KCN,[104] 10 to 20 μ*M* ethidium bromide,[49,104] 5 *M* urea or 20 to 25% ethanol or acetone,[97] and phosphonoacetic acid.[105]

Chang and Bollum[106] demonstrated that both DNA polymerase α and β could use oligodeoxyribonucleotides to initiate polydeoxyribonucleotide synthesis. However, only DNA polymerase α can use oligoribonucleotides to initiate the DNA synthesis. A similar observation was made by de Recondo et al.,[107] who demonstrated the ability of a purified DNA polymerase α to utilize the polyribonucleotide strands of a synthetic RNA-DNA hybrid as an initiator for the synthesis of the complementary polydeoxyribonucleotide strand. However, there is a considerable amount of evidence implicating the ability of DNA polymerase β to utilize a polyribonucleotide template if provided with a complementary oligodeoxyribonucleotide as an initiator. Haines et al.[108] demonstrated that the incorporation of dTTP into a poly(A) poly(dT) template by purified rat liver DNA polymerase β was threefold higher than that of activated calf thymus DNA as template. Similar activities associated with DNA polymerase β were also reported in chicken embryo,[62] rabbit bone marrow,[86] normal human blood lymphocytes,[57] and BALB/3T3 cells.[109] However, limited incorporation of dATP substrate into poly(A) oligo(dT) primer template was observed by DNA polymerase β isolated from HeLa cells[87] and KB cells.[49] It has often been demonstrated that DNA polymerase β shows a high ability to incorporate a single substrate (deoxynucleoside trihotriphosphate) into an acid-insoluble form in the presence of an activated DNA template.[10,84] This could be due to a limited DNA repair synthesis occurring at the large number of 3′-OH ends in the template.

The finding that DNA polymerase β is absent from the nuclei of yeast was first made by Wintersberger.[110] Since then, several investigators[111-113] confirmed and extended the observations to a number of other lower eukaryotes, plants, and protozoa as shown in Table 2. Yeast cells contain a major DNA polymerase activity resembling DNA polymerase α in higher eukaryotes called DNA polymerase A. This enzyme fluctuates in its activity during the cell cycle, sensitivity to SH-blocking reagents, and its ability to use RNA primer-DNA template for DNA synthesis. Polymerase α-like enzymes were also found in all other phyla examined, from lower vertebrates to protozoa, fungi, and plants, while β-like polymerase is present in all multicellular organisms but not in fungi or plants. Even though there is no DNA polymerase β in yeast, this organism contains a second molecular weight enzyme called DNA polymerase B which shares some but not all properties with DNA polymerase β in animal cells. It is possible that DNA polymerase B fulfills a function similar to that of DNA polymerase β.

C. DNA Polymerase γ

DNA polymerase γ was first found in HeLa cells[114,115] and is now known to present in a variety of animal cells as well as yeast (Table 2 and Reference 11). As compared to DNA polymerase α and β, DNA polymerase γ represents only a minor fraction of total DNA polymerase activity. It is present in both cytoplasmic and nuclear fractions. In addition to natural or synthetic DNA templates, DNA polymerase γ can also copy a variety of synthetic ribohomopolymers such as poly(A), poly(C), etc.[116-118] DNA polymerase γ, including those isolated from various sources,[114-120] is characteristically different from DNA polymerase α and β by its preference in utilization of template. DNA polymerase γ uses ribohomopolymers as template better than deoxyribohomopolymers. For example, it copies poly(A) five to ten times better than poly(dA). Because of its ability to copy ribohomopolymer templates, properties of DNA polymerase γ were extensively compared with those of the RNA-direct DNA polymerase (reverse transcriptase) of type C RNA tumor viruses. Both DNA polymerase γ and viral reverse transcriptase can copy polyribonucleotide with the same efficiency. However, DNA polymerase γ is not able to copy natural RNA. Immunological studies have shown that antibodies against viral reverse transcriptase did not inhibit DNA polymerase γ activity.[116,121] Gerard's recent experiment[118,122] revealed characteristic differences in utilization of synthetic polyribonucleotide between DNA polymerase γ and reverse transcriptase. He found that a poly (2′-*O*-methylcytidylate) template cannot be copied by HeLa cell or murine (3T6 cells) DNA polymerase, γ, whereas it is an effective template for reverse transcriptase.

Due to the low level of DNA polymerase γ present in the cell and difficulty in isolation, the enzyme has not yet been purified to homogeneity. Like DNA polymerase α and β, polymerase γ has a heterogeneous nature. Sedimentation analyses indicate that DNA polymerase γ, isolated from human, murine, and bovine cells has a molecular weight range from 110,000 to 300,000.[116,117,123,124] Some characteristics of DNA polymerase γ distinguish it from the DNA polymerases α and β, such as 50 m*M* phosphate in the presence of 100 to 300 m*M* KCl stimulated DNA polymerase γ but strongly inhibits both DNA polymerases α and β. Like DNA polymerase α, DNA polymerase γ is also sensitive to sulfhydryl blocking

Table 2
PHYLOGENY OF EUKARYOTIC DNA POLYMERASES

Phyla	Class	DNA polymerase α	DNA polymerase β	DNA polymerase γ
Vertebrates	Mammals	+	+	+
	Birds	+	+	+
	Reptiles	+	+	+
	Amphibians	+	+	+
Arthropods	Insects (embryos)	+	−	+
Mollusks	Cephalopods	+	+	
Coelenterates	Hydrozoas	+	+	
Protozoa	Ciliates	+	−	
Thallophytes	Ascomycetes	+	−	+
Spermaphytes	Monocotyledones	+	−	+

reagents: 1 m*M* *N*-ethylmaleimide inhibits DNA polymerase γ activity to 50 to 80%; 2 μg/mℓ aphidicolin,[125] 100 m*M* butylanilinouracil,[126] and 0.1 μg/mℓ heparin[127] strongly inhibit DNA polymerase α, but do not affect the activities of DNA polymerases β and γ; 20 μ*M* ethidium bromide has little effect on DNA polymerases α and β activity, whereas it inhibits DNA polymerase γ activity up to 60%.[128]

Since the first studies of Kalf and Ch'ih[129] and Meyer and Simpson,[56] it has been assumed that mitochondria possess a unique polymerase that differs from the other cellular DNA polymerases. However, a recent report[130] indicates that mitochondrial DNA polymerase is DNA polymerase γ.

D. Functions of DNA Polymerases α, β, and γ

Since there are no mutants with altered DNA polymerase genes in animal cells, a definite role for each of these enzymes is difficult to determine. Many investigations have been correlative and concerned with measuring the activity of each of the DNA polymerases in cells or tissues under varying conditions of growth or quiescence. A close correlation between the activity of the DNA polymerase α and the in vivo rate DNA replication has been shown during the developmental stages of the rat brain.[83,84,131] Chiu and Sung[131] have shown that DNA synthetic activity in rat brain rises to a peak approximately 6 days after birth, decreasing rapidly thereafter. DNA polymerases α and β were extracted from rat brain at different stages of development and analyzed on sucrose gradient centrifugation, clearly revealing a close correlation between the maximum rate of DNA synthesis in vivo and the maximum levels of activity of DNA polymerase α. As shown in Figure 2, DNA polymerase α is very active in the cerebella from younger rats, peaks at around the sixth day after birth, then decreases rapidly to very low activity in the adult rat cerebella. When the in vivo DNA synthesis was at minimum as in the adult brain, the activity DNA polymerase α accounted for only 4% of the total activity. Throughout this maturation process the activity of the DNA polymerase β remained relatively constant.[83,84,131]

Similar experiments using regenerating rat liver indicated that DNA synthesis was also correlated with increased activity of DNA polymerase α. Lynch et al.[132,133] showed that rat liver nuclei contain both DNA polymerases α and β. DNA polymerase α was shown to increase tenfold after partial hepatectomy. Chang and Bollum[134] also found that DNA polymerase α increased six- to sevenfold in regenerating rat liver while polymerase β activity was unchanged. Moreover, experiments with stationary phase mouse L cells, stimulated to proliferate by dilution into fresh growth medium, showed that the activity of DNA polymerase α declined substantially when the cells were in the stationary phase and subsequently increased 5- to 12-fold, in parallel with DNA synthesis, measured in vivo by thymidine

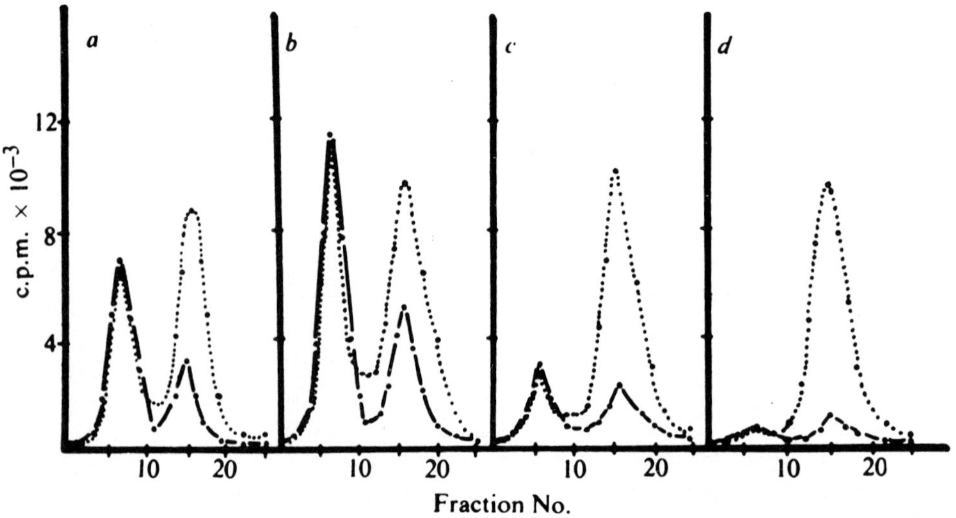

FIGURE 2. Developmental changes in DNA polymerase α and β of rat cerebellum. The profile of DNA polymerases α and β was analyzed by sucrose gradient centrifugation. The bottom of gradient is on the left. DNA polymerases were extracted from rat cerebella of different ages with 0.01 *M* Tris-HCl buffer (●—●) or 0.2 *M* phosphate buffer (●----●). a: 2 day old; b: 10 day old; c: 17 day old; d: adult. (From Chiu, J. F. and Sung, S. C., *Biochim. Biophys. Acta,* 269, 364, 1972. With permission.)

incorporation, as the cells proceeded from the stationary phase into the logarithmic phase of growth. In contrast, there were only relatively minor changes in the levels of DNA polymerase β activity during the cycle of events.[135]

Observations with synchronized cultured cells have shown a positive correlation between the rate of DNA synthesis and the activity of the DNA polymerase α. Spadari and Weissbach[115] measured the levels of DNA polymerase α, β, and γ in the S phase of synchronized HeLa cells. They found a rise in the DNA polymerase α activity while the cells progressed through the S phase. A concomitant increase in the DNA polymerase activity was also noted in the early part of the S phase, whereas the activity of DNA polymerase β remained constant throughout the S phase. Chiu and Baril[136] extended this work by examining the complete cell cycle in synchronized HeLa cells. They observed that the DNA polymerase α activity started to increase in the G_1 period before the S phase, and that this marked rise in the DNA polymerase α level was abolished by addition of cyclohexamide, indicating that protein synthesis was necessary for the increase in enzymatic activity.

It is a general observation that proliferating tissues contain a high level of DNA polymerase α that may represent as much as 90% of the total cellular DNA polymerase activity. In quiescent tissues or cells in the G_o state, the DNA polymerase β is the predominant activity.[82,84,137] A rise in the DNA polymerase α level can also be seen in the experimental neoplasia of rat liver, which is induced by a hepatocarcinogen.[138] Recently, Chiu et al.[139] demonstrated that a small level of DNA polymerase α was found in normal colon nuclei. On the other hand, extremely high activity of DNA polymerase α was found in nuclei of transplantable colon adenocarcinoma. The activity of DNA polymerase β was the same in both tissues. The effect of administration of a carcinogen (1,2-dimethylhydrazine) on large bowel nuclear DNA polymerase was studied. As shown in Figure 3, DNA polymerase α activity increased dramatically, while DNA polymerase β did not exhibit any significant change. The increase of DNA polymerase α activity paralleled the increase of in vivo DNA synthesis activity. Suzuki et al.[139a] have examined changes in the DNA polymerases α and β of HeLa cells during incubation after treatment with the carcinogen, *N*-methyl-*N′*-nitro-

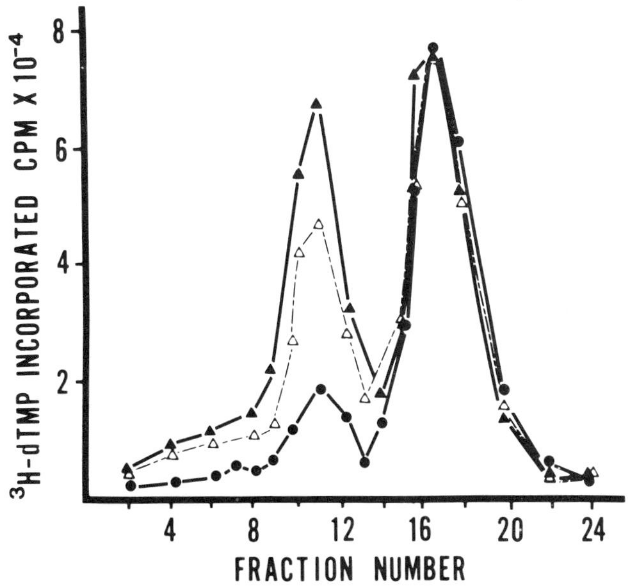

FIGURE 3. Nuclear DNA polymerase activities in colorectal epithelia of rats with 1,2-dimethylhydrazine (DMH). ●—●: normal control; △—△: 10 weeks on DMH; ▲—▲: 20 weeks on DMH.

N-nitrosoguanidine (MNNG). They found that DNA polymerase α activity increased by treatment of the cells with 10 μ*M* MNNG, while DNA polymerase β activity did not change. However, a concentration of 60 μ*M* of MNNG inhibited DNA polymerase α activity while inducing high levels of DNA polymerase β activity.

In addition to the physiological experimental data, several laboratories have recently used SV40 and polyoma DNA as model systems to study the role of DNA polymerases in DNA synthesis. A three- to tenfold increase in DNA polymerase α activity appears concomitant with the onset of DNA replication in papovavirus-infected mammalian cells.[140-142] Although DNA polymerase γ activity may vary, the activity of DNA polymerase β remains constant

Differential response of DNA polymerases α, β, and γ to chemical inhibitors, NEM, ddTTP, and aphidicolin provides another approach to perform a meaningful physiological experiment. The DNA polymerase γ can be distinguished from DNA polymerase α by its aphidicolin resistance, and from DNA polymerase β by its sensitivity to NEM.[12] In all physiological experiments studied, replicative DNA synthesis is completely inhibited by low concentrations of aphidicolin,[13,125,143,143a] whereas unscheduled DNA synthesis and the replication of mitochondrial DNA appear to be insensitive to aphidocolin. The latter two processes are affected by appropriate concentrations of ddNTP.[13] DNA polymerase γ is 250 to 520 times more sensitive to ddTTP than DNA polymerase α.[144-151] SV40 and cellular DNA replication is not affected by ddTTP, whereas it strongly inhibits both Ad_2, Ad_5, and mitochondrial DNA replication,[144-152] and DNA polymerase γ activity. Therefore, DNA polymerase α is exclusvely required for cellular and papovavirus DNA replication, and DNA polymerase γ is required in mitochondrial DNA replication.

Recently, McLennan has reported that DNA polymerase α in addition to DNA polymerase γ, is involved in mitochondrial DNA replication.[152a] However, Zimmerman et al.[152b] have further analyzed the synthesis of HeLa cell nuclear and mitochondrial DNA in vivo and in vitro with isolated mitochondria in the presence of specific inhibitors of DNA polymerase activity. They have shown that DNA polymerase α is not involved in mitochondrial DNA replication, a process which requires only DNA polymerase γ.

Table 3
THE FIDELITY OF DNA POLYMERASES FROM 1,2-DIMETHYLHYDRAZINE-TREATED RAT COLON MUCOSAE

	Incorporation (pmol)		
Polymerase	dTMP	dGMP	Level of misincorporation
Normal colon			
DNA polymerase α	56	0.019	1/2947
DNA polymerase β	468	0.102	1/4588
DMH-treated colon			
DNA polymerase α	224	0.183	1/1224
DNA polymerase β	496	0.119	1/4168

Note: DMH-treated colon DNA polymerases were extracted from colon mucosae of rats treated with 1,2-dimethylhydrazine for 20 weeks. For determination of correct incorporation, the reaction mixture (0.1 mℓ in final volume) contained: 2 nmol Tris-HCl (pH 7.4), 0.02 μmol $MnCl_2$, 0.2 μmol $MgCl_2$, 10 μmol KCl, 8 μmol dithiothreitol, 2 nmol each of dATP, dTTP, and dGTP, 1 μCi ^{3}H-dTTP (sp. act. = 1 Ci/mmol), and 20 μg of poly (dA-dT) (dA-dT). For measurement of incorrect incorporation, identical reaction mixtures were used except that they contained ^{3}H-dGTP (final sp. act. = 10 Ci/mmol, 10 μCi/assay) instead of ^{3}H-dTTP.

E. Fidelity of DNA Synthesis

The accumulation of DNA sequence errors has been proposed as a basis for aging[153] and for carcinogenesis and tumor progression.[154,155] One mechanism for the introduction of such errors would be a decreased fidelity of DNA polymerases during DNA synthesis, or DNA repair. Infidelity of DNA synthesis by mutagenized or altered DNA polymerases has been proposed as a mechanism for the accumulation of somatic mutations and subsequent alteration in cellular control. The infidelity of DNA polymerases in DNA synthesis was demonstrated in malignant cells[154] or cells which were induced by carcinogens.[156,157]

Sirover and Loeb[156] examined the effect of metal cations on the fidelity of DNA synthesis. Using AMV DNA polymerase, they measured the incorporation of complementary and noncomplementary nucleotides using Co^{2+}, Mn^{2+}, and Ni^{2+} as metal activators. The results suggest that the fidelity of in vitro DNA synthesis is a function of the divalent metal cation used during catalysis. As AMV DNA polymerase lacks a detectable proofreading exonuclease activity, any changes in fidelity could not be mediated by differences in the excision of noncomplementary nucleotides but must reflect alterations in base selection during catalysis. Cobalt and nickel are known carcinogens.[158] Manganese is a potent mutagen[159] and has been recently reported to be a carcinogen.[160] The relationship between decreases in the fidelity of DNA synthesis and metal mutagenesis, or carcinogenesis or both, has been discussed.[156] Recently, Chan and Becker[157] studied the fidelity of DNA polymerase activity from rat liver under *N*-2-fluorenylacetamide hepatocarcinogenesis and from 24-hr regenerating liver. They found that the DNA polymerase α from carcinogen-exposed rat livers were strikingly error-prone whereas the nuclear and cytoplasmic DNA polymerase β retained their fidelity throughout the feeding cycles. Similar results were also demonstrated by Chiu[139] on the fidelity of nuclear DNA polymerases during dimethylhydrazine (DMH)-induced colon carcinogenesis. As shown in Table 3, nuclear DNA polymerase β from normal and from DMH-treated rat colon had low error rates. However, high error rates were observed with nuclear DNA polymerase α from DMH-treated rat colons.

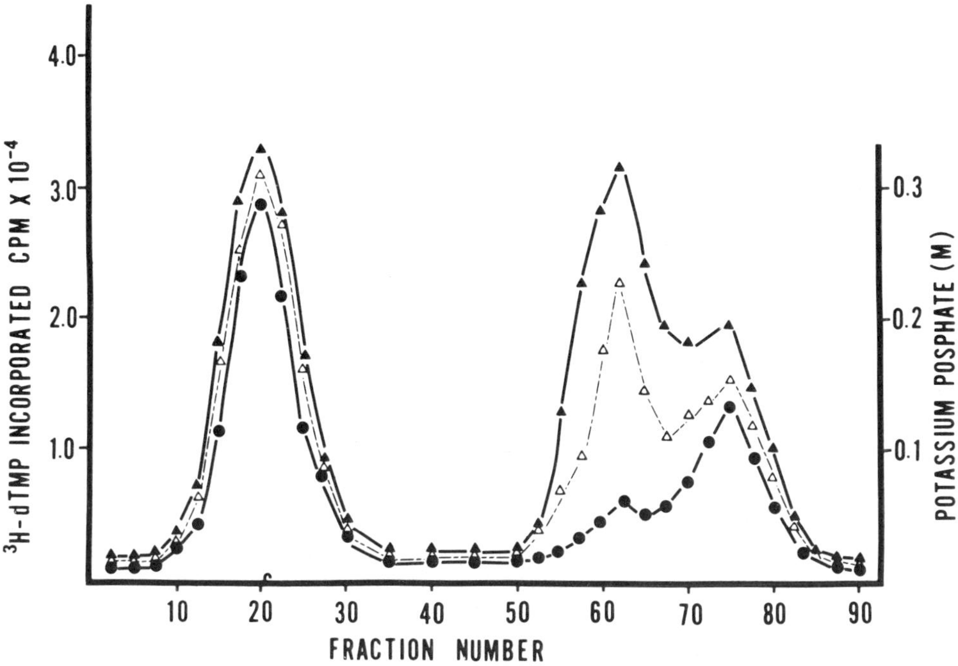

FIGURE 4. DEAE-cellulose chromatograms of nuclear extracts from normal or DMH treated rat large bowel mucosae and transplantable large bowel adenocarcinoma. ●—●: normal control; △----△: 20 weeks on DMH; ▲—▲: transplantable large bowel adenocarcinoma.

Chiu and Sung[161] have reported the distribution of DNA polymerase activities between cytoplasm and nuclei in developing brain. They found that there are at least two forms of DNA polymerase α; one form is easily washed from nuclei and the other is tightly bound to DNA or chromatin. Several other investigators have reported the same observation.[49,79,133] Multiple forms of DNA polymerase α have been demonstrated in calf thymus,[47,75,78] mouse myeloma,[53,162] rat ascites hepatoma,[163,164] and HeLa cells[165] by ion-exchange column chromatography. In order to see if DMH treatment affects one of the subspecies of DNA polymerase α which appears during colon carcinogenesis and to see if this subspecies of DNA polymerase α contributes to the infidelity of DNA synthesis during carcinogenesis, Chiu isolated and fractionated colon nuclear DNA polymerases of rat treated with DMH for 10, 15, and 20 weeks.[139] As shown in Figure 4, three DNA polymerase activities were separated on a DEAE-cellulose column, similar to data reported previously.[165,166] In addition to the DNA polymerase β in the nonadsorbed fractions, and the expected DNA polymerase α which eluted at 0.19 *M* potassium phosphate (designated as DNA polymerase α_2), a consistent and discrete peak of DNA polymerase α activity was found eluting at 0.14 *M* potassium phosphate (designated as DNA polymerase α_1). DNA polymerase α_1 is at a low level in normal colon (Figure 4), however, it represents the major DNA polymerase activity in the nuclear extracts of DMH-treated rat colon. DNA polymerase α_1 represents almost 50% of the total nuclear bound DNA polymerase activity detected in rat colon mucosae at the late stage of large bowel carcinogenesis.

The fidelity of DNA polymerases α_1, α_2, and β purified through DEAE-cellulose column chromatography was studied. The levels of infidelity of DNA polymerases α_1, α_2, and β were 1/769, 1/3653, and 1/7741, respectively. DNA polymerase α_1 fractions gave a high level of infidelity in incorporation of deoxyribonucleoside triphosphates into DNA, while

DNA polymerases α_2 and β appeared to be normal in the fidelity of DNA synthesis when compared to enzymes obtained from normal control colons.

III. DNA-BINDING PROTEINS AND ENZYMES INVOLVED IN DNA REPLICATION

A. DNA Ligases

DNA ligases are DNA-joining enzymes that join single-strand breaks in double-stranded DNA by catalyzing the formation of phosphodiester bonds. These enzymes were first identified in uninfected and phage-infected *E. coli* in 1967.[167-171] The discovery of ligases also coincided closely with the proposal of a discontinuous mechanism for DNA replication, a model which relies heavily on the existence of a ligase-like activity to join discontinuously synthesized DNA segments.[172] Following their discovery in prokaryotes, DNA ligase activities have also been detected in a variety of animal tissues and in plants.[173] Both prokaryotic and eukaryotic DNA ligases have been reviewed.[15,173-175]

Ligases are specific for duplex polynucleotides and catalyze the synthesis of phosphodiester bonds at single-stranded breaks between opposed 3′-hydroxyl and 5′-phosphoryl termini. Phosphodiester bond synthesis by the *E. coli* enzyme is coupled to the cleavage of the pyrophosphate bond of DNP, whereas the energy for phosphodiester bond synthesis by bacteriophage-induced or eukaryotic ligases is provided by the hydrolysis of the α, β-pyrophosphate bond of ATP.[15,173-175] The mammalian DNA ligases also differ in many respects from bacterial DNA ligases.

DNA joining activities have been measured by a variety of assay methods: (1) the change in sedimentation coefficient after covalently joining the DNA fragments, (2) conversion of internally located 5′-^{32}p-labeled phosphoryl groups to a form resistant to alkaline phosphatase, (3) linkage of one polynulceotide chain to a second one attached to cellulose, and (4) the conversion of ^{3}H-labeled d(A-T) copolymer, chain length of approximately 1000 nucleotides, with 3′-hydroxyl and 5′-phosphoryl termini, to a circular molecule resistant to exonuclease III.

The mechanism of the reaction catalyzed by eukaryotic DNA ligases can be illustrated in Figure 5. The first step consists of a nucleophilic attack of the ϵ-amino group of a lysine residue of the enzyme on the adenylyl phosphorus of ATP to form ligase-adenylate with the elimination of pyrophosphate (PPi). The second step involves transferring the adenylyl group from the enzyme to the 5′-phosphoryl terminus of single-stranded interruption (a "nick") in duplex DNA. At the final step, the 5′-phosphoryl group is attacked by the opposing 3′-hydroxyl terminus to form a phosphodiester bond, and AMP is eliminated. The proof for this mechanism rests on the evidence of isolation of covalent intermediates. A covalent intermediate, enzyme adenylate complex, has been isolated after incubation of purified mammalian DNA ligase with radioactive ATP.[176] This ligase-adenylate complex can join single-strand breaks in DNA in the absence of ATP, and one adenylate residue is released in each repair event. When DNA ligase I is purified from calf thymus, part of the enzyme is found in the form of such an enzyme-AMP complex, and a small amount of ATP-independent DNA ligase activity can consequently be observed with the enzyme preparations.[176] Similar results have been obtained with a DNA ligase in the chromatin fraction of rabbit-bone marrow.[177] A second reaction intermediate, a covalent DNA-AMP complex has also been recovered after brief incubation at 0°C of isolated DNA ligase-AMP complex with DNA containing single-strand breaks. Free AMP was released on further incubation of the isolated DNA-AMP complex with DNA ligase and Mg^{2+}, in agreement with the notion that this complex is a reaction intermediate.[178]

The physiological function of DNA ligase on DNA replication was demonstrated with *E. coli* mutant.[179,180] In mutants defective in DNA ligase, it can be shown that the joining of

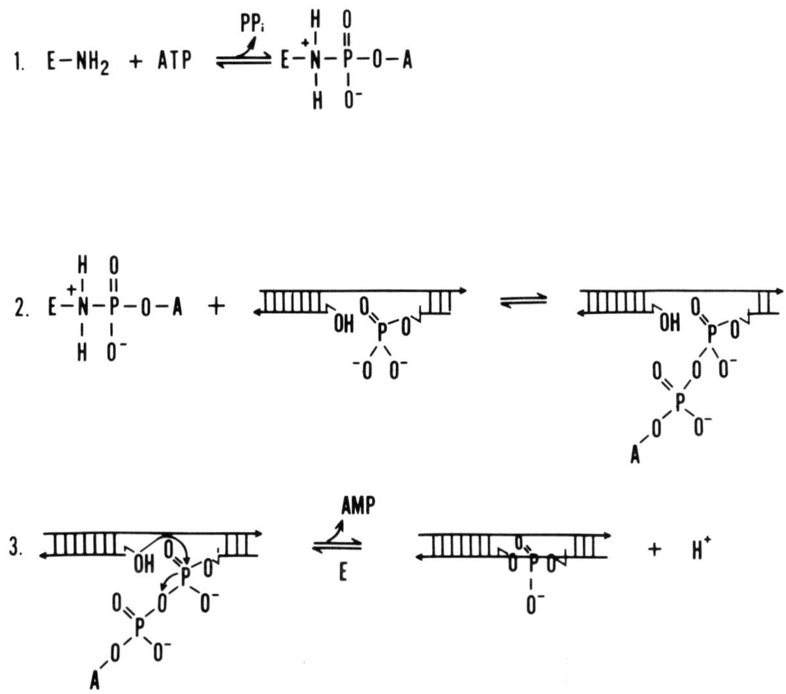

FIGURE 5. Mechanism of the reaction catalyzed by eukaryotic DNA ligases.

Okazaki fragments to high molecular weight DNA is severely impaired. When such mutants are observed at the nonpermissive temperature, essentially no joining of Okazaki fragments is seen; DNA synthesis, on the other hand, continues for several hours. Figure 6 shows the results of pulse-labeling experiments with strains carrying lig$^+$, lig 4, lig ts 7, and poly A_1 lig 4. Both lig 4 and lig ts 7 have less than 1% of wild-type (lig$^+$) ligase activity at 42°C. *E. coli* strains poly A_1 lig 4 are double mutant constructed by Gellert and associates.[179] As shown in Figure 6, *E. coli* strains carrying lig 4 alone seal Okazaki fragments (10S newly synthesized DNA fragments) slowly at 42°C (Figure 6b), but normally at 30°C (data not shown). By comparing the time required to produce similar size distributions in lig$^+$ and lig 4 strains, one can estimate that joining is about tenfold slower in lig 4. Similarly, strains carrying lig ts 7 have a very marked defect in sealing of Okazaki fragments at 42°C (Figure 6c), so that they still have only a small percentage of pulse-labeled DNA in high molecular weight pieces, even 5 min after the pulse. The rate of sealing by mutant strain ligts 7 is reduced by a factor of about 40. Double mutant poly A_1 lig 4 has a drastic defect in ability to seal pulse-labeled fragments (Figure 6d). Sealing of pulse-labeled fragments in this mutant is slowed more than 50-fold compared to the wild type and 5-fold compared to a lig$^+$ poly A_1 strain. In the double mutant, [^{3}H]thymidine is incorporated at a normal rate for more than 30 min at 42°C.

Conditionally lethal mutant cell lines with defective DNA synthesis have also been isolated from mouse cell lines.[181,182] One such mutant, ts A1S9, shows a defect in joining of newly synthesized DNA fragments to a higher molecular DNA while DNA repair ability remains normal under the nonpermissive conditions. It was proposed that ts A1S9 cells might have a defect in a DNA ligase specifically involved in DNA replication.[181]

Bacterial cells only contain a single DNA ligase, however, mammalian cells have two different DNA ligases which are serologically unrelated and respond in a different fashion

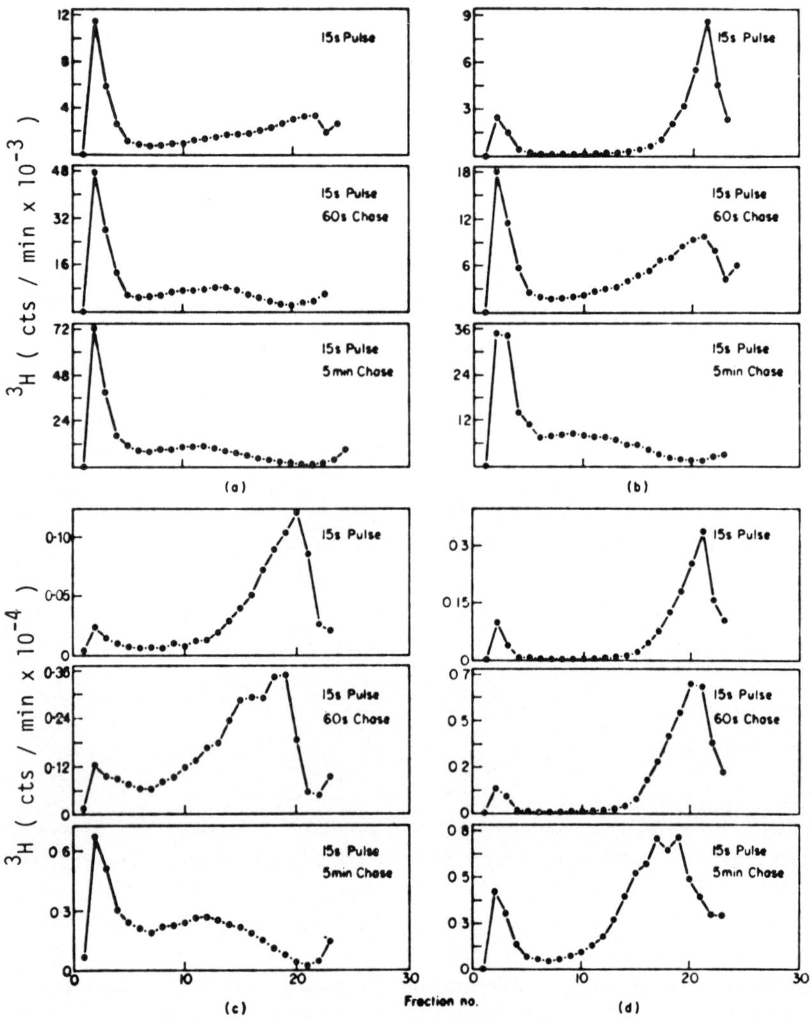

FIGURE 6. Alkaline sucrose density gradients on cultures which were ^{3}H-thymidine pulse-labeled at 42°C. Experiments were carried out as described in Reference 179. Fractions were collected from the bottoms of the gradients: (a) strain N1615 (lig$^+$); (b) strain N1618 (lig 4); (c) strain N2672 (lig ts7); (d) strain N1662 (pol A_1 lig 4). (From Gottsman, M. M., Hicks, M. L., and Gellert, M., *J. Mol. Biol.*, 77, 531, 1973. With permission.)

to cell proliferation.[15] These two mammalian enzymes have been named DNA ligase I (cytoplasmic DNA ligase) and DNA ligase II (nuclear DNA ligase). Whether DNA ligase I is of nuclear origin is still unknown. DNA ligase I has a large molecular weight of 175,000 to 220,000, while DNA ligase II has a molecular weight of about 85,000. Actively growing mammalian cells contain more DNA ligase activity than nongrowing cells. Regenerating rat livers contain about five times more ligase activity than normal rat liver.[218,219] It was found that the increased ligase activity in the regenerating tissue was due to a large increase in DNA ligase I activity, while DNA ligase II levels did not change significantly.[185]

Mitochondria contain little DNA ligase activity. This DNA ligase appears to be identical with DNA ligase II. Mitochondrial ligase behaves like DNA ligase II during gradient chromatography on phosphocellulose, and both activities have the same pH dependence and optimal $MgCl_2$ concentrations.[186]

B. DNA-Binding Proteins

This group of proteins is also called "DNA-unwinding proteins" or "DNA-extending proteins", i.e., the proteins that destabilize the double helix by binding more strongly to single-stranded DNA. The first such protein described is the gene 32 protein isolated from bacteriophage T_4-infected cells.[187] The gene 32 protein binds strongly and cooperatively to single-stranded DNA and thus promotes denaturation (unwinding) of double-helical DNA. Genetic, physiological, and biochemical experiments suggests that the gene 32 protein is involved in the recombination and replication.[187] DNA binding proteins have also been isolated from *E. coli,* bacteriophage-infected *E. coli,* and mammalian cells.[20,188]

Another important property of these proteins, besides their interaction with DNA, is their capacity to specifically stimulate the homologous DNA polymerase activity. For example, the T_4 32 protein stimulates only T_4 DNA polymerase and not any host enzyme. Similar stimulation of DNA binding proteins of various DNA polymerases have been observed with the *E. coli* protein and *E. coli* DNA polymerase, and with the phage T7 protein and phage T7-induced DNA polymerase.[189-191] This stimulation cannot be explained simply in terms of a generalized destabilization of the double helix but must imply a direct physical and functional interaction between the binding proteins and the DNA polymerases. The data result from biochemical and genetic studies support this conclusion.[192]

Similar DNA-binding proteins were also observed in eurkaryotes.[20] Table 4 lists a number of such proteins from various sources. As shown in the table, most of these proteins specifically stimulate DNA polymerase α activity. As mentioned previously, DNA polymerase α is responsible for replicative DNA synthesis, therefore these DNA-binding proteins are likely to be involved in DNA replication.[22,193-197] However, Chiu and Sung[101] and Blue and Weissbach[198] have described another protein which specifically stimulates polymerase β. Recently, Chang et al.[198b] purified a DNA-binding protein from *Saccharomyces cerevisiae* and named this protein "protein C". The purified protein C is essentially free of any known enzyme activity, however, it appears to stimulate yeast DNA polymerase I strongly.

The mechanism of these proteins involving in DNA replication is not known. However, a detailed analysis of the interaction of the *E. coli* DNA binding protein with DNA polymerase II has suggested a model to account for these various observations. *E. coli* DNA binding protein interacts with *E. coli* DNA polymerase II to form a protein-protein complex in the absence of DNA. DNA-binding protein, when bound to DNA, retains its ability to interact with DNA polymerase II and to form a ternary complex between DNA-binding protein, DNA polymerase II, and DNA. Neither *E. coli* DNA polymerase I nor DNA polymerase III nor phage T4-induced DNA polymerase interacts with DNA-binding protein. Thus, it appears that the specific stimulation is due to complex formation. The stimulation rate and extent of synthesis seen when the DNA-binding protein is added can be explained by destabilizing the double helix of DNA template as well as by binding to the polymerase during DNA synthesis, such that the polymerase does not dissociate from the template during processing.

C. DNA-Dependent ATPase

DNA-dependent ATPases are ATP-dependent unwinding enzymes that actively unwind DNA in a reaction driven by the hydrolysis of ATP or some other nucleoside or deoxynucleoside triphosphate. DNA unwinding ATPase have been found in bacteria, phage-infected bacteria, and plant and animal cells.[20,188] The ATPase activity of protein has an absolute requirement for single-stranded DNA.[198] No activity is demonstrated with duplex DNA that is free of single strands. In the presence of single-stranded DNA they hydrolyzed ATP to ADP and Pi. The other ribonucleoside triphosphates are also hydrolyzed, but at a slower rate, while dATP is as effective as ATP.

A DNA-dependent ATPase, has been reported in *E coli* and characterized more exten-

Table 4
DNA-BINDING PROTEINS IN EUKARYOTES

Source	DNA-binding protein	Binding specificity	Helix ⇌ coil →	Helix ⇌ coil ←	Effect on DNA polymerase activity	Ref.
HeLa cells	CI factor (33,000)	SS-DNA	No	No	Strongly stimulate (>10×) DNA polymerase α	194
	(85,000)	SS-DNA,DS-DNA	No	No	Strongly stimulate (>10×) DNA polymerase β	198
Calf thymus	UP1 (mixed) (24,000)	SS-DNA	Yes	No	Strongly stimulate (>10×) DNA polymerase α	193
	Basic protein (33,000)	SS-DNA	No	No	Inhibit DNA polymerase	193
	Acidic protein (33,000—40,000)	SS-DNA	Yes	No	Strongly stimulate (>10×) DNA polymerase α	193
Rat brain	Acidic protein	DS-DNA	—	—	Stimulate DNA polymerase β	101
Rat liver	Tetramer (24,000)	SS-DNA	Yes	—	Stimulate (3×) DNA polymerase α	195
Rat liver	(25,000)	SS-DNA	No	No	Inhibit DNA polymerase	195b
Mouse cells	Phosphoprotein (30,000—35,000)	SS-DNA, DS-DNA	—	—	Stimulate DNA polymerase α	22
Yeast	Protein C (37,000)	SS-DNA, DS-DNA	—	—	Strongly stimulate (50×) yeast DNA polymerase I	198b

sively. This protein has a monomer molecular weight of 180,000. It binds tightly to single-stranded DNA or to unpaired regions in DNA-duplexes or DNA-RNA hybrids. Several experimental data suggest a model according to which the enzyme first interacts with a single-stranded region adjacent to the duplex DNA and then processively invades and unwinds the adjoining helix using the free energy of ATP hydrolysis as the driving force.[199] The complete unwinding of a duplex region 2000 base pairs in length requires approximately 80 enzymes in this process and suggests that the protein may, in addition to the separation of the fork, have to coat the single strands to prevent renaturation. The cooperative interaction of several molecules could serve to stabilize the denatured sites. It was suggested that the DNA-dependent ATPase might function in a number of biological processes such as replication, transcription, and recombination.

D. DNA-Untwisting Enzymes

An enzyme with the capacity to relax a closed circular superhelical DNA and to alter the topological winding number (α) of DNA was first discovered in *E. coli* and called the "ω protein".[200] Subsequently, the eukaryotic DNA untwisting enzyme was discovered in nuclear extracts of mouse embryo cells and was referred to as the DNA-untwisting enzyme.[201] Similar proteins which have been shown to be enzymes that are capable of reducing the number of superhelical turns in circular DNA, also have been identified from many sources. They have been referred to as "ω proteins" in *Drosophila* eggs,[202] calf thymus,[203] and *Bacillus megaterium;*[204] as untwisting enzymes in rat liver,[205] and yeast;[206] as DNA relaxing enzymes in

KB cells,[207] *Xenopus laevis* eggs,[208,209] chicken erythrocytes,[210] and monkey cells,[211] and as DNA nicking-closing enzymes in HeLa and mouse cells,[212] duck cells,[213] and *Vaccinia* virions.[214] There is no uniform nomenclature for these proteins and none of these names is completely satisfactory.

DNA-untwisting enzymes relax the closed circular superhelical DNA and still retain the product DNA covalently closed. Since product DNA is permanently relaxed even after the enzyme is removed, these enzymes must change the topological winding number (α) rather than the helix winding number (β). The topological states of circular double-helical DNA follow the relationship $\tau = \alpha - \beta$,[215] where τ is the number of superhelical turns (positive or negative), α is the number of turns, and β is the number of duplex turns. The only way the topological winding can be changed is to break one strand, wind one end of that strand relative to the helix axis, and reseal the break. The remarkable feature of these enzymes is that they act by concerted single-strand nicking and ligasing reactions on the polynucleotide without any known cofactor or external source of chemical energy. This indicates that there must be some mechanism for conserving the energy released from the broken strand and using it for resealing the break.

The untwisting enzymes which were purified from rat liver[205] and KB cells[207] are single polypeptide chain proteins with molecular weights between 60,000 and 70,000. The enzymes are found in nuclei in association with chromatin.[212,216] The purified rat liver untwisting enzyme has been used to study the mechanism of the nicking-ligasing reaction. At high enzyme DNA ratio, the nicked intermediate has been trapped by rapidly adjusting the pH of the reaction mixture to either alkali (pH $>$12.5) or to acidic (pH 4.5).[217,218] Presumably the DNA or enzyme is denatured under these conditions and fails to reseal the breaks. High salt concentration, i.e., 0.5 *M* KCl inhibits the enzyme to make the breaks, yet it cannot stop the enzyme to seal the break. Using alkaline CsCl gradient centrifugation, Champoux[218] demonstrated that DNA strands which have been broken by rat liver untwisting enzyme exhibit a reduced buoyant density due to the covalent attachment of this enzyme to one of the ends of the broken strands.[218] It confirmed the notion that DNA strand breakage involves not hydrolysis of a phosphodiester bond, but transfer of one end of the broken DNA strand to a site on the enzyme.[200,201] The nature of the covalent linkage is not known.

Although the reaction mechanism is different, the physiological function of DNA-untwisting enzyme is similar to the functions of DNA-binding protein (DNA-unwinding protein) and DNA-dependent ATPase (as shown in Table 5). Several experimental results propose a model suggesting how these enzymes may play biological roles in the opening of the DNA duplex leading to DNA replication, transcription, and DNA recombination.[20] With respect to the replication fork, it is tempting to postulate that DNA-unwinding protein, DNA-dependent ATPase, and DNA-untwisting enzyme act in concert to locally separate the strands of the helix, stabilize the opened state, and release the resulting topological strain, respectively.

IV. SUMMARY AND CONCLUSION

Mammalian cells contain at least three distinct DNA polymerases, i.e., DNA polymerases α, β, and γ. Experimental evidence suggests that DNA polymerase α is responsible for DNA replication and DNA polymerase β seems to be involved in DNA repair. DNA polymerase γ is required in mitochondrial DNA replication. However, the role of nuclear DNA polymerase γ is still unknown. The infidelity of DNA polymerase α in DNA synthesis was demonstrated in malignant cells and cells induced by carcinogens. Chromosome duplication is discontinuous in nature. DNA ligases join the intermediate fragments to high molecular DNA. DNA-binding proteins unwind DNA helix, while others bind to the single-stranded DNA and even stimulate homologous DNA polymerase activities.

Although a great deal is being learned about how DNA replication is accomplished, little

Table 5
PROTEINS AFFECTING DNA HELIX STRUCTURE

Protein	Enzymatic activity	Function
DNA-binding proteins (DNA-unwinding proteins)	None	Stabilizes single-stranded regions
DNA helicase (DNA-dependent ATPase)	ATPase	Actively denatures helical DNA or DNA-RNA hybrids
DNA-untwisting enzyme (protein)	Nicking-closing	Relaxes supercoiled DNA

is known about the control of replication. The fractionation of entire DNA replicating systems into their accessory protein components, the characterization of these protein components, and the reconstitution of these accessory proteins into a DNA replicating system to define their specific roles in DNA replication are immediate approaches to understanding the control mechanism of DNA replication.

ACKNOWLEDGMENTS

I thank my colleagues Drs. Joseph H. Gans and Douglas P. Quick for their constructive criticism. Previously unreported work from this laboratory was supported by grants CA 25136 from the National Cancer Institute. I am grateful to Mr. Robert Shaw for his devoted and patient secretarial efforts in preparing this manuscript.

REFERENCES

1. **Kornberg, R. D.,** Structure of chromatin, *Annu. Rev. Biochem.*, 46, 931, 1977.
2. **McGhee, J. D. and Felsenfeld, G.,** Nucleosome structure, *Annu. Rev. Biochem.*, 49, 1115, 1980.
3. **Williamson, D. H.,** Protein synthesis and the control of nuclear DNA replication in yeast and other eukaryotes, in *Cell Cycle Controls,* Padilla, G. M., Cameron, I. L., and Zimmerman, A., Eds., Academic Press, New York, 1974, 143.
4. **Taylor, J. H.,** Units of DNA replication in chromosomes of eukaryotes, *Int. Rev. Cytol.*, 37, 1, 1974.
5. **Edenberg, H. J. and Huberman, J. A.,** Eukaryotic chromosome replication, *Annu. Rev. Genet.*, 9, 245, 1975.
6. **Hand, R.,** Eukaryotic chromosome replication and its regulation, in *Cell Biology, A Comprehensive Treatise,* Vol. 2, Goldstein, L. and Prescott, D. M., Eds., Academic Press, New York, 1979, 389.
7. **Prescott, D. M.,** The cell cycle and the control of cellular reproduction, *Adv. Genet.*, 18, 99, 1976.
8. **Wickner, S.,** DNA replication proteins of *Escherichia coli, Annu. Rev. Biochem.*, 47, 1163, 1978.
9. **Champoux, J. J.,** Proteins that affect DNA conformation, *Annu. Rev. Biochem.*, 47, 449, 1978.
10. **Weissbach, A.,** Eukaryotic DNA polymerases, *Annu. Rev. Biochem.*, 46, 25, 1977.
11. **Wintersberger, E.,** DNA-dependent DNA polymerases from eukaryotes, *Trends Biochem. Sci.*, 2, 58, 1977.
12. **DePamphilis, M. L. and Wasserman, P. M.,** Replication of eukaryotic chromosomes: a close-up of the replication fork, *Annu. Rev. Biochem.*, 49, 627, 1980.
13. **Kornberg, A.,** *DNA Replication,* Freeman and Co., San Francisco, 1980.
14. **Scovassi, A. I., Plevani, P., and Bertazzoni, U.,** Eukaryotic DNA polymerases, *Trends Biochem. Sci.*, 5, 335, 1980.

14a. **Chang, L. M. S. and Bollum, F. J.,** Immunological reagent for comparisons of DNA polymerase-α and DNA polymerase-β, *J. Biol. Chem.*, 256, 494, 1981.

15. **Soderhall, S. and Lindahl, T.,** DNA ligases of eukaryotes, *FEBS Lett.*, 67, 1, 1976.

16. **Lehman, I. R.,** DNA joining enzymes (ligases), in *The Enzymes,* Vol. 10, Boyer, P. D., Ed., Academic Press, New York, 1974, 237.
17. **Hackman, H. J. and Lezuis, A. G.,** An ATPase depending on the presence of single-stranded DNA from mouse myeloma, *Eur. J. Biochem.,* 61, 325, 1976.
18. **Cobianchi, F., Riva, S., Mastromi, G., Spardari, S., Pedrali-Noy, G., and Falschi, A.,** Enhancement of the rate of DNA polymerase α activity on duplex DNA by a DNA-binding protein and a DNA-dependent ATPase of mammalian cells, *Cold Spring Harbor Symp. Quant. Biol.,* 43, 639, 1979.
19. **Otto, B.,** DNA-dependent ATPase in concanvalin A stimulated lymphocytes, *FEBS Lett.,* 79, 175, 1977.
20. **Falachi, A., Cobianchi, F., and Riva, S.,** DNA-binding proteins and DNA-unwinding enzymes in eukaryotes, *Trends Biochem. Sci.,* 5, 154, 1980.
21. **Hoch, S. O. and McVey, E.,** Purification and characterization of two major DNA-binding proteins in human serum, *J. Biol. Chem.,* 252, 1881, 1977.
22. **Otto, B.,** A single-strand-specific DNA-binding protein from mouse cells that stimulate DNA polymerase. Its modification by phosphorylation, *Eur. J. Biochem.,* 73, 17, 1977.
23. **Mechali, M. and DeRecondo, A. M.,** Co-fractionation of an endonuclease activity during the purification of DNA polymerase γ from regenerating rat liver. Properties and separation from DNA polymerase, *Eur. J. Biochem.,* 58, 461, 1975.
24. **Pedrini, A. M., Ranzani, G., Pedrali-Noy, G. C. F., Spadari, S., and Falaschi, A.,** A novel endonuclease of human cells specific for single-stranded DNA, *Eur. J. Biochem.,* 70, 275, 1976.
25. **Otto, B. and Knippers, R.,** An endonuclease from mouse cells specific for single-stranded DNA, *Eur. J. Biochem.,* 71, 617, 1976.
26. **Lavin, M. F., Kikuchi, T., Counsilman, C., Jenkins, A., Winzox, D. J., and Kidson, C.,** A mammalian nicking endonuclease, *Biochemistry.,* 15, 2409, 1976.
27. **Wang, E. C. and Furth, J. J.,** Mammalian endonuclease, DNase V. Purification and properties of enzyme of calf thymus, *J. Biol. Chem.,* 252, 116, 1977.
28. **Byrnes, J. J., Downey, K. M., Black, V. L., and So, A. G.,** A new mammalian DNA polymerase with 3′ to 5′ exonuclease activity: DNA polymerase δ, *Biochemistry,* 15, 2817, 1976.
29. **Banks, G. R.,** A ribonuclease H from *Ustilago maydis.* Properties, mode of action and substrate specificity of the enzyme, *Eur. J. Biochem.,* 47, 499, 1974.
30. **Tashiro, F., Mita, T., and Higashinakagawa, T.,** Multiple forms of nuclear ribonuclease H from *Tetrahymena pyriformis, Eur. J. Biochem.,* 65, 123, 1976.
31. **Modak, M. J. and Marcus, S. L.,** Specific inhibition of DNA polymerase-associated RNase H by DNA, *J. Virol.,* 22, 243, 1977.
32. **Roeder, R. G.,** in *RNA Polymerase,* Losick, R. and Chamberlin, M., Eds., Cold Spring Harbor Laboratory, N.Y., 1976, 285.
33. **Elgin, S. C. R. and Weintraub, H.,** Chromosomal proteins and chromatin structure, *Annu. Rev. Biochem.,* 44, 725, 1975.
34. **Chiu, J. F. and Hnilica, L. S.,** Nuclear nonhistone proteins: chemistry and function, in *Chromatin and Chromosome Strucutre,* Li, H. J. and Eckhard, R. A., Eds., Academic Press, New York, 1977, 193.
35. **Gefter, M. L.,** DNA replication, *Annu. Rev. Biochem.,* 44, 45, 1975.
36. **Byrnes, J. J. and Black, V. L.,** Comparison of DNA polymerases α and δ from bone marrow, *Biochemistry,* 17, 4226, 1978.
37. **Foster, D. N. and Gurney, T.,** Sizes of polymerases from nuclei isolated by a non-aqueous method, *J. Cell Biol.* 63, 103a, 1974.
38. **Bollum, F. J.,** Deoxynucleotide-polymerizing enzymes in mammalian cells: immunofluorescence, in *Antiviral Mechanisms In the Control of Neoplasia,* Chandra, P., Ed., Plenum Press, New York, 1979, 587.
39. **Belden, A., Pedrali-Noy, G., and Weissbach, A.,** DNA polymerase of mitochondria is a γ-polymerase, *J. Biol Chem.,* 252, 3351, 1977.
40. **Bertazzoni, U., Scovassi, A. I., and Brun, G. M.,** Chick-embryo DNA polymerase γ. Identity of γ-polymerases purified from nuclei and mitochondria, *Eur. J. Biochem.,* 81, 237, 1977.
41. **Hubscher, U., Kuenzle, C. C., and Spadari, S.,** Identity of DNA polymerase from synaptosomal mitochondria and rat brain nuclei, *Eur. J. Biochem.,* 81, 249, 1977.
42. **Sheinin, R., Humbert, J., and Pearlman, R. E.,** Some defects of eukaryotic DNA replication, *Annu. Rev. Biochem.,* 47, 277, 1978.
43. **Herrick, G., Spear, B. B., and Veomett, G.,** Intracellular localization of mouse DNA polymerase α, *Proc. Natl. Acad. Sci. U.S.A.,* 73, 1136, 1976.
44. **Siebert, G., Schlatterer, B., and Shakoori, R. A.,** Organic solvent mixtures as zonal centrifugation media: separation of nuclei and of nucleoli, in *Methodological Developments in Biochemistry,* Vol. 4, Reid, E., Ed., Longman Group, Edinburgh, 1976, 13.

45. **Yoneda, M. and Bollum, F. J.,** Deoxyribonucleotide-polymerizing enzymes of calf thymus gland. I. Large scale purification of terminal and replicative deoxynucleotidyl transferases, *J. Biol. Chem.*, 240, 3385, 1965.
46. **Holmes, A. M., Hesslewood, I. P., and Johnston, I. R.,** Evidence that DNA polymerase α of calf thymus contains a subunit of molecular weight 155,000, *Eur. J. Biochem.*, 62, 229, 1976.
47. **Yoshida, S., Kondo, T., and Ando, T.,** Multiple molecular species of cytoplasmic DNA polymerase from calf thymus, *Biochim. Biophys. Acta,* 353, 463, 1974.
48. **Wang, T. Y.,** Nonhistone chromatin proteins from calf thymus and their role in DNA biosynthesis, *Arch. Biochem. Biophys.*, 122, 629, 1967.
49. **Sedwick, W. D., Wang, T. S-F., and Korn, D.,** Purification and properties of nuclear and cytoplasmic deoxyribonucleic acid polymerases from human KB cells, *J. Biol. Chem.*, 241, 5026, 1972.
50. **Masui, H. and Garren, L. D.,** On the mechanism of action of adrenocorticotropic hormone. Stimulation of deoxyribonucleic acid polymerase and thymidine kinase activities in adrenal glands, *J. Biol. Chem.*, 245, 2627, 1970.
51. **Chiu, J. F. and Sung, S. C.,** Separation and properties of DNA polymerases from developing rat brain, *Biochim. Biophys. Acta,* 246, 44, 1971.
52. **Wallace, P. G., Hewish, D. R., Venning, M. M., and Burgoyne, L. A.,** Multiple forms of mammalian deoxyribonucleic acid polymerase. An attempt to relate their interactions with nuclei and free deoxyribonucleic acid *in vitro* with their possible functions *in vivo, Biochem. J.*, 125, 47, 1971.
53. **Matsukage, A., Bohn, E. W., and Wilson, S. H.,** Multiple forms of DNA polymerase in mouse meyloma, *Proc. Natl. Acad. Sci. U.S.A.*, 71, 578, 1974.
54. **Tsuruo, T., Satoh, H., and Ukita, T.,** DNA polymerases of ascites hepatoma cells. I. Purification and properties of a DNA polymerase from soluble fraction, *Biochem. Biophys. Res. Commun.*, 48, 769, 1972.
55. **Wang, T. Y.,** Isolation of a terminal DNA-nucleotidyl transferase from calf thymus non-histone chromatin proteins, *Arch. Biochem. Biophys.*, 127, 235, 1968.
56. **Meyer, R. R. and Simpson, M. V.,** DNA biosynthesis in mitochondria: partial purification of a distinct DNA polymerase from isolated rat liver mitochondria, *Proc. Natl. Acad. Sci. U.S.A.*, 61, 130, 1968.
57. **Smith, R. G. and Gallo, R. C.,** DNA-dependent DNA polymerases I and II from normal human-blood lymphocytes, *Proc. Natl. Acad. Sci. U.S.A.*, 69, 2879, 1972.
58. **Roufa, D. J., Moses, R. G., and Reed, S. J.,** The DNA polymerases of chinese hamster cells. Subcellular distribution and properties of two DNA polymerases, *Arch. Biochem. Biophys.*, 167, 547, 1975.
59. **Craig, R. K. and Keir, H. M.,** Deoxyribonucleic acid polymerases of BHK-21/C13 cells. Partial purification and characterization of the enzymes, *Biochem. J.*, 145, 215, 1975.
60. **Brun, G., Rougeon, F., Lauber, M., and Chapeville, F.,** Purification and properties of DNA polymerases from chick embryo, *Eur. J. Biochem.*, 41, 241, 1974.
61. **Williams, A. F.,** DNA polymerase in avian erythroid cells, *J. Cell Sci.*, 11, 785, 1972.
62. **Stavrianopoulos, J. G., Karkas, J. D., and Chargaff, E.,** Nucleic acid polymerases of the developing chicken embryo: a DNA polymerase preferring a hybrid template, *Proc. Natl. Acad. Sci. U.S.A.*, 68, 2207, 1971.
63. **Fansler, B. and Loeb, L. A.,** Sea urchin nuclear DNA polymerase. II. Changing localization during early development, *Exp. Cell Res.*, 57, 305, 1969.
64. **Loeb, L. A.,** Purification and properties of deoxyribonucleic acid polymerase from nuclei of sea urchin embryos, *J. Biol. Chem.*, 244, 1672, 1969.
65. **Westergaard, O.,** Separation of two DNA polymerase fractions from tetrahymena cells after excision-repairable damage to DNA, *Biochim. Biophys. Acta,* 213, 36, 1970.
66. **Wintersberger, U. and Wintersberger, E.,** Studies on deoxyribonucleic acid polymerases from yeast. I. Partial purification and properties of two DNA polymerases from mitochondria-free cell extracts, *Eur. J. Biochem.*, 13, 11, 1970.
67. **Wintersberger, U. and Wintersberger, E.,** Studies on deoxyribonucleic acid polymerases from yeast. II. Partial purification and characterization of mitochondrial DNA polymerase from wild type and respiration-deficient yeast cells, *Eur. J. Biochem.*, 13, 20, 1970.
68. **Schiebel, W. and Raffael, A.,** Two groups of deoxyribonucleic acid polymerases from *Physarum polycephalum* classified by differential sensitivity to N-ethylmaleimide, heparin, cytosine arabinoside triphosphate and ethidium bromide, *FEBS Lett.*, 121, 81, 1980.
69. **Bollum, F. J.,** Mammalian DNA polymerases, *Prog. Nucleic Acid Res. Mol. Biol.*, 15, 109, 1975.
70. **Sedwick, W. D., Wang, T. S-F., and Korn, D.,** "Cytoplasmic" deoxyribonucleic acid polymerase. Structure and properties of the highly purified enzyme from human KB cells, *J. Biol. Chem.*, 250, 7045, 1975.
71. **Spadari, S. and Weissbach, A.,** The dissimilitude of the low and high molecular weight deoxyribonucleic acid-dependent deoxyribonucleic acid polymerases of HeLa cells, *J. Biol. Chem.*, 249, 2991, 1974.

72. **Byrnes, J. J., Downey, K. M., and So, A. G.,** Bone marrow cytoplasmic deoxyribonucleic acid polymerase. Variation of pH and ionic environment as a possible control mechanism, *Biochemistry,* 12, 4378, 1973.

72a. **Mechali, M., Abdiedebat, J., and de Recondo, A. M.,** Eukaryotic DNA polymerase α. Structural analysis of the enzyme from regenerating rat liver, *J. Biol. Chem.,* 255, 2114, 1980.

72b. **Grosse, F. and Krauss, G.,** Purification and partial characterization of a DNA polymerase α species from calf thymus, *Nucleic Acids Res.,* 8, 5703, 1980.

73. **Karkas, J. D., Margulies, L., and Chargaff, E.,** A DNA polymerase from embryos of *Drosophila melanogaster.* Purification and properties, *J. Biol. Chem.,* 250, 8657, 1975.

74. **Crerar, M. and Pearlman, R. E.,** Deoxyribonucleic acid polymerase from *Tetrahymena pyriformis.* Purification and properties of the major activity in exponentially growing cells, *J. Biol. Chem.,* 249, 3123, 1974.

75. **Holmes, A. M., Hesslewood, I. P., and Johnston, I. R.,** The occurrence of multiple activities in the high-molecular-weight DNA polymerase fraction of mammalian tissues. A preliminary study of some of their properties, *Eur. J. Biochem.,* 43, 487, 1974.

76. **Herrick, G., Delius, H., and Alberts, B.,** Stingle-stranded DNA structure and DNA polymerase activity in the presence of nucleic acid helix-unwinding proteins from calf thymus, *J. Biol. Chem.,* 251, 2142, 1976.

77. **Weissbach, A., Spadari, S., and Knopf, K. W.,** HeLa cell DNA polymerases: properties and possible functions, in *ICN-UCLA Symp. Mol. Cell. Biol.,* Vol. 3, ICN Biologicals, Cleveland, 1975, 64.

78. **Momparler, R. L., Rossi, M., and Labitan, A.,** Partial purification and properties of two forms of deoxyribonucleic acid polymerase from calf thymus, *J. Biol. Chem.,* 248, 285, 1973.

79. **Weissbach, A., Schlaback, A., Fridlender, B., and Bolden, A.,** DNA polymerases from human cells, *Nature (London) New Biol.,* 231, 167, 1971.

80. **Baril, E. F., Brown, O. E., Jenkins, M. D., and Laszlo, J.,** Deoxyribonucleic acid polymerase with rat liver ribosomes and smooth membranes. Purification and properties of the enzymes, *Biochemistry,* 10, 1981, 1971.

81. **Haines, M. E., Holmes, A. M., and Johnston, I. R.,** Distinct cytoplasmic and nuclear DNA polymerases from rat liver, *FEBS Lett.,* 17, 63, 1971.

82. **Chiu, J. F. and Sung, C. S.,** Solubilization and characterization of particulate form of DNA polymerase from adult rat liver nuclei. *Biochem. Biophys. Res. Commun.,* 46, 1830, 1972.

83. **Chiu, J. F. and Sung, C. S.,** Solublization and characterization of a particulate form of DNA polymerase from adult rat brain nuclei, *Biochim. Biophys. Acta,* 262, 397, 1972.

84. **Chiu, J. F. and Sung, C. S.,** Particulate form of DNA polymerase in rat brain, *Nature (London) New Biol.,* 239, 176, 1972.

85. **Adams, R. L. P., Henderson, M. A. L., Wood, W., and Lindsay, J. G.,** Multiple forms of nuclear deoxyribonucleic acid polymerases and their relationship with the soluble enzyme, *Biochem. J.,* 131, 237, 1973.

86. **Chang, L. M. S. and Bollum, F. J.,** Low molecular weight deoxyribonucleic acid polymerase from rabbit bone marrow, *Biochemistry,* 11, 1264, 1972.

87. **Schlaback, A., Fridlender, B., Bolden, A., and Weissbach, A.,** DNA-dependent DNA polymerases from HeLa cell nuclei. II. Template and substrate utilization, *Biochem. Biophys. Res. Commun.,* 44, 879, 1971.

87a. **Korn, D., Fisher, P. A., Battey, J., and Wang, T. S. F.,** Structural and enzymological properties of human DNA polymerases α and β, *Cold Spring Harbor Symp. Quant. Biol.,* 43, 613, 1978.

87b. **Fisher, P. A., Chen. J. T., and Korn, D.,** Enzymological characterization of KB cell DNA polymerase α. Regulation of template binding by nucleic acid base composition, *J. Biol. Chem.,* 256, 133, 1981.

88. **Sugino, A., Hirose, S., and Okazaki, R.,** RNA-linked nascent DNA fragments in *Escherichia coli, Proc. Natl. Acad. Sci. U.S.A.,* 69, 1863, 1972.

89. **Sugino, A. and Okazaki, R.,** RNA-linked DNA fragments, *in vitro, Proc. Natl. Acad. Sci. U.S.A.,* 70, 88, 1973.

90. **Lark, K. G.,** Evidence for the direct involvement of RNA in the initiation of DNA replication in *Escherichia coli* 15T$^-$, *J. Mol. Biol.,* 64, 47, 1972.

91. **Schekman, R., Wickner, W., Westergaard, O., Brutlag, D., Geider, K., Bertsch, L. L., and Kornberg, A.,** Initiation of DNA synthesis: synthesis of ϕX174 replicative form requires RNA synthesis resistant to rifampicin, *Proc. Natl. Acad. Sci. U.S.A.,* 69, 2691, 1972.

92. **Wickner, W., Brutlag, D., Schekman, R., and Kornberg, A.,** RNA synthesis initiates *in vitro* conversion of M13 DNA to its replicative form, *Proc. Natl. Acad. Sci. U.S.A.,* 69, 965, 1972.

93. **Sato, S., Ariake, S. and Saito, M.,** RNA bound to nascent DNA in Ehrlich ascites tumor cells, *Biochem. Biophys. Res. Commun.,* 49, 827, 1972.

94. **Magnusson, G., Pigiet, V., Winnacker, E. L., Abrams, A., and Reichard, P.,** RNA-linked short DNA fragments during polyoma replication, *Proc. Natl. Acad. Sci. U.S.A.*, 70, 412, 1973.
95. **Keller, W.,** RNA-primed DNA synthesis *in vitro, Proc. Natl. Acad. Sci. U.S.A.*, 69, 1560, 1972.
96. **Stavrianopoulos, J. G., Karkas, J. D., and Chargaff, E.,** Mechanism of DNA replication by highly purified DNA polymerase of chicken embryo, *Proc. Natl. Acad. Sci. U.S.A.*, 69, 2609, 1972.
97. **Chang, L. M. S.,** Low molecular weight deoxyribonucleic acid polymerase from calf thymus chromatin. I. Purification of homogeneous enzymes, *J. Biol. Chem.*, 248, 3789, 1973.
98. **Wang, T. S-F., Sedwick, W. D., and Korn, D.,** Nuclear deoxyribonucleic acid polymerase. Further observations on the structure and properties of the enzyme from human KB cells, *J. Biol. Chem.*, 250, 7040, 1975.
99. **Brun, G., Rougeon, F., Lanber, M., and Chapeville, F.,** Purification and properties of DNA polymerases from chick embryo, *Eur. J. Biochem.*, 41, 241, 1974.
100. **Tsuruo, T., Hirayama, K., Kawaguchi, M., Satoh, H., and Ukita, T.,** Low molecular weight DNA polymerase of rat ascites hepatoma cells, *Biochim. Biophys. Acta*, 366, 270, 1974.
101. **Chiu, J. F. and Sung, S. C.,** Effect of spermidine on the activity of DNA polymerases, *Biochim. Biophys. Acta*, 281, 535, 1972.
102. **Probst, G. S., Stalker, D. M., Mosbaugh, D. W., and Meyer, R. R.,** Stimulation of DNA polymerase by factors isolated from Novikoff hepatoma, *Proc. Natl. Acad. Sci. U.S.A.*, 72, 1171, 1975.
103. **Tsuro, T., Tomita, Y., Satoh, H., and Ukita, T.,** DNA polymerases of ascites hepatoma cells. II. Purification and properties of DNA polymerases from nuclear membrane-chromatin fraction, *Biochem. Biophys. Res. Commun.*, 48, 776, 1972.
104. **Berger, H., Huang, R. C. C., and Irvin, J. L.,** Purification and characterization of a dexoyribonucleic acid polymerase from rat liver, *J. Biol. Chem.*, 246, 7275, 1971.
105. **Bolden, A., Aucker, J., and Weissbach, A.,** Synthesis of *Herpes Simplex* virus, Vaccinia virus, and Adenovirus DNA in isolated HeLa cell nuclei. I. Effect of viral-specific antisera and phosphonoacetic acid, *J. Virol.*, 16, 1584, 1975.
106. **Chang, L. M. S. and Bollum, F. J.,** A chemical model for transcriptional initiation of DNA replication, *Biochem. Biophys. Res. Commun.*, 46, 1354, 1972.
107. **de Recondo, A. M., Lepesant, J. A., Fichot, O., Grasset, L., Rossignol, J. M., and Cazillis, M.,** Synthetic template specificity of a deoxyribonucleic acid polymerase from regenerating rat liver, *J. Biol. Chem.*, 248, 131, 1973.
108. **Haines, M. E., Wcikremasinghe, R. G., and Johnston, I. R.,** Purification and partial characterization of rat liver nuclear DNA polymerase, *Eur. J. Biochem.*, 31, 119, 1972.
109. **Ross, J., Scolnick, E. M., Todaro, G. J., and Aaronson, S. A.,** Separation of murine cellular and murine leukemia virus DNA polymerases, *Nature (London) New Biol.*, 231, 163, 1971.
110. **Wintersberger, U.,** Absence of a low-molecular-weight DNA polymerase from nuclei of the yeast, *Saccharomyces cerevisiae, Eur. J. Biochem.*, 50, 197, 1974.
111. **McLennan, A. G. and Keir, H. M.,** Deoxyribonucleic acid polymerases of *Euglena gracilis.* Purification and properties of two distinct deoxyribonucleic acid polymerases of high molecular weight, *Biochem. J.*, 151, 227, 1975.
112. **Loomis, L. W., Rossomando, E. F., and Chang, L. M. S.,** DMA polymerase of *Dictyostelium discoideum, Biochim. Biophys. Acta*, 425, 469, 1976.
113. **Chang, L. M. S.,** Phylogeny of DNA polymerase β, *Science*, 191, 1183, 1976.
114. **Fridlender, B., Fry, M., Bolden, A., and Weissbach, A.,** A new synthetic RNA-dependent DNA polymerase from human tissue culture cells, *Proc. Natl. Acad. Sci. U.S.A.*, 69, 452, 1972.
115. **Spadari, S. and Weissbach, A.,** The interrelation between DNA synthesis and various DNA polymerase activities in synchronized HeLa cells, *J. Mol. Biol.*, 86, 11, 1974.
116. **Spadari, S. and Weissbach, A.,** HeLa cell R-deoxyribonucleic acid polymerases. Separation and characterization of two enzymatic activities, *J. Biol. Chem.*, 249, 5809, 1974.
117. **Knopf, K. W., Yamada, M., and Weissbach, A.,** HeLa cell DNA polymerase γ: further purification and properties of the enzyme, *Biochemistry*, 15, 4540, 1976.
118. **Gerard, G. F.,** Poly(2′-O-methylcytidylate)-oligo deoxyguanylate. I. Template-primer specific for reverse transcriptase, it is not utilized by HeLa cell γ DNA polymerases, *Biochem. Biophys. Res. Commun.*, 63, 706, 1975.
119. **Lewis, B. J., Abrell, J., Smith, R. G., and Gallo, R. C.,** DNA polymerases in human lymphoblastoid cells infected with simian sarcoma virus, *Biochim. Biophys. Acta*, 349, 148, 1974.
120. **Tarrago-Litvak, L., Castroviejo, M., and Litvak, S.,** Studies on a DNA polymerase γ-like enzyme from wheat embryos, *FEBS Lett.*, 59, 125, 1975.
121. **Gallagher, R. E., Todaro, G. J., Smith, R. G., Livingston, D. M., and Gallo, R. C.,** Relationship between RNA-directed DNA polymerase (reverse transcriptase) from human acute leukemia blood cells and primate type-c virus, *Proc. Natl. Acad. Sci. U.S.A.*, 71, 1309, 1974.

122. **Gerard, G. F., Rottman, F., and Green, M.,** Poly(2′-O-methylcytidylate). Oligodeoxyguanylate as a template for the ribonucleic acid directed deoxyribonucleic acid polymerase in ribonucleic acid tumor virus particles and a specific probe for the ribonucleic acid directed enzyme in transformed murine cells, *Biochemistry*, 13, 1632, 1974.
123. **Matsukage, A., Bohn, E. W., and Wilson, S. H.,** On the DNA polymerase III of mouse myeloma: partial purification and characterization, *Biochemistry*, 14, 1006, 1975.
124. **Yoshida, Y., Ando, T., and Kondo T.,** Synthetic RNA-dependent DNA polymerase from calf thymus, *Biochem. Biophys. Res. Commun.*, 60, 1193, 1974.
125. **Ikegami, S., Taguchi, T., Ohashi, M., Oguro, H., Nagano, H., and Mano, Y.,** Aphidicolin prevents mitotic cell division by interfering with the activity of DNA polymerase α, *Nature (London)*, 275, 458, 1978.
126. **Wright, G. E., Baril, E. F., and Brown, N. C.,** Butylanilinouracil: a selective inhibitor of HeLa cell DNA synthesis and HeLa cell DNA polymerase α, *Nucleic Acids Res.*, 8, 99, 1980.
127. **Di Cioccio, R. A. and Srivastava, B. I. C.,** Inhibition of deoxynucleotide-polymerizing enzyme activities of human cells and of simian sarcoma virus by heparin, *Cancer Res.*, 38, 2401, 1978.
128. **Tarrago-Litvak, L., Viratelle, O., Darriet, D., Dalibart, R., Graves, P. V., and Litvak, S.,** The inhibition of mitochondrial DNA polymerase from animal cells by intercalating drugs, *Nucleic Acids Res.*, 5, 2197, 1978.
129. **Kalf, G. F. and Ch'ih, J. J.,** Purification and properties of deoxyribonucleic acid polymerase from rat liver mitochondria, *J. Biol. Chem.*, 243, 4904, 1968.
130. **Bolden, A., Pedrali-Noy, G., and Weissbach, A.,** DNA polymerase of mitochondria is a γ-polymerase, *J. Biol. Chem.*, 252, 3351, 1977.
131. **Chiu, J. F. and Sung, S. C.,** Pattern of developmental changes in two DNA polymerases of rat brain, *Biochim. Biophys. Acta*, 269, 364, 1972.
132. **Lynch, W. W. and Lieberman, I.,** A DNA polymerase in liver nuclei whose activity rises with DNA synthesis after partial hepatectomy, *Biochem. Biophys. Res. Commun.*, 52, 843, 1973.
133. **Lynch, W. E., Surrey, S., and Lieberman, I.,** Nuclear deoxyribonucleic acid polymerases of liver, *J. Biol. Chem.*, 250, 8179, 1975.
134. **Chang, L. M. S. and Bollum, F. J.,** Variation of deoxyribonucleic acid polymerase activities during rat liver regeneration, *J. Biol. Chem.*, 247, 7948, 1972.
135. **Chang, L. M. S., Brown, M., and Bollum, F. J.,** Induction of DNA polymerase in mouse L cells, *J. Mol. Biol.*, 74, 1, 1973.
136. **Chiu, R. W. and Baril, E. F.,** Nuclear DNA polymerases and the HeLa cell cycle, *J. Biol. Chem.*, 250, 7951, 1975.
137. **Craig, R. K., Costello, P. A., and Keir, H. M.,** Deoxyribonucleic acid polymerases of BHK-21/C13 cells. Relationship to the physiological state of the cells, and to synchronous induction of synthesis of deoxyribonucleic acid, *Biochem. J.*, 145, 233, 1975.
138. **Chiu, J. F., Craddock, C., Morris, H. P., and Hnilica, L. S.,** Nuclear DNA polymerases during experimental neoplasia of rat liver, *Cancer Biochem. Biophys.*, 1, 13, 1974.
139. **Chiu, J. F.,** manuscript submitted for publication, 1982.

139a. **Suzuki, K., Miyaki, M., Akamatsu, N., and Ono, T.,** Increase of DNA polymerase activities in HeLa cells treated with N-methyl-N′-nitro-N-nitrosoguanidine, *FEBS Lett.*, 119, 150, 1980.

140. **Falaschi, A. and Spadari, S.,** in *DNA Synthesis: Present and Future*, Molineux, I. and Kohiyama, M., Eds., Plenum Press, New York, 1978, 487.
141. **Mechali, M., Girard, M., and de Recondo, A. M.,** DNA polymerase activities in growing cells infected with simian virus 40, *J. Virol.*, 23, 117, 1977.
142. **Riva, S., Cline, M. A., and Laipis, P. J.,** in *DNA Synthesis: Present and Future*, Molineux, I. and Kohiyama, M., Eds., Plenum Press, New York, 1978, 641.
143. **Habaro, A., Kano, K., Nagano, H., Mano, Y., Ikegami, S., and Yamashita, T.,** Inhibition of DNA synthesis in the adenovirus DNA replication complex by aphidicolin and 2′,3′-dideoxythymidine triphosphate, *Biochem. Biophys. Res. Commun.*, 92, 8, 1980.

143a. **Oguro, M., Shioda, M., Nagano, H., Mano, Y., Hanaoka, F., and Yamada, M.,** The mode of action of aphidicolin on DNA synthesis in isolated nuclei, *Biochem. Biophys. Res. Commun.*, 82, 13, 1980.

144. **Edenberg, H. J., Anderson, S., and De Pamphilis, M. L.,** Involvement of DNA polymerase α in simian virus 40 DNA replication, *J. Biol. Chem.*, 253, 3273, 1978.
145. **Abboud, M. M. and Horwitz, M. S.,** The DNA polymerases associated with the adenovirus type 2 replication complex: effect of 2′,3′-dideoxythymidine-5′-triphosphate on viral DNA synthesis, *Nucleic Acids Res.*, 6, 1025, 1979.
146. **Krokan, H., Schaffer, P., and DePamphilis, M. L.,** Involvement of eukaryotic deoxyribonucleic acid polymerase α and γ in the replication of cellular and viral deoxyribonucleic acid, *Biochemistry*, 18, 4431, 1979.

147. **Van der Vliet, P. C. and Kwant, M. M.**, Roleof DNA polymerase γ in adenovirus DNA replication, *Nature (London)*, 276, 532, 1978.
148. **Waqar, M. A., Evans, M. J., and Huberman, J. A.**, Effect of 2′,3′-dideoxythymidine-5′-triphosphate on HeLa cell *in vitro* DNA synthesis: evidence that DNA polymerase α is the only polymerase required for cellular DNA replication, *Nucleic Acids Res.*, 5, 1933, 1978.
149. **Ono, K., Ogasawara, M., and Matsukage, A.**, Inhibition of the activity of DNA polymerase α by 2′,3′-dideoxythymidine-5′-triphosphate, *Biochem. Biophys. Res. Commun.*, 88, 1255, 1979.
150. **Wist, E.**, The role of DNA polymerases α, β and γ in nuclear DNA synthesis, *Biochim. Biophys. Acta*, 562, 62, 1979.
151. **Fisher, P. A., Wang, T. S., and Korn, D.**, Enzymological characterization of DNA polymerase α. Basic catalytic properties, processivity, and gap utilization of the homogeneous enzyme from human KB cells, *J. Biol. Chem.*, 254, 6128, 1979.
152. **Hubscher, U., Kuenzle, C. C., and Spadari, S.**, Functional roles of DNA polymerases β and γ, *Proc. Natl. Acad. Sci. U.S.A.*, 76, 2316, 1979.
152a. **McLennan, A. G.**, Effect of aphidicolin and 2′,3′-dideoxythymidine on mitochondrial DNA replication, *Biochem. Biophys. Res. Commun.*, 94, 116, 1980.
152b. **Zimmerman, W., Chen, S. M., Bolden, A., and Weissbach, A.**, Mitochondrial DNA replication does not involve DNA polymerase α, *J. Biol. Chem.*, 255, 11847, 1980.
153. **Linn, S., Kairis, M., and Holliday, R.**, Decreased fidelity of DNA polymerase activity isolated from aging human fibroblasts, *Proc. Natl. Acad. Sci. U.S.A.*, 73, 2818, 1976.
154. **Loeb, L. A., Battula, N., Springgate, C. F., and Seal, G.**, On mutagenic DNA polymerases and malignancy, in *Fundamental Aspects of Neoplasia*, Gottlieb, A. A., Plescie, O. J., and Bishop, D. H. L., Eds., Springer Verlag, New York, 1975, 243. pp. 243—254.
155. **Lieberman, M. W.**, Approaches to the analysis of fidelity of DNA repair in mammalian cells, *Int. Rev. Cytol.*, 45, 1, 1976.
156. **Sirover, M. A., and Loeb, L. A.**, On the fidelity of DNA replication. Effect of metal activation during synthesis with avian myeloblastosis virus DNA polymerase, *J. Biol. Chem.*, 252, 3605, 1977.
157. **Chan, J. Y. H. and Becker, F. F.**, Decreased fidelity of DNA polymerase activity during N-2-fluorenolacetamide hepatocarcinogenesis, *Proc. Natl. Acad. Sci. U.S.A.*, 76, 814, 1979.
158. **Sunderman, F. W., Jr.**, Carcinogenic effects of metals, *Fed. Proc.*, 37, 40, 1978.
159. **Orgel, A. and Orgel, L. A.**, Induction of mutations in bacteriophage T_4 with divalent manganese, *J. Mol. Biol.*, 14, 453, 1965.
160. **Stoner, G. D., Shimkin, M. B., Troxell, M. C., Thompson, T. L., and Terry, L. S.**, Test for carcinogenicity of metallic compounds by the pulmonary tumor response in strain A mice, *Cancer Res.*, 36, 1744, 1976.
161. **Chiu, J. F. and Sung, S. C.**, Intracellular distribution of DNA polymerase activities in developing and adult rat brain, *J. Neurochem.*, 20, 617, 1973.
162. **Hackmann, H. J. and Lezius, A. G.**, High-molecular-weight DNA polymerases from mouse Myeloma. Purification and properties of three enzymes, *Eur. J. Biochem.*, 50, 357, 1975.
163. **Tanabe, K. and Takahashi, T.**, Studies on the molecular species of DNA polymerase extracted from rat Ascites hepatoma cells, *J. Biochem. (Tokyo)*, 79, 85, 1976.
164. **Tsuruo, T., Hirayama, K., Satoh, H., and Ukita, T.**, A new form of high molecular weight DNA polymerase in the nuclei of rat Ascites hepatoma cells, *J. Biochem. (Tokyo)*, 78, 401, 1975.
165. **Noy, G. P. and Weissbach, A.**, HeLa cell DNA polymerases: the effect of cycloheximide *in vivo* and detection of a new form of DNA polymerase α, *Biochim. Biophys. Acta*, 477, 70, 1977.
166. **Matsukage, A., Sivarajan, M., and Wilson, S. H.**, Studies on DNA α-polymerase of mouse myeloma: partial purification and composition of three molecular forms of the enzyme, *Biochemistry*, 15, 5305, 1976.
167. **Gellert, M.**, Formation of covalent circles of lambda DNA by *E. coli* extracts, *Proc. Natl. Acad. Sci., U.S.A.*, 57, 148, 1967.
168. **Olivera, B. M. and Lehman, I. R.**, Linkages of poly nucleotides through phosphodiester bonds by an enzyme from *Escherichia coli*, *Proc. Natl. Acad. Sci. U.S.A.*, 57, 1426, 1967.
169. **Weiss, B. and Richardson, C. C.**, Enzymatic breakages and joining of deoxyribonucleic acid. I. Repair of single-strand breaks in DNA by an enzyme system from *Escherichia coli* infected with T_4 bacteriophage, *Proc. Natl. Acad. Sci. U.S.A.*, 57, 1021, 1967.
170. **Gefter, M. L., Becker, A., and Hurwitz, J.**, The enzymatic repair of DNA_1. I. Formation of circular λ DNA, *Proc. Natl. Acad. Sci. U.S.A.*, 58, 240, 1967.
171. **Cozzarelli, N. R., Melechen, N. E., Jovin, T. M., and Kornberg, A.**, Polynucleotide cellulose as a substrate for polynucleotide ligase induced by phage T_4, *Biochem. Biophys. Res. Commun.*, 28, 578,1967.
172. **Okazaki, R. Okazaki, T., Sakabe, K., Sugimoto, K., and Sugino, A.**, Polynucleotide ligase from myeloid and lymphoid tissues, *Proc. Natl. Acad. Sci. U.S.A.*, 61, 680, 1968.

173. **Lehman, I. R.,** DNA joining enzymes (ligases), in *The Enzymes,* Vol. 10, Boyer, P. D., Ed., Academic Press, New York, 1974, 237.
174. **Richardson, C. C.,** Enzymes in DNA metabolism, *Annu. Rev. Biochem.,* 38, 795, 1969.
175. **Gellert, M.,** DNA ligase from *E. coli,* in *Procedures in Nucleic Acid Research,* Vol. 2, Cantoni, G. L. and Davies, D. R., Eds., Harper & Row, New York, 1971, 875.
176. **Soderhall, S. and Lindahl, T.,** Mammalian deoxyribonucleic acid ligase. Isolation of an active enzyme-adenylate complex, *J. Biol. Chem.,* 248, 672, 1973.
177. **Gaziev, A. I. and Kuzin, A.,** The localization of DNA ligase in the chromatin of animal cells, *Eur. J. Biochem.,* 37, 7, 1973.
178. **Soderhall, S.,** Properties of DNA-adenylate complex formed in the reaction between mammalian DNA ligase I and DNA containing single-strand breaks, *Eur. J. Biochem.,* 51, 129, 1975.
179. **Gottsman, M. M., Hicks, M. L., and Gellert, M.,** Genetics and function of DNA ligase in *Escherichia coli, J. Mol. Biol.,* 77, 531, 1973.
180. **Konrad, E. B., Modrich, P., and Lehman, I. R.,** Genetic and enzymatic characterization of a conditional lethal mutant of *Escherichia coli* K_{12} with a temperature-sensitive DNA ligase, *J. Mol. Biol.,* 77, 519, 1973.
181. **Shenin, R.,** Preliminary characterization of the temperature sensitive defect in DNA replication in mutant mouse L cell, *Cell,* 7, 49, 1976.
182. **Slater, M. L. and Ozer, H. L.,** Temperature-sensitive mutants of Balb/3T3 cells: description of a mutant affected in cellular and polyoma virus DNA synthesis, *Cell,* 7, 289, 1976.
183. **Tsukada, K.,** Changes in polynucleotide ligase during rat liver regenerating, *Biochem. Biophys. Res. Commun.,* 57, 758, 1974.
184. **Austin, G. E.,** Purification of DNA ligase from rat liver, *Fed. Proc.,* 34, 701, 1975.
185. **Soderhall, S.,** DNA ligases during rat liver regeneration, *Nature (London),* 260, 640, 1976.
186. **Levin, C. J. and Zimmerman, S. B.,** A DNA ligase from mitochondria of rat liver, *Biochem. Biophys. Res. Commun.,* 69, 514, 1976.
187. **Alberts, B. and Frey, L.,** T_4 bacteriophage gene 32: a structural protein in the replication and recombination of DNA, *Nature (London),* 227, 1313, 1970.
188. **Abdel-Monem, M. and Hoffmann-Berling, H.,** DNA unwinding enzymes, *Trends Biochem. Sci.,* 5, 128, 1980.
189. **Molineux, I. J., Friedman, S., and Gefter, M. L.,** Purification and properties of the *Escherichia coli* deoxyribonucleic acid-unwinding protein. Effects on deoxyribonuclcic acid synthesis *in vitro, J. Biol. Chem.,* 249, 6090, 1974.
190. **Reuben, R. C. and Gefter, M. L.,** A DNA-binding protein induced by bacteriophage T_7, *Proc. Natl. Acad. Sci. U.S.A.,* 70, 1846, 1973.
191. **Scherzinger, E., Litfin, F., and Jost, E.,** Stimulation of T_7 DNA polymerase by a new phage-coded protein, *Mol. Gen. Genet.,* 123, 247, 1973.
192. **Kornberg, A.,** Aspects of DNA replication, *Cold Spring Harbor Symp. Quant. Biol.,* 43, 1, 1979.
193. **Herrick, G., Delius, H., and Alberts, B.,** Single-stranded DNA structure and DNA polymerase activity in the presence of nucleic acid helix-unwinding proteins from calf thymus, *J. Biol. Chem.,* 251, 2142, 1976.
194. **Novak, B. and Baril, E. F.,** HeLa DNA polymerase α activity *in vitro*: specific stimulation by a non-enzymatic protein factor, *Nucleic Acids Res.,* 5, 221, 1978.
195. **Duquet, M. and De Recondo, A. M.,** A deoxyribonucleic acid unwinding protein isolated from regenerating rat liver. Physical and functional properties, *J. Biol. Chem.,* 253, 1660, 1978.

195a. **Bonne, C., Duquet, M., and de Recondo, A.,** Single-strand DNA binding protein from rat liver interactions with supercoiled DNA, *Nucleic Acids Res.,* 8, 9955, 1980.

196. **Carrara, G., Gattoni, S., Mercanti, D., and Tocchini-Valenti, G. P.,** Purification of a DNA-binding protein from *Xenopus laevis* unfertilized eggs, *Nucleic Acids Res.,* 4, 2855, 1977.
197. **Otto, B., Fanning, E., and Richter, A.,** DNA polymerases and a single-strand-specific DNA-binding protein associated with simian virus 40 nucleoprotein complexes, *Cold Spring Harbor Symp. Quant. Biol.,* 43, 705, 1979.
198. **Blue W. T. and Weissbach, A.,** Detection of a DNA polymerase β stimulating protein in HeLa nuclear extracts, *Biochem. Biophys. Res. Commun.,* 84, 603, 1978.

198b. **Chang, L. M. S., Lurie, K., and Plevani, P.,** A stimulatory factor for yeast DNA polymerase, *Cold Spring Harbor Symp. Quant. Biol.,* 43, 587, 1979.

199. **Abdel-Monem, H. and Hoffman-Berling, H.,** Enzymatic unwinding of DNA. I. Purification and characterization of a DNA-dependent ATPase from *Escherichia coli, Eur. J. Biochem.,* 65, 431, 1976.
200. **Wang, J. C.,** Interaction between DNA and an *Escherichia coli* protein ω, *J. Mol. Biol.,* 55, 523, 1971.
201. **Champoux, J. J. and Dulbecco, R.,** An activity from mammalian cells that untwists super-helical DNA — a possible swivel for DNA replication, *Proc. Natl. Acad. Sci. U.S.A.,* 69, 143, 1972.

202. **Brase, W. A. and Wang, J. C.,** An ω protein from *Drosophila melanogaster, Biochemistry,* 13, 4299, 1974.
203. **Pulleyblank, D. E. and Morgan, A. R.,** Partial purification of "ω" protein from calf thymus, *Biochemistry,* 14, 5205, 1975.
204. **Burrington, M. G. and Morgan, A. R.,** The purification from *Escherichia coli* of a protein relaxing superhelical DNA, *Can. J. Biochem.,* 54, 301, 1976.
205. **Champoux, J. J. and McConaughy, B. L.,** Purification and characterization of the DNA untwisting enzyme from rat liver, *Biochemistry,* 15, 4638, 1976.
206. **Durnford, J. M. and Champoux, J. J.,** The DNA untwisting enzyme from *Saccharomyces cerevisiae, J. Biol. Chem.,* 253, 1086, 1978.
207. **Keller, W.,** Characterization of purified DNA-relaxing enzyme from human tissue culture cells, *Proc. Natl. Acad. Sci. U.S.A.,* 72, 2550, 1975.
208. **Mattoccia, E., Attardi, D. G., and Tocchini-Valentini, G. P.,** DNA-relaxing activity and endonuclease activity in *Xenopus laevis* oocytes, *Proc. Natl. Acad. Sci. U.S.A.,* 73, 4551, 1976.
209. **Lasky, R. A., Mills, A. D., and Morris, N. R.,** Assembly of SV40 chromatin in a cell-free system from *Xenopus* eggs, *Cell,* 10, 237, 1977.
210. **Bina-Stein, M., Vogel, T., Singer, D. S., and Singer, M. F.,** H5 histone and DNA-relaxing enzyme of chicken erythrocytes. Interaction with superhelical DNA, *J. Biol. Chem.,* 251, 7363, 1976.
211. **DeLeys, R. J. and Jackson, D. A.,** Electrophoretic analysis of covalently closed SV40 DNA: Boltzman distributions of DNA species, *Nucleic Acids Res.,* 3, 641, 1976.
212. **Vosberg, H. P., Grossman, L. I., and Vinograd, J.,** Isolation and partial characterization of the relaxation protein from nuclei of cultured mouse and human cells, *Eur. J. Biochem.,* 55, 79, 1975.
213. **Camerini-Otero, R. D. and Felsenfeld, G.,** Supercoiling energy and nucleosome fraction: the role of the arginine-rich histone kernel, *Nucleic Acids Res.,* 4, 1159, 1977.
214. **Bauer, W. R., Ressener, E. C., Kates, J., and Patzke, J. V.,** A DNA nicking-closing enzyme encasidated in vaccinia virus: partial purification and properties, *Proc. Natl. Acad. Sci. U.S.A.,* 74, 1841, 1977.
215. **Bauer, W. and Vinograd, J.,** Circular DNA, in *Basic Principles in Nucleic Acid Chemistry,* Vol. 2, Ts'o, P., Ed., Academic Press, New York, 1974, 265.
216. **Champoux, J. J. and Durnford, J. M.,** Assay and partial purification of the DNA untwisting activity from rat liver, in *ICN-UCLA Symp. Mol. Cell. Biol.,* Vol. 3, ICN Biologicals, Cleveland, 1975, 83.
217. **Champoux, J. J.,** Evidence for an intermediate with a single-strand break in the reaction catalyzed by the DNA untwisting enzyme, *Proc. Natl. Acad. Sci. U.S.A.,* 73, 3488, 1976.
218. **Champoux, J. J.,** Strand breakage by the DNA untwisting enzyme results in covalent attachment of the enzyme to DNA, *Proc. Natl. Acad. Sci. U.S.A.,* 74, 3800, 1977.

INDEX

A

B

C

D

E

F

G

H

I

J

K

L

M

N

O

Q

R

S

T